Manual of Horticultural Techniques

园艺技术手册

刘文星　杨　明　编　译
罗大强　副编译

哈爾濱工業大學出版社

图书在版编目（CIP）数据

园艺技术手册 = Manual of Horticultural Techniques：英文 / 刘文星，杨明编译. — 哈尔滨：哈尔滨工业大学出版社，2024. 5
ISBN 978-7-5767-1449-4

Ⅰ. ①园… Ⅱ. ①刘… ②杨… Ⅲ. ①园艺-设施农业-手册-英文 Ⅳ. ①S62-62

中国国家版本馆 CIP 数据核字（2024）第 103631 号

YUANYI JISHU SHOUCE

策划编辑　闻　竹　常　雨
责任编辑　闻　竹　常　雨
封面设计　童越图文
出版发行　哈尔滨工业大学出版社
社　　址　哈尔滨市南岗区复华四道街 10 号　邮编 150006
传　　真　0451-86414749
网　　址　http://hitpress.hit.edu.cn
印　　刷　哈尔滨市石桥印务有限公司
开　　本　787mm×1092mm　1/16　印张 22.5　字数 668 千字
版　　次　2024 年 5 月第 1 版　2024 年 5 月第 1 次印刷
书　　号　ISBN 978-7-5767-1449-4
定　　价　148.00 元

编译委员会

Foreword

The revitalization of rural areas, the construction of a new socialist countryside, and the achievement of agricultural modernization are major decisions made by the central government. At the current stage, China will continue to consolidate and expand the achievements of poverty alleviation and ensure effective integration with rural revitalization. Implementing agricultural scientific and technological innovation, improving the quality of agricultural technical personnel, and enhancing the skills of farmers are important guarantees for implementing the strategy of rural revitalization.

To meet the needs of the rural revitalization development strategy, promote and facilitate the transformation of agricultural scientific and technological achievements, strengthen the construction of agricultural technical personnel, and enhance the vocational skills of farmers, Liaoning Vocational College has organized relevant teachers and agricultural experts to compile the *Manual of Horticultural Techniques*, which includes "Q&A on Practical Techniques for Strawberry Cultivation" "Q&A on Practical Techniques for Grape Cultivation" "Q&A on Practical Techniques for Fresh Tomato Production" "Q&A on Practical Techniques for Sweet Melon Production" "Q&A on Practical Techniques for Vegetable Disease and Pest Control" "Q&A on Practical Techniques for Cut Flower Production" and "Q&A on Practical Techniques for Edible Mushroom Production".

This book closely combines practical production, emphasizes pertinence, applicability, and popularization, and applies modern agricultural technologies and achievements. The language is easy to understand and close to practical production. The aim is to provide new knowledge, new technologies, and new information for agricultural technical personnel, farmers, and other users, solve practical problems encountered in the development of modern agriculture, and improve the ability of agricultural technical personnel and farmers to acquire new knowledge, master new technologies, and obtain new information. It is believed that the publication of this book will play an important role in improving the scientific and technological literacy of farmers, promoting industrial restructuring, increasing agricultural production, increasing farmers' income, accelerating modernization, and supporting rural revitalization.

Editor

2021.10

Contents

Chapter 1 Q&A on Practical Techniques for Strawberry Cultivation

1. Overview

001 Why develop strawberry production?

Strawberries have the advantages of fast fruiting, early maturity, high yield, easy cultivation management, low incidence of pests and diseases, low production costs, and high economic benefits. They are highly favored by producers and consumers. In recent years, the average yield per hectare in the northern strawberry production areas of China has reached 15 000 – 30 000 kg. Currently, protected cultivation allows strawberries to be available in the market from November to December, with a selling price of over 40 yuan per kilogram. Strawberries available from March to April are priced at over 20 yuan per kilogram, and those available from May to June are priced at over 10 yuan per kilogram. The income from a half-year period of cultivation on an area of 667 square meters can exceed 8 000 yuan for open-field cultivation and reach over 50 000 yuan for protected cultivation.

Strawberries have strong adaptability and a wide cultivation range. They can be grown in open fields using conventional methods or in various forms of protected cultivation. The fruits can be harvested in stages. Frozen storage allows for long-term preservation, which helps regulate the supply during the off-season and ensures a balanced year-round availability.

Strawberries are rich in nutrients and are known as the "Queen of Fruits". According to analysis, the sugar content of strawberry ranges from 6% to 12%, organic acid content ranges from 0.6% to 1.6%, vitamin C content ranges from 50 to 120 milligrams per 100 grams, vitamin A content is 60 IU, vitamin B1 content is 0.3 milligrams per 100 grams, vitamin D content is 0.07 milligrams per 100 grams, vitamin K content is 0.1 milligrams per 100 grams, fiber content ranges from 1.0% to 1.6%, and aromatic substance content ranges from 0.16% to 0.25%. Additionally, the fruit contains abundant mineral elements such as calcium and phosphorus. In recent years, a substance called "strawberry amine" has been extracted from the fruit, roots, stems, and leaves of strawberries, which has remarkable effects on treating leukemia and hypochromic anemia. The health and medical functions of strawberries have gained attention. Strawberry plants have an elegant appearance, a long flowering and fruiting period, and can be grown in pots or as vertical tower plantings. They have both greening and ornamental value.

002 What are the prospects for strawberry development?

In recent years, strawberry cultivation in China has developed rapidly, with nearly 30 provinces and cities engaged in strawberry cultivation. Regions with early cultivation, large cultivation areas, and rapid development include Baoding in Hebei Province, Dandong in Liaoning Province, Yantai and Linyi in Shandong Province, Zhenjiang in Jiangsu Province, and the Guanzhong area in Shaanxi Province. Significant development has also been observed in Beijing, Tianjin, Shanghai, and other areas. Currently, the main production method for strawberries is protected cultivation, with the increasing popularity of plastic greenhouses, medium-sized greenhouses, small greenhouses, and mulching. China has imported over a hundred strawberry varieties from abroad, and domestic breeding has yielded more than a dozen excellent varieties. Many places have established strawberry variety resource nurseries and conducted research on high-yield cultivation. A series of research achievements have been made in protected cultivation, variety breeding, tissue culture, virus-free seedling production, pest and disease control, and storage and processing, which have positively contributed to the development of strawberry production in China. The prospects for strawberry production are becoming increasingly promising.

003 What key technologies should be emphasized to improve strawberry production?

To achieve successful strawberry production, it is crucial to focus on the following five key technical aspects:

First, breeding and utilizing high-quality varieties with large fruits suitable for fresh consumption. It has been reported that foreign countries are working on developing giant strawberries with individual fruit weights of nearly 100 grams, and some initial achievements have been made. A small number of these varieties have been introduced domestically. At the same time, emphasis should also be placed on breeding varieties with a very short dormancy period and low chilling requirements for off-season cultivation.

Second, implementing effective protected cultivation techniques. Utilizing modern technology, selecting suitable varieties with short dormancy periods, breaking dormancy using hormone applications, and employing protective structures for early or delayed cultivation can achieve year-round harvest and supply.

Third, actively promoting virus-free seedlings and popularizing non-toxic cultivation techniques for high-yield and high-quality production.

Fourth, developing the strawberry processing industry by adopting new processing techniques and packaging methods to produce nutritionally rich strawberry products with certain therapeutic effects. For example, extracting medicinal components like "strawberry amine" from strawberry roots, stems, and leaves can further enhance the efficacy of strawberries.

Fifth, ensuring proper cold storage and preservation techniques for strawberries to extend the availability of fresh fruits.

2. Growth and Yield Characteristics

004 What are the characteristics of strawberry roots?

Strawberries are perennial evergreen herbaceous plants. They have short stature and grow in clusters, with stolons that creep along the ground. Although they can live for over ten years, the peak fruiting period occurs at around 2-3 years of age.

The root system of strawberries is composed of adventitious roots that grow from the stems. New plants formed from stolons develop primary roots, which are approximately 1 millimeter in diameter and emerge at the even-numbered nodes. The number of primary roots typically ranges from a dozen to several dozen, sometimes even exceeding a hundred. Secondary roots (lateral roots) can also develop from the primary roots. The new roots are initially milky white and gradually age to light yellow, brown, and eventually black as they die off. New adventitious roots also emerge from the upper part of the plant's stem. As the position of the new stem gradually rises, the sites of root formation move upward, sometimes even becoming exposed above the ground. If the conditions, mainly humidity, are not suitable after emergence, it can affect the normal production and growth of new roots. Therefore, measures such as soil covering can be taken to promote root growth and ensure safe overwintering. After covering the perennial stems with soil, new roots can also be induced. In addition to their absorptive function, new roots can store nutrients, allowing the rhizomes to penetrate into the soil, resulting in robust and well-nourished mother plants that produce more flowers and fruits. The lifespan of new roots is usually around one year.

Strawberry roots are shallow and primarily distributed in the soil layer above 30 centimeters from the surface. In sandy loam soil, more than 70% of the root mass is concentrated within a 20-centimeter depth. The growth dynamics of strawberry roots throughout the year are as follows: they begin to grow in early spring, approximately 10 days earlier than the aboveground parts. First, the non-aging roots (autumn roots) elongate, followed by the emergence of new roots from the upper rhizomes. There are usually three root growth peaks in a year. The first peak occurs in regions with high summer temperatures or severe winter cold, where cultivars tolerant to high and low temperatures should be considered. For home gardens and container cultivation, everbearing strawberries can be selected. The second peak occurs from late February to mid-April, and the third peak has the highest number of new roots and takes place from early July to mid-August. The third peak occurs from mid-September to the end of November. In late autumn and early winter, the root system enters a vigorous growth period and ceases growth relatively late.

005 How many types of stems does a strawberry plant have, and what are their characteristics?

The stalk of the strawberry includes three types: the new stalk, the rhizome and the stolon. The first two types of stalks grow above or below the ground, while the stolon is a special above-ground stem that extends along the ground. The characteristics of these stems are briefly described as follows.

(1) New stems. New stems are short, compact stems that grow in the current year. They have a slightly arched shape and exhibit slow elongation and growth. Each year, they only grow 0.5–2 centimeters in length. However, they undergo strong thickening growth. New stems are densely covered with leaves that have long petioles. Axillary buds are formed between the leaf axils, and in autumn, the terminal bud can develop mixed flower buds. In the second year, they become the first inflorescence of the main stem. The inflorescence faces the concave side of the new stem.

The axillary buds on the new stems, also known as lateral buds, exhibit early maturity. Some lateral buds can develop into shortened new stem branches in the same year, while others can produce long, slender stolons that give rise to new plants. Some lateral buds remain dormant as latent buds and only sprout and regenerate when stimulated. The occurrence of new stems is influenced by factors such as variety, plant age, seedling vigor, and planting time. The peak period of new stem generation generally occurs from July to September. After the temperature drops in October, new stem formation ceases.

New stems in the lower and middle parts sprout adventitious roots, which become rhizomes in the second year.

(2) Rhizomes. After the second year, the leaves on the new stems wither and fall off, and the stems take on a similar appearance to roots. These stems are called rhizomes (in fact, they are aged new stems). Rhizomes have nodes and annual rings and serve as storage organs for nutrients. In the plant's third year of growth, the lower portion of the old rhizomes gradually dies off from the bottom up. The older the rhizomes, the weaker the aboveground growth. The internal structure of rhizomes differs from that of new stems. Rhizomes have annual rings and a higher degree of lignification, while the phloem of new stems has well-developed spiral vascular bundles with numerous living cells and strong vitality. The undeveloped buds on the new stems are latent buds of the rhizomes. When the aboveground parts of strawberry plants are damaged, these latent buds can produce new stems. The base of the new stems generates adventitious roots, promoting plant growth.

(3) Stolons. Stolons, also known as runners, are a special type of aboveground stem

that develops from lateral buds on new stems. Stolons have long and slender internodes and serve as the reproductive organs of strawberries. Sometimes, flowers can also form at the tips of stolons or on the nodes of the inflorescence stalk.

During the initial stage of stolon development, they grow upward until they exceed the height of the leaves. Then, they spread along the ground towards areas with fewer plant clusters and better sunlight. The occurrence of stolons is mainly influenced by the variety and plant age. Varieties that produce more new stems usually have fewer stolons. Plants that are 2–3 years old exhibit the strongest ability to produce runners. Stolons usually have two nodes. The first node does not produce stolon plantlets but may give rise to stolon branches. The second node can produce stolon plantlets, which develop normal leaves upward and adventitious roots downward, eventually becoming independent plants. Stolon plantlets can continue to produce secondary stolons from lateral buds, generating the next generation of stolon plantlets. This continuous process forms a long stolon, and stolon plantlets are produced at even-numbered nodes. Generally, stolon plantlets closer to the mother plant form earlier and exhibit better growth and development.

006 What are the internal and external factors that influence the occurrence of strawberry stolons, and how can they be regulated?

The timing of stolon formation in strawberries is related to variety characteristics. The initiation of stolon formation for the first generation generally occurs during the period of fruit enlargement. Early-maturing varieties exhibit earlier stolon formation. The majority of stolon formation occurs after berry harvest. However, even with 16 hours of long daylight exposure, stolon formation does not occur at low temperatures. Increasing the intensity of light is beneficial for stolon formation.

The occurrence of stolons is also closely related to the duration of the low-temperature period experienced by the mother plants. Stolon formation is vigorous when the low-temperature requirements for plant dormancy are fully met. Therefore, in protected cultivation, it is important not to start warming too early. Otherwise, the plants' low-temperature requirements will not be met, resulting in a lack of stolon plantlets.

After meeting the low-temperature requirements, high temperatures and long daylight exposure promote stolon formation. Spraying 0.005% gibberellin twice between June and July can promote stolon formation. However, it may have a certain inhibitory effect on flowering. Spraying gibberellin combined with thinning flowers significantly increases the propagation coefficient.

Although stolon plantlets are the main means of strawberry propagation, excessive and vigorous stolon formation can lead to overcrowding of seedlings, poor nutrient conditions, increased susceptibility to pests and diseases, lower quality of plantlets, and impaired flower bud differentiation in mother plants.

007 What is the relationship between the occurrence of strawberry stolon plantlets and the mother plant?

A strawberry mother plant can produce 3-5 generations of stolon plantlets (runners) within a year. There are significant differences in the timing and quantity of stolon plantlet formation among different generations. Four varieties, including Baobao Early, Hongyi, Chunxiang, and Dana, can produce 4 generations of plantlets in a year. The peak period of the first-generation plantlet formation occurs from late June to early August, accounting for 52.3%, 66.7%, 61.6%, and 94.8% of the total plantlet quantity, respectively. Early-maturing varieties like Baobao Early exhibit earlier initiation and peak periods of plantlet formation, while late-maturing varieties like Dana exhibit later periods. Other generations of plantlets start to form successively during the mid to late peak period of the first-generation plantlets, without a distinct peak. The first and second generations of plantlets usually account for more than half of the total plantlet number. The lower the generation, the closer to the mother plant, the earlier the occurrence, the thicker the new stems of the plantlets, the more robust the plantlets, and the higher the rate of flowering. Plantlets with new stems thicker than 0.8 centimeters can all flower and bear fruit after transplanting.

During the initial stage of plantlet formation, the mother plant provides nutrients and water to the plantlets. Once the plantlets develop roots and touch the ground, they begin to directly absorb water and nutrients from the soil. The better the development of the mother plant, the faster the root growth of the plantlets. After being separated from the mother plant, the plantlets can grow and develop independently.

008 How do strawberry flower buds differentiate?

Strawberry flower buds are mixed buds. During the annual cycle, as long as the temperature and daylight conditions are suitable, strawberry plants can undergo flower bud differentiation. Along with the apical bud becoming a flower bud, some axillary buds can also differentiate into flower buds. Within a floral cluster, after the primary flowers differentiate, secondary inflorescences form from the axils of the primary inflorescence, and tertiary inflorescences form from the axils of the secondary inflorescence.

The process of strawberry flower bud differentiation can be divided into the following three periods.

(1) Initial differentiation stage. This stage has a fast differentiation speed, taking about

6 days. The growing point becomes round and prominent, and the floral primordia form.

(2) Inflorescence differentiation stage. The formation of the apical inflorescence meristem takes about 11 days. The apical inflorescence meristem continuously differentiates and develops, and in the later stage, the sepals of the apical flower become prominent, while the second inflorescence(from the first axillary bud) forms floral primordia.

(3) Floral organ differentiation stage. This stage requires a longer time, around 16 days. The apical inflorescence forms, and most of the small sepals extend inward to form the receptacle. The inner floral corolla, stamens, and pistils gradually become prominent and develop to maturity. In the middle and late stages, the floral primordia of the third inflorescence(from the second axillary bud) form.

009 How does sunlight and temperature affect strawberry flower bud differentiation?

Strawberry flower bud differentiation occurs under relatively low temperatures and short daylight conditions. Generally, the temperature should be below 17 ℃, and the daylight should be less than 12 hours. Flower bud differentiation can occur when the temperature is between 10 ℃ and 24 ℃ and under short daylight conditions of less than 12 hours. When the temperature reaches 30 ℃, neither long nor short daylight conditions can induce flower bud differentiation. When the temperature is below 10 ℃, flower bud differentiation can occur regardless of long or short daylight conditions. However, when the temperature drops below 5 ℃, flower bud differentiation is inhibited. Therefore, for strawberry flower bud differentiation, temperature has a greater influence than daylight conditions. The required temperature range for strawberry flower bud differentiation is 5 – 27 ℃, with an optimal temperature around 15 ℃. Within the required temperature range, low temperatures promote flower bud differentiation, while high temperatures delay it. During the late stage of flower bud differentiation, if the temperature drops below 5 ℃, after a period of physiological dormancy, strawberries can be induced to flower and fruit by providing protective facilities and increasing temperature or supplementing light during winter and spring.

3. Requirements of Strawberries for Environmental Conditions

010 What are the soil requirements for strawberries?

Strawberries prefer well-drained, loose, and fertile sandy soils with good organic matter content that can retain water and nutrients. The pH of the soil is crucial, with an optimal range of 5.5–6.5. When the organic matter content is high(above 1.0%), strawberries can also grow well in soils with a pH range of 5–7. However, when the pH exceeds 8, plant growth is poor, and the plants may develop leaf scorching and eventually die. The ideal groundwater level is below 1 meter. Strawberries are not suitable for cultivation in saline-alkali soils, marshy areas, waterlogged depressions, calcareous soils, heavy clay soils, or sandy loam soils. On heavy clay soils, strawberries tend to have darker fruit color, sour taste, lower quality, and a maturity period that is 2–3 days later than that of sandy soils.

011 What are the water requirements for strawberries?

Strawberries have large leaves, frequent leaf turnover between old and new leaves, high transpiration rates, shallow root systems, high water content in the berries, and abundant runner and sucker formation. These factors indicate that strawberries have high water requirements. Insufficient water during the transplanting stage can affect seedling survival and establishment. Water deficiency during the seedling and bud initiation stages can hinder normal stem and leaf growth. Water shortage during the flowering stage can shorten the flowering period and affect normal flower opening. Water scarcity during the fruiting stage can hinder fruit development and enlargement, leading to reduced yield and quality. Water shortage during the propagation stage can affect runner formation and young plant rooting, resulting in reduced quantity and quality of seedlings. The water requirements of strawberries vary depending on the phenological stages. In early spring, when strawberries are sprouting, the soil moisture content should be around 80% of the field's maximum water-holding capacity. During the fruit ripening period, water should be controlled appropriately. After harvesting, additional irrigation is needed to promote runner and daughter plant formation. In hot summer, when plants are in a dormant or slow growth phase, excessive watering should be avoided, and the soil should be kept from drying out. During the vigorous growth phase in early autumn, sufficient water supply should be ensured. During the flower bud differentiation period, water should be moderately controlled, with the soil moisture content around 60% of the field's maximum water-holding capacity. Before winter, sufficient water should be irrigated to prepare for freezing and combined with cold protection measures.

Strawberries are not tolerant of waterlogging. Excessive soil moisture and poor drainage can lead to lignification of the primary roots, yellowing of leaves, diseases, and fruit rot. Prolonged waterlogging can even cause plant suffocation and death. Therefore, drainage should be considered during the rainy season. Excessive irrigation should be avoided, and techniques such as furrow planting, raised beds, drip irrigation, or seepage irrigation can be employed to prevent waterlogging.

012 What are the temperature requirements for strawberries?

Strawberries have a strong adaptability to temperature and exhibit a certain degree of cold tolerance. However, they are not tolerant of extremely cold or hot conditions and prefer a relatively warm climate. The root system of strawberries becomes active at 2 ℃, while the aboveground growth begins at 5 ℃. The optimal temperature for root growth is between 15 ℃ and 23 ℃, and for overall plant growth, it is between 20 ℃ and 25 ℃. When exposed to temperatures as low as −7 ℃, strawberries may suffer from frost damage, and at −10 ℃, most plants may be killed. Strawberries that have undergone cold acclimation during the autumn can tolerate temperatures as low as −8 ℃ for the roots and −15−−10 ℃ for the buds. Even in regions with temperatures as low as −40 ℃, strawberries can be cultivated with winter cover protection. Early-maturing varieties generally have less cold tolerance than late-maturing ones in early winter, while the situation is reversed. The temperature during the flowering period should be above 5 ℃, with around 25 ℃ being the most suitable. Temperatures below 0 ℃ or above 40 ℃ can affect pollination, fertilization, and result in malformed fruits. The temperature should be between 5 ℃ and 17 ℃ for proper flower bud differentiation. In the summer, when temperatures exceed 30 ℃, combined with drought and intense sunlight, strawberry growth is severely inhibited, making it difficult to develop new leaves, and older leaves are prone to scorching or browning. Measures such as watering, shading, or planting tall-stemmed crops on ridges can be employed to reduce the temperature. In autumn, after experiencing multiple light frosts or cold spells, the plants enter a dormant state with enhanced cold resistance.

013 What are the light requirements for strawberries?

Strawberries are photophilic plants but can tolerate partial shading. Even the overwintering leaves under cover can maintain their green color and photosynthetic function in the following spring. Interplanting strawberries with other crops in young orchards, despite partial shading, can still promote vigorous growth and flower bud formation, resulting in good yields. Excessive planting density, excessive shading, or insufficient sunlight in the field can lead to elongated petioles, lighter leaf color, elongated flower stalks, smaller or non-opening flowers, smaller fruits, increased acidity, delayed maturation, lighter color, and poor quality. Insufficient light in autumn can affect flower bud differentiation, weaken the plants, reduce nutrient reserves, decrease cold resistance, and even cause winter death. However, excessive light combined with drought and high temperatures can result in poor plant growth, smaller leaves, and weakened root systems. In severe cases, plant death can occur. Under open-field conditions, strawberry plants are shorter, sturdier, and produce smaller, deep red, shiny fruits with higher sugar content and a strong sweet aroma. Strawberry runners exhibit clear positive phototropism and tend to extend towards areas with better sunlight. The light requirements of strawberries vary during different growth stages. During the flowering and fruiting period, they require 12–15 hours of long-day photoperiods. During the flower bud formation period, shorter photoperiods of 10 – 12 hours and lower temperatures are required. Artificially providing more than 16 hours of long-day photoperiods can lead to poor flower bud formation and even prevent flowering and fruiting. After flower bud differentiation, long-day photoperiods are needed.

014 What are the different methods of strawberry propagation, and what are their characteristics?

There are five methods of strawberry propagation: runner propagation, transplanting and root division, mother plant division, seed propagation, and tissue culture propagation. Runner propagation is the most commonly used method in production due to its simplicity, high propagation rate, and the ability to produce 50 000–100 000 runner plants per 667 square meters of land per year. Runner plants are vegetative offspring that maintain the characteristics of the original variety and can start fruiting as early as the second year when planted in the autumn. Protected cultivation can result in even earlier harvests, potentially yielding profits in the same year.

Transplanting and root division is an important method for producing high-quality seedlings in facilities.

Mother plant division, also known as "stooling", involves separating new stems with new roots from older plants for planting. This method is less commonly used in production and is mainly employed when changing planting locations or when propagating varieties with low runner production. The propagation rate is relatively low.

Seed propagation involves sowing strawberry seeds to produce seedlings. This method is rarely used in production and is primarily employed for breeding new varieties or when introducing varieties from distant or foreign sources or when accelerating the propagation of superior varieties.

Tissue culture propagation involves using the apical meristem of strawberry stems (or other tissues) to produce tissue culture plantlets in vitro. This method has a high propagation rate, with a single strawberry plant capable of producing tens of thousands or even hundreds of thousands of tissue culture plantlets within a year. It allows for industrial-scale production and the cultivation of disease-free strawberry plants. This method has emerged in recent years and is employed to varying degrees in different locations for strawberry propagation.

015 How can strawberry seedlings be propagated using runner propagation?

Propagating strawberry seedlings through runner propagation can help avoid soil-borne diseases, produce robust seedlings, and achieve high yields. Runners generally start to emerge when the fruits are ripe or after fruit harvest. The ideal conditions for runner production are 12 – 16 hours of sunlight, temperatures above 14 ℃, and the plants being completely out of dormancy. To promote runner production and increase the propagation rate, it is recommended to remove flower buds and spray a 0.001% concentration of gibberellic acid solution on the plants when they have developed three to four new leaves (typically in mid-May). This treatment significantly increases the number and quality of runners.

Currently, people often turn the fruit garden into a nursery field for propagating seedlings after the berry harvest. They remove rows alternately, dig out sick, weak, mixed, and dense seedlings, fertilize, loosen the soil, rake it flat, and leave a certain space to facilitate the growth and rooting of stolons.

To ensure the quality of the stolon seedlings, it is best to establish a dedicated strawberry parent garden and propagation nursery, so that fruiting and seedling cultivation do not affect each other. The mother plant field and propagation bed should be established on loose and fertile soil, and high-quality and vigorous seedlings should be planted. The spacing between plants and rows is generally 100 cm × 100 cm or 50 cm × 100 cm. Proper management, including timely removal of flower clusters, fertilization, irrigation, loosening of the soil, and weed control, is necessary. When the runners hang down to the ground, they should be guided to spread evenly in the rows between the mother plants, and the stems should be fixed in the soil using straw or other materials. Alternatively, soil can be pressed onto the second and fourth nodes of the runners to promote the formation of adventitious roots and the early development of robust seedlings. Excessive or weak runners should be removed in a timely manner to concentrate nutrients. After the daughter plants have developed three to four compound leaves, they should be separated from the mother plants. The mother plant field should be relocated after approximately three years. Generally, only five to six good runners and 2–3 seedlings near the mother plant are retained.

In situations where there is a shortage of high-quality mother plants, propagation can be achieved by pressing the runners onto nutrient bowls. During the peak runner production period, flower pots with a diameter of 15–20 cm filled with fertile soil are buried around the mother plants. The runners' leaf clusters are pressed into the potting soil, kept moist, and once new roots have formed, the pots are transferred to the mother plant field. This method allows for the continued production of runners and expansion of propagation.

To accelerate the propagation of superior seedlings, leaf clusters with at least two normal leaves but without roots can be cut and inserted into water. The base of the leaf cluster should touch the water surface, and the water should be changed every other day. After the leaf cluster develops 5 – 6 roots, it can be planted in the field. In locations with suitable conditions, mist propagation can be used. This involves inserting the rootless leaf clusters into a sand bed or tray in a greenhouse or plastic tunnel equipped with misting equipment. The temperature should be maintained between 14 ℃ and 20 ℃. After approximately 10 days, when the leaf clusters have developed roots, they can be transferred to small pots or containers with a diameter of 10 cm for further growth before transplantation or direct planting. Shading should be provided during the rooting period.

016 What are the benefits of transplanting and root division, and how is it performed?

Transplanting and root division are important methods for producing high-quality seedlings in facilities.

Transplanting and root division involve cutting a certain number of seedling roots to reduce nutrient and water absorption, suppress vegetative growth, promote early flower bud differentiation, and accelerate flower bud development. However, excessive or improper root division can lead to poor plant growth, weak vigor, and reduced yields, even if flower bud differentiation occurs earlier.

Transplanting and root division include seedling collection and false planting. Seedling collection and false planting are typically performed in mid-July. Healthy runner seedlings with three leaves are selected, graded, and cleaned. Disease-free and old leaves are removed, and the seedlings are cut from the mother plants, leaving a 3 cm segment of the runner attached. The seedlings are then immersed in water for root soaking. After soaking, they are false planted on raised beds with a spacing of 15 cm × 15 cm. During false planting, the larger seedlings should have the runner segment exposed, while smaller seedlings have the runner segment inserted into the soil at an angle. The seedlings are covered with cold-resistant cloth or similar materials. Regular watering is done before evening to maintain a certain level of humidity. After about 10 days, the covering can be removed. During the false planting period, underground pests should be controlled.

After the seedlings have successfully established, root division can be performed. Using a small shovel, the soil around the seedlings is cut, and the seedlings with soil clumps are removed. Then they are transplanted at a spacing of 15 cm, one after another. The gaps between the seedlings are filled with soil, and old leaves and runners are removed. The seedlings should be watered thoroughly the day before transplantation to prevent soil clumps from breaking apart. Transplanting and root division can be performed every 10 days from late August to early September, for a total of 2–3 times. In late September, the seedlings with soil clumps are planted in a greenhouse for accelerated cultivation, which often yields good results. This method is more suitable for varieties with deep dormancy. Mulching for insulation can begin in mid-October, and the ripe fruits can be harvested in late December. Transplanting and root division can increase the yield by more than 85% compared to conventional seedling production methods on an area of 667 square meters.

017 How can strawberry seedlings be propagated using mother plant division?

Mother plant division is suitable for renewing strawberry fields or for varieties that do not produce runners easily. Generally, after fruit harvest, when the old plants have developed new leaves and new roots underground, they are dug out. The old and blackened rhizomes and black roots are pruned, and 1–2 years old rhizomes with 5–8 healthy leaves and shallow yellowish robust adventitious roots are separated and planted individually. This propagation method has the advantages of fast establishment, high survival rate, and the ability to produce normal yields in the second year. It also saves costs and labor by not requiring dedicated propagation fields. The disadvantage is a relatively low propagation rate, with 3-year-old mother plants typically producing only 8–14 viable daughter plants for transplantation. During the separation of daughter plants, larger wounds may occur, increasing the risk of soil-borne diseases. If some daughter plants have leaves but no roots, only a few leaves should be retained, and they can be inserted into a shaded rooting bed for careful management to promote root development and leaf growth. These plants can still be used for planting in the autumn. Another option is to dig out the mother plants with soil clumps after fruit harvest, plant them with wider spacing or two plants per hole, and allow them to produce new stems and runners. The number of flower clusters generally increases by one-third, resulting in significantly higher yields. After the second year's harvest, the 3-year-old plants are removed, and the 2-year-old plants that have produced fruits can be used for seedling production.

018 How can strawberry seedlings be propagated using seed propagation?

Seed propagation of strawberry seedlings is not commonly used in production. However, it can be employed when introducing varieties from abroad or distant locations, accelerating the propagation of superior seedlings, or for hybrid breeding purposes.

In natural conditions, strawberries primarily undergo self-pollination, and their offspring inherit the characteristics of the mother plants. Seedlings produced from seeds have well-developed root systems, are less prone to aging, and exhibit strong adaptability and resistance. They are generally free from viruses or carry very few viruses, enhancing the vitality of the seedlings. Seedlings typically begin fruiting within 10 – 15 months. Approximately 10 grams of seeds can be obtained from every kilogram of fresh strawberry fruits. The seeds can be collected by scraping them from the fruit surface or by placing the whole fruit in a cloth bag, squeezing out the juice, rinsing the seeds with water, and drying them on paper. Alternatively, the ripe fruits can be rubbed in a basin to remove the flesh and impurities, leaving only the seeds, which can then be dried. Strawberry seeds do not have a distinct dormancy period, and they can germinate well immediately after collection or after a certain period of time. As long as the conditions are suitable, seeds can be sown throughout the year. However, the highest germination rate is achieved when seeds are sown immediately after collection. Prior to sowing, the seeds can be soaked in water for a day and then subjected to a low-temperature treatment at 0–3 ℃ in a refrigerator for about two weeks. This treatment improves the germination rate. Strawberry seeds should be stored in a cool, dry place under room temperature conditions, and their germination capacity can be maintained for up to two years.

Strawberry seeds are small and should be sown finely. Potting soil that has been sieved finely should be used. Before sowing, the soil should be thoroughly watered, and the seeds should be evenly scattered on the soil surface. A layer of fine sand, about 2–3 mm thick, should be spread over the seeds, and a plastic film should be placed on top to maintain a certain temperature and humidity. If the potting soil becomes dry, it can be sprayed with water using a sprayer, or the sowing container can be placed in a water tank for absorption to keep it moist. Germination usually occurs within 1–2 weeks. When the seedlings have grown 1–2 true leaves, they can be separated. Each nutrient bowl should contain only one seedling. After separation, the seedlings can be placed on a seedbed and transplanted when they have 3–4 true leaves. They can be transplanted to the field or propagation field. If sown in spring or summer, they can be transplanted in autumn, while those sown in autumn can be transplanted in the following spring.

019 How can strawberry seedlings be propagated using tissue culture propagation?

Tissue culture propagation involves using the apical meristem of strawberry stems to induce the growth of young shoots under sterile conditions on a culture medium. The shoots are then multiplied through axillary bud proliferation to rapidly increase the number of plantlets in vitro. After undergoing a certain period of acclimatization and conditioning, the plantlets can be transferred to the field. The main advantages of tissue culture propagation are its fast propagation rate, high multiplication coefficient, ability to facilitate large-scale production, and the production of disease-free plantlets. The plantlets produced through tissue culture have well-developed root systems and vigorous growth, resulting in an average yield increase of over 15%. Tissue culture propagation does not occupy land and is not affected by environmental conditions. It can also be used to produce disease-free seedlings, achieving the goals of stronger seedlings, higher yields, and better quality.

Tissue culture propagation of strawberries involves preparing a culture medium, sterilization and inoculation, subculture, and plantlet transfer. The method starts by inoculating the apical meristem of the strawberry stem onto a culture medium under sterile conditions. Under specific temperature, humidity, and light conditions, the culture is induced, and the shoots are continually transferred and multiplied. Once a large number of rootless plantlets are obtained, they can be removed from the sterile environment and planted in pots containing moist perlite. The pots should be covered with plastic film, and the temperature should be maintained above 10 ℃. After about two weeks, the plantlets will root and become seedlings. Alternatively, the induced rooting plantlets can be washed to remove the culture medium and planted in pots or containers filled with sandy loam or humus soil. The pots are placed in a greenhouse and covered to maintain humidity for cultivation, which improves the survival rate.

020 How can we prevent the re-infection of viruses when planting virus-free strawberry seedlings?

If virus-free strawberry seedlings are planted in an area infected with viruses, there is a risk of re-infection. Therefore, the following measures can be taken to prevent it.

(1) Keep a distance from virus-infected old strawberry fields. Newly established virus-free strawberry nurseries should be at least 1 500 meters away from old strawberry fields.

(2) Avoid consecutive cropping or continuous cropping with plants from the Solanaceae family to prevent soil contamination and the transmission of pathogens through crop residues or nematodes.

(3) Control aphids effectively, especially the strawberry aphid. The period when stolons are vigorously growing is the main time for aphid transmission, so aphid control should be taken seriously.

(4) Cultivate virus-free mother plants in a greenhouse or cover them with a mesh net (20–25 mesh) to prevent aphids. Covering the ground with silver reflective film can also deter aphids.

In addition, it is recommended to conduct soil disinfection in the virus-free strawberry nursery before planting. The soil should be loose and fertile, and the planting density should not exceed 1 500 plants per 667 square meters. After three years of field cultivation, the seedlings should be replaced. After planting, it is important to ensure adequate irrigation and maintain an air humidity of around 80% and a temperature of around 22 ℃. The plastic film can be removed after 7 days, and the survival rate can reach nearly 100%.

4. Orchard Establishment and Planting

021 How many planting systems are there for strawberries, and what are their characteristics?

Strawberries are perennial herbaceous plants with a lifespan of usually over 5 years, with some plants living up to 10 years. The highest yield is generally achieved in the first or second year. After flowering and fruiting, stolons are produced, which can be used for propagation. In European and American countries, strawberries are typically renewed after 2–3 years of fruiting. In Japan, stolons are taken every year for replanting. In China, the commonly used planting systems for strawberries are the 3-year planting system and the 1-year planting system.

The choice of planting system should be based on factors such as cultivation type, site conditions, and economic benefits. With the 3-year planting system, the strawberry yield usually decreases significantly in the third year. However, both the 2-year planting system and the 1-year planting system can achieve high yields and profits with proper management.

In some areas, the 2-year-old strawberries may exhibit smaller leaves, smaller fruits, and lower yields. This is often due to poor management of the 1-year-old strawberries during the period between fruit harvest and winter.

022 How to choose a strawberry production site?

When planting strawberries, it is important to choose a site that is flat, sunny, with fertile and well-drained soil, and convenient for irrigation and drainage. In areas with a low water table, planting can be done in trenches or ditches. On sloping terrain, terraces or contour planting can be used. When intercropping or rotating with other crops, the previous crop should not have common diseases with strawberries. Grape vineyards infested with nematodes and orchards with removed old trees without soil disinfection are not suitable for strawberry cultivation. Due to the concentrated harvest period of strawberries, when establishing a commercial base, the planting area should be reasonably arranged according to local labor conditions to avoid unnecessary economic losses.

023 How to manage the soil in a strawberry field?

Before planting strawberries, the soil should be deeply plowed to a depth of about 30 centimeters. The soil should be compact and level to prevent the seedlings from sinking after watering, which could affect their survival. If the field has excessive weeds, about half a month before plowing, spray 0. 5 kilograms of 10% glyphosate mixed with 50 kilograms of water per 667 square meters to control the weeds. After the weeds have withered, the soil can be plowed. During plowing, incorporate base fertilizer. Generally, apply more than 5 000 kilograms of well-rotted organic fertilizer per 667 square meters, along with 50 kilograms of calcium superphosphate and 50 kilograms of potassium chloride or a compound fertilizer containing nitrogen, phosphorus, and potassium. If the soil is deficient in certain nutrients, appropriate micronutrient fertilizers should also be applied. Since strawberries have a high planting density and a short growth cycle, with sufficient base fertilizer, additional fertilization in the second spring can meet their growth and fruiting requirements. For strawberry fields with consecutive years of cultivation, it is important to ensure sufficient base fertilizer. The base fertilizer should be evenly spread across the entire field and then plowed to ensure thorough mixing of the fertilizer with the soil.

Early plowing is recommended, preferably before the winter, to allow the soil to mature. After plowing, create ridges according to the planting requirements. In northern regions, flat planting is generally used, with ridge lengths of 10－20 meters, ridge widths of 1. 2－1. 5 meters, and ridge heights of about 15 centimeters. Flat planting has the advantage of convenient irrigation, cultivation, fertilization, and cold protection. However, the disadvantage is that the ridge surface is not easy to level, and uneven watering may lead to excessive humidity and poor ventilation in certain areas, causing fruit rot due to waterlogging. In areas with abundant rainfall and high water tables, raised bed cultivation is recommended. The ridge height should be 30 centimeters, ridge surface width should be 1－1. 3 meters, ridge base width should be 1. 4－1. 7 meters, and the trench width should be 40 centimeters. When using plastic mulch, the ridge width should be reduced to 70－75 centimeters. Raised bed cultivation has the advantages of convenient irrigation and drainage, maintaining soil looseness, good ventilation, better fruit coloring, high fruit quality, and reduced soil contamination. The disadvantage is that it is susceptible to wind and frost damage, and sometimes inadequate water supply may occur. In northern regions, raised bed cultivation can also be practiced, with slightly lower and narrower ridges. After creating the ridges, water them lightly or compact them appropriately to make the soil compact.

024 How to select strawberry varieties and arrange their combination in a strawberry field?

Strawberries can undergo self-pollination and fertilization, but cross-pollination tends to increase yields. Therefore, in addition to the main cultivated varieties, it is important to select suitable pollination varieties in a strawberry field. For example, when the main cultivated variety is "Baojiao Early", suitable pollination varieties can include "Chunxiang" "Mingbao", and "Mingjing". Generally, for each main cultivated variety, 2-3 pollination varieties should be selected. The main cultivated variety should not exceed 70% of the total planting area, with the remaining area dedicated to pollination varieties. To extend the supply period of strawberries, it is important to pay attention to the combination of early, mid-season, and late-maturing varieties. When planting on a large scale, it is recommended to have at least 3-4 different varieties. The distance between the main cultivated variety and the pollination varieties should generally not exceed 30 meters. The same variety should be planted together for easier management and harvesting. In large planting areas with uneven terrain, early and mid-early varieties should be planted in higher areas. Higher ground temperatures in the spring promote early root activity and reduce the risk of late spring frosts during the flowering period.

When selecting strawberry varieties, factors to consider include:

(1) Regional adaptability. Varieties suitable for cold regions in the north generally have a longer dormancy period. When grown in warmer regions, these varieties may exhibit characteristics similar to everbearing strawberries, with fewer stolons and a longer flowering and fruiting period. Therefore, in southern regions where the latitude is further south, it is important to select varieties that require a shorter chilling time and have a shallower physiological dormancy. In the north, varieties resistant to late spring frosts should be chosen, while in the south, varieties resistant to high summer temperatures and drought should be selected. Additionally, the chosen varieties should have strong resistance to common local diseases and pests.

(2) Cultivation purpose. Different varieties have different requirements for fresh consumption and processing. Even for varieties suitable for both fresh consumption and processing, different processing products may have specific requirements for the variety. Therefore, the selection of strawberry varieties should be based on the cultivation purpose.

(3) Planting method. Different planting methods, such as protected cultivation and open-field cultivation, have different requirements for varieties. For potted ornamental strawberries, it is preferable to choose smaller plant types, such as everbearing varieties.

025 How to prepare strawberry seedlings for planting?

Strawberry seedlings for planting, also known as stolon seedlings, should be free from diseases and pests, have plenty of new roots, and have a stem thickness of at least 1 centimeter. They should have at least 4 fully expanded leaves, a well-developed central bud, short and sturdy petioles, dark green leaves, and a fresh weight of about 30 grams per plant, with the underground root system accounting for about one-third of the total plant weight. Seedlings with excessively long petioles should not be used for planting. New stolon seedlings propagated from division of old plants must have plenty of new roots.

Before planting, remove yellow, old, and diseased leaves, leaving 2–3 crown leaves. If the planting location is nearby, it is best to plant the seedlings immediately after digging them up. For dug-up seedlings, keep the roots moist by lightly sprinkling water to prevent them from drying out. However, the roots should not be soaked in water for a long time. For seedlings transported over long distances, remove the soil, trim the lower leaves, bundle them in groups of 50 plants, soak the roots in water, and then place them in wet sacks, grass bags, or plastic bags for transportation using baskets or boxes. When seedlings are supplied from other regions, it is necessary to prepare the planting site in advance. After the seedlings arrive, soak the roots in water or dip them in mud slurry, and then place them in a shaded area for immediate planting.

026 How to determine the planting period for strawberries?

The planting period for strawberries should be determined based on factors such as the harvest time of the previous crop, the growth status of the seedlings, local temperature and humidity conditions, and whether the seedlings have sufficient time for growth and development after planting. Generally, planting is done in the autumn. Autumn planting has a longer duration, with a large supply of stolon seedlings in the same year, good soil moisture, high air humidity, and a short establishment period, resulting in high survival rates. The recommended temperature for planting is around 15–20 ℃. In regions such as Shandong, Hebei, and southern Liaoning, planting is recommended in mid to late August. In the Shanghai and Hangzhou area, planting is suitable in early to mid-October. Planting too late can result in a short growth period, preventing the seedlings from becoming robust before winter and affecting the yield in the following year. Spring planting often utilizes overwintered seedlings, which may suffer root damage and have lower individual plant yields compared to autumn-planted seedlings. Spring planting should be carried out as early as possible after the soil thaws. For cold-stored seedlings, the planting period can be calculated based on the expected harvest time, usually around 60 days before that time.

027 What are the main planting systems for strawberries?

The commonly used planting systems for strawberries include the following three aspects.

(1) Planting with fixed plants. This method involves planting with a specific plant and row spacing. Before the fruits ripen, new stolons that emerge from the plants are continuously removed to concentrate nutrients, thereby increasing yield and quality. After fruit harvest, the old plants are retained while the newly grown stolons are removed. After the second year of fruiting, the stolon seedlings are retained, and the old plants are removed, renewing the plants in the same location. This system ensures stable yields without changing the planting site.

(2) Carpet planting. In this method, larger plants are planted with wider plant and row spacing. The stolon seedlings that emerge from the plants root and grow between the rows until the entire field is covered, forming a carpet-like cover. Alternatively, the stolons can be allowed to root and grow within a predetermined area, forming a strip-like carpet. This planting method can be used when there is a shortage of seedlings or limited labor availability. In the first year, the number of seedlings may be insufficient, resulting in lower yields, but higher yields can be obtained in the second year.

(3) Ridge planting. This method involves planting on raised ridges, which facilitates the use of plastic mulch. In some areas, the seedlings are planted in furrows, and irrigation and fertilization are carried out in the furrows. Organic fertilizer can be applied in the furrows in spring or autumn, and the effect is also good. However, planting in furrows makes the fruits more prone to soil contamination.

028 How to determine the planting density and orientation of strawberries?

The planting density of strawberries should be determined based on the planting method, soil fertility, and variety characteristics. For the 1-year planting system, smaller plant and row spacing are suitable, while for the multi-year planting system, the plant and row spacing should be increased appropriately. Varieties with smaller plant sizes, such as "Gorella", can be planted at higher densities. In general, for flat planting with a width of 1.2–1.5 meters, there should be 4–6 rows per bed, with a row spacing of 20–25 centimeters and a plant spacing of 15–20 centimeters. In northern regions, low ridges are used for planting, with a ridge height of 15 centimeters, ridge width of 50–55 centimeters, and trench width of 20–25 centimeters. The plant spacing can be reduced appropriately, with a planting density of around 10 000 plants per 667 square meters. In protected cultivation, the planting density can be increased accordingly.

The inflorescences of strawberries emerge from the new stems in a certain pattern. Generally, the new stems are slightly arched, and the inflorescences extend from the side facing the arch. To facilitate fruit setting and harvesting, the inflorescences of each strawberry plant should extend in the same direction. When planting, the arch of the new stem should face a fixed direction. For flat planting, the direction of the inflorescence in the border rows should face inward to avoid the inflorescences extending onto the edge of the bed, which would affect operations. For ridge planting, the arch of the new stem can face the trench to facilitate harvesting.

029 How to determine the planting depth of strawberries, and how to plant them?

The planting depth of strawberries is crucial for their survival. Planting too deep can bury the crown of the seedling, leading to rotting. Planting too shallow can expose the roots, making it difficult for new roots to form and resulting in the drying out of the seedling. The appropriate planting depth is when the stem of the seedling is level with the ground. Uneven ridge surfaces or loose soil can cause the seedlings to sink or become waterlogged after watering, reducing their survival rate. Therefore, special attention should be paid to soil preparation quality before planting. During planting, strive to achieve a depth that is neither burying the crown nor exposing the roots. It is especially important to prevent excessive planting depth.

When planting strawberries, place the roots in the planting hole and fill it with fine soil. Firmly press the soil around the roots and gently lift the seedling to ensure good contact between the roots and the soil. Immediately after planting, water the seedlings thoroughly to help them establish roots. For seedlings with exposed roots, sinking crowns, or inflorescences not aligned with the planned direction, adjust or replant them promptly. If any seedlings are missed during planting, they should be replanted in a timely manner to ensure full and robust seedling establishment.

030 How to improve the survival rate of strawberry seedlings?

To improve the survival rate of strawberry seedlings, the following measures can be taken.

(1) Treatment with rooting agents. Before planting, soak the seedlings in a rooting agent for 2–6 hours to promote root formation.

(2) Removal of withered leaves and black roots. Before planting, remove some of the old leaves and black roots from the seedlings to reduce leaf surface area, decrease plant transpiration, and promote the development of new roots.

(3) Planting on cloudy days or during the morning or evening. Planting on cloudy or rainy days with high humidity and low leaf transpiration can accelerate seedling establishment and improve survival rates. In cases of excessive rainfall or heavy rain, proper drainage should be ensured to prevent the seedlings from being waterlogged. When planting on sunny days, it is advisable to do so in the early morning or evening.

(4) Shade cover. On sunny days with intense sunlight after planting, besides supplementing water, shade can be provided using reed screens, plastic mesh nets, or branches with leaves. If conditions permit, temporary small shelters can be created using shade nets or green or silver plastic films. After the seedlings have established and acclimated, they should be exposed to sunlight in a timely manner and proper ventilation should be maintained to prevent scorching of the seedlings when removing the shade covers.

(5) Timely watering. Immediately after planting, thoroughly water the seedlings. In the first three days after planting, water lightly every day. After 4-5 days, switch to watering every 2 - 3 days. Avoid excessive watering, as it can lead to poor soil aeration, root respiration problems, root rot, and seedling death. After the seedlings have established, they can be allowed to dry out appropriately.

(6) Soil clump planting. For close-distance planting, the seedlings can be planted with soil clumps to shorten the establishment period and improve survival rates. Clump planting is an effective measure to improve survival rates in areas with limited seedling resources, poor irrigation conditions, or sticky or acidic/alkaline soil.

5. Management of Open-field Strawberry Plantations

031 How to perform intertillage, mulching, and soil cultivation in a strawberry field?

Strawberries have shallow roots and prefer moist and loose soil. After planting and removing winter protection in early spring, multiple intertillage operations should be carried out. Intertillage is generally done at a depth of 3 – 4 cm after rainfall. Intertillage can be combined with topdressing, soil cultivation, removal of runners, and removal of weak and diseased plants. In areas with dry climate and poor irrigation conditions, ground cover can be used during the strawberry growing season. In areas with good irrigation conditions, dense planting, and plastic mulch, ground cover is generally not necessary. It has been proven that covering strawberries with plastic mulch can increase early yields by 1.7 times. Black plastic mulch is preferred for ground cover. In northern regions prone to late spring frosts, covering with plastic mulch can advance the flowering period, making the plants more susceptible to frost damage. In southern regions, where the fruit ripening period coincides with the hot season, plastic mulch cover can lead to fruit scorching and rot. Therefore, when using ground cover, attention should be paid to avoiding unfavorable conditions.

032 How to fertilize strawberries?

Strawberries have high nutrient requirements and require substantial amounts of fertilizer. Nitrogen promotes the growth of strawberry stems and leaves and facilitates the development of flower buds, inflorescences, and fruits. Phosphorus promotes flower bud formation and improves fruiting capacity. Potassium promotes fruit development and ripening, increases sugar content, and improves fruit quality. Since strawberries undergo both vegetative growth and abundant flowering and fruiting within a year, they require sufficient supplies of nitrogen, phosphorus, potassium, and other nutrients. Specific fertilization methods are described as follows.

(1) Base fertilizer. Applied before planting, it includes the application of 2 500 kg or more of chicken manure or 5 000 kg of high-quality stable manure per 667 square meters.

Phosphorus and potassium fertilizers can be added together with organic fertilizers.

For fields with a two-year or three-year planting system, after fruit harvest and before the second vigorous growth of stems and leaves, combined with weeding and soil cultivation, apply well-rotted organic fertilizers.

(2) Topdressing. On the basis of sufficient organic fertilizers, topdressing is generally performed during the early growth period, vigorous growth period after harvest, and flower bud differentiation period. From late March to early April, apply 10 – 15 kg of urea or compound fertilizer per 667 square meters to accelerate plant growth, promote leaf expansion, and facilitate flowering and fruit set. From late August to early September, apply 10–20 kg of urea or compound fertilizer per 667 square meters to promote flower bud differentiation, ensure vigorous growth, and enhance winter hardiness. Post-harvest topdressing mainly aims to rejuvenate growth and propagate runner plants.

(3) Foliar fertilization. It has been proven that spraying 0.4% urea solution several times during the flowering period can increase fruit yield. In practice, spraying 0.3%–0.5% potassium dihydrogen phosphate solution, 2% calcium superphosphate soaking solution, and 0.3% borax solution all have significant yield-increasing effects. Foliar fertilization is particularly important during the bud differentiation, flowering, and flower bud differentiation periods. The optimal time for spraying is around 4:00 p.m. –5:00 p.m.

033 How to irrigate strawberries?

The soil in strawberry fields should be kept consistently moist. In addition to combining irrigation with topdressing, timely watering should be carried out during periods of drought. During the fruit enlargement period, strawberries require the most water, and maintaining sufficient water supply is crucial for high yields. In early spring, if the weather is not excessively dry, watering in the evening is appropriate. The amount of water should not be excessive to avoid lowering the soil temperature. In late autumn, water should be controlled appropriately to prevent excessive vegetative growth. Before the soil freezes, one round of freeze-proof water should be applied to facilitate safe overwintering. The best time for watering is in the morning and evening. Drip irrigation or seepage irrigation, if available, is more beneficial for the growth and development of strawberries and can reduce fruit rot. During the rainy season or in low-lying areas with high water tables, special attention should be paid to drainage and flood prevention.

034 What are the plant management practices for strawberries? How are they carried out?

(1) Thinning. For strawberries, the first flowers to open at lower positions usually

produce the best fruits, which are larger and ripen earlier. As the flower order increases, later flowers often fail to set fruits. Even if they do, the fruits are too small to be of value, resulting in ineffective fruit production. Therefore, it is necessary to thin out these higher-order flowers(i. e., ineffective flowers) to reduce nutrient consumption.

(2) Leaf removal. Since new leaves continuously emerge throughout the year for strawberries, it is necessary to remove old leaves, residual leaves, and diseased leaves to reduce nutrient consumption.

(3) Fruit elevation. Strawberry plants are short, and after flowering and fruit set, the fruits can hang down and touch the ground, leading to soil contamination and susceptibility to diseases and pests. It can also result in uneven fruit coloring and affect fruit quality. Therefore, it is best to use plastic mulch or spread chopped straw or wheat straw around the plants to elevate the fruits.

(4) Removal of runners. The purpose of removing runners is to prevent excessive nutrient consumption by runners, improve light conditions, promote flower bud differentiation, enhance winter hardiness, and increase fruit yield and quality. Removing runners can increase yields by an average of about 40%. Runners usually develop in large numbers after fruit harvest, and unnecessary runners should be removed in a timely manner according to the requirements for propagating new plants. Manual removal can be labor-intensive, so growth regulators such as ethephon, daminozide, or paclobutrazol can be used to inhibit runner formation. Removing runners can be combined with intertillage, soil cultivation, and removal of old(diseased) leaves.

(5) Plant spacing. For fields with a 2-year or 3-year planting system, after fruit harvest, thinning should be carried out to leave a certain number of new stem branches. Remove or transplant the excess seedlings. It is advisable to leave 3 – 5 new stem branches per plant.

035 Why is it necessary to cover strawberries with winter protection materials? How should it be done?

Strawberries are evergreen plants but are not cold-tolerant. In northern regions of China, winter protection measures must be taken during cultivation to preserve a sufficient number of green leaves for safe overwintering.

Practical experience has shown that covering strawberries with winter protection materials is an important measure for achieving high yields in open-field cultivation. Observations have indicated that covering strawberries with straw during winter results in more green leaves and inflorescences in early spring, leading to a significant increase in yield, generally more than 40% compared to the control group. In various regions of Shandong, strawberries can survive without straw cover during winter, but the green leaves mostly wither, which severely affects the yield in the following year.

Common materials for covering include wheat straw, rice straw, corn stalks, tree leaves, well-rotted horse manure, etc. The effect is even better when combined with plastic mulch. The covering period generally starts after soil freezing and sealing with water and when the soil has just frozen. In regions with severe cold and strong winds, windbreaks can be installed. The thickness of the cover is about 5 cm, and it should be slightly thicker in areas north of Shandong. In early spring, when the average temperature is above 0 ℃, the winter protection materials can be gradually removed, and the ground should be cleaned and loosened to retain moisture. In Shandong, freeze-proof water is usually applied in early November, straw cover is applied in late November, and timely removal is done in early March. For fields with plastic mulch cover, the covering materials above the mulch should be removed in a timely manner, and the mulch should be punctured to allow the seedlings to emerge.

036 What are the hazards of late spring frosts to strawberries? How to prevent frost damage in spring?

Strawberry plants are short and highly sensitive to late spring frosts. When the newly emerged but unexpanded young leaves are frozen, the leaf tips and margins turn black. Open flowers are more severely affected, with the pistils being the most vulnerable. The center of the flowers turns black and fails to develop into fruits. When the damage is mild, only some pistils change color due to freezing, resulting in the development of deformed fruits. Young fruits appear as oil stains after freezing. At −1 ℃, the damage to plants is mild; at −3 ℃, the damage is severe. If low temperatures persist for several hours during the flowering period, the damage becomes more severe, resulting in significant yield losses. Preventive measures should be taken in areas where strawberries are susceptible to late spring frosts. For example, selecting locations with good ventilation for strawberry cultivation, delaying the removal of winter protection materials to postpone the flowering period, and using frost-resistant varieties. If conditions permit, measures such as smoking or sprinkler irrigation can be employed to reduce frost damage.

Additionally, in certain regions, continuous high temperatures or drought during the summer can cause strawberries to wilt or even dry up. Therefore, protecting strawberries during the summer is also an important issue in production. Measures to prevent heat damage include keeping the soil moist, intercropping in young orchards, or interplanting tall-stem crops in strawberry rows, implementing a one-year planting system, and selecting heat-tolerant varieties.

037 How to clear and renew a strawberry field and perform simple soil disinfection?

After harvesting in a one-year planting system, the strawberry stems and leaves can be directly plowed into the soil as green manure. Generally, there are about 500 kg of fresh strawberry stems and leaves per 667 square meters. Fresh stems and leaves contain approximately 0.59% nitrogen, equivalent to applying 2.95 kg of pure nitrogen, as well as other nutrients such as phosphorus and potassium.

For fields with a multi-year planting system, due to heavy soil depletion, high incidence of diseases and pests, and a large amount of weed growth, it is necessary to collect and burn the strawberry stems and leaves. If switching to dryland crops, all strawberry root segments in the soil should be removed, and soil disinfection is recommended before sowing to eliminate pathogens.

Simple soil disinfection can be achieved through solarization, which involves creating long ridges in a curved shape (ridge width determined by the width of the plastic film) after soil plowing. Cover the ridges with plastic film, and solar radiation will raise the soil temperature above 60 ℃. After 3-4 weeks, high-temperature disinfection can be achieved. Chemical disinfection can be done by spraying a 1 000-fold dilution of methyl tolyl fluanid solution on the ground and then plowing to kill pathogens in the soil.

038 What are the main growth regulators used for strawberries?

The main growth regulators used for strawberries include the following three aspects.

(1) Gibberellins. Spraying 0.01% gibberellins on strawberries can inhibit dormancy and promote early ripening. The same concentration of gibberellins, when sprayed twice during the early growth period of strawberries, can increase the occurrence of runners. According to experiments conducted by Northwest Agricultural University, spraying 0.01% gibberellins during the flowering and fruiting period can increase strawberry yield and sugar content. Additionally, gibberellins can promote inflorescence formation, induce the development of parthenocarpic fruits, reduce losses caused by poor pollination, and advance the market availability of fruits.

(2) Paclobutrazol (PP333). This is a growth retardant that mainly inhibits the occurrence of strawberry runners and vegetative growth while promoting reproductive growth. When applied at appropriate concentrations and timing, it has a significant yield-increasing effect. The recommended concentration is 0.25%. The application period is during the early stage of runner development. Improper application that leads to excessive inhibition can result in yield reduction.

(3) Others. Spraying 0.02%-0.05% cytokinins during the flowering and fruit set period is beneficial for fruit set and early fruit enlargement. Root soaking with naphthylacetic acid (NAA) promotes rooting. Spraying with daminozide (MH) controls plant growth, and other growth regulators can be used as well. Ethephon can be used to induce concentrated ripening of strawberry fruits for mechanical harvesting.

6. Protected Cultivation

039 What is the significance of protected cultivation? What are the contents and types of protected cultivation for strawberries?

Protected cultivation refers to the cultivation of plants using cold protection and insulation facilities such as windbreaks, raised beds, hotbeds, plastic greenhouses, and tunnels during the cold winter and spring seasons when plants cannot grow in open fields. The purpose is to achieve early maturity, high yield, and extended supply periods. Protected cultivation for strawberries utilizes the short growth cycle, dwarf stature, and easy control of growth and development of strawberry plants based on the study of their physiological and ecological characteristics. It involves covering strawberry strong seedlings that have developed flower buds with plastic film, setting up plastic tunnels, constructing greenhouses, or even directly cultivating them in sunlight greenhouses. Measures such as temperature control, supplemental lighting, and application of gibberellins or other growth regulators can be employed to advance the harvest period of strawberries significantly and extend the supply period, laying the foundation for year-round strawberry supply.

Protected cultivation for strawberries has the characteristics of easy material availability, low investment, simple operation, high yield, and good economic benefits. The yield per 667 square meters is generally above 1 200 kg, and it can exceed 1 500 kg in high-yielding cases. Different methods of protected cultivation can advance the harvest period by 1 to 4 months.

Protected cultivation can be divided into forcing cultivation and semi-forcing cultivation based on the purpose and application of technical measures. Forcing cultivation aims to promote flower bud formation, while semi-forcing cultivation mainly breaks plant dormancy and promotes normal growth.

040 What are semi-forcing cultivation, forcing cultivation, and inhibitory cultivation for strawberries?

(1) Semi-forcing cultivation refers to the cultivation technique that breaks the dormancy of strawberries under artificial conditions and promotes their early growth and development.

There are various methods for semi-forcing cultivation, but they all involve providing artificial insulation to meet the low-temperature requirements for strawberry dormancy and advance the maturity period. Although semi-forcing cultivation increases costs, it significantly advances the harvest period (usually before March) and greatly increases economic benefits, with an income per 667 square meters reaching over 50 000 yuan.

(2) Forcing cultivation refers to the technique of artificially promoting the early formation of flower buds in strawberries and preventing them from entering dormancy, thus promoting their continuous growth and fruiting. Forcing cultivation has high yields, good fruit quality, and an earlier maturity period compared to semi-forcing cultivation. However, it requires higher costs and more advanced management techniques. The key to forcing cultivation is to promote flower formation and inhibit dormancy.

(3) Inhibitory cultivation refers to the technique of keeping strawberry plants in a suppressed state under artificial refrigeration conditions for a longer period and promoting their growth and development at a planned time. This method is widely used in Japan and has also been tested in China. Inhibitory cultivation can extend the harvest period, with an even earlier maturity compared to forcing cultivation, allowing for a large harvest before winter. However, this cultivation method has higher costs and more complicated management.

041 What are the characteristics of strawberry dormancy?

After late autumn and early winter, when the daylight hours become shorter, the temperature drops, and the weather gradually becomes colder, strawberries enter a dormant state. During dormancy, the new leaves have short petioles, small leaf area, and a parallel angle of leaf attachment to the ground. As dormancy deepens, the plants become shorter and closer to the ground. Strawberry leaves do not fall off during dormancy and can remain green during the winter when suitable environmental conditions or artificial protection are provided. However, in northern regions, if not covered and protected during winter, the leaves will wither, affecting the yield of strawberries in the following year.

042 What is the relationship between the duration and depth of strawberry dormancy and the variety?

According to the physiological characteristics and activities during strawberry dormancy, it can be divided into two stages: natural dormancy and forced dormancy. Natural dormancy is determined by the physiological characteristics of strawberries and requires a certain amount of low temperature to pass through smoothly. Even if suitable environmental conditions for plant activity are provided, the plants still remain dormant during this stage. Forced dormancy refers to the dormancy caused by unsuitable growth conditions after natural dormancy has been broken. During this stage, the plants can resume normal growth and development when appropriate conditions are provided. After flower bud differentiation, strawberries gradually enter dormancy, which deepens over time. Generally, dormancy is at its deepest state in mid to late November.

The duration and depth of natural dormancy in strawberries vary depending on the variety's requirement for low temperatures. According to experiments conducted in Japan, the time required for strawberries to experience temperatures below 5 ℃ during natural dormancy is 20-50 hours for the variety "Chunxiang" and 400-500 hours for the variety "Baojiao Early".

043 How to artificially break strawberry dormancy?

The external conditions to break strawberry dormancy are low temperatures and long daylight hours. Artificially providing long daylight treatments is more effective in breaking dormancy. The main measures to artificially break dormancy are as follows.

(1) Cold storage of plants: Place strawberry plants under low-temperature conditions to quickly satisfy their low-temperature dormancy requirements. For varieties like "Dana" and "Baojiao Early," the temperature is generally maintained at 0 - 2 ℃ during cold storage.

(2) Photoperiod extension: After subjecting the plants to a certain degree of low temperatures, move them into plastic greenhouses and provide long daylight treatments while maintaining insulation.

(3) Spraying with gibberellins: Spraying strawberry plants with a 0.005%-0.01% gibberellin solution has a significant effect in breaking dormancy. The signs of broken dormancy include the unfolding of new leaves, upright leaf petioles, longer leaf blades, and elongation of flower peduncles. After strawberry flowering ends, runner formation begins. Runner formation is an important indicator to determine whether strawberry dormancy has ended.

044 What are the technical points of using plastic film mulching for strawberries?

In northern regions, it is necessary to protect strawberries from cold during winter. Using plastic film mulching not only ensures the safe overwintering of plants but also maintains a high rate of green leaf retention, laying the foundation for early sprouting and vigorous growth in spring. Plastic film mulching for strawberries advances the harvest period by at least 7 days compared to open-field cultivation and extends the harvest duration by 3 - 4 days. The proportion of marketable fruits significantly increases, with a 20% increase in total yield and a greater increase in market value. The fruits have good quality, and the occurrence of gray mold disease is reduced.

The following technical points should be noted for plastic film mulching of strawberries.

(1) Selection of plastic film. Choose plastic film with a thickness of 0. 008–0. 015 mm. In northern regions, transparent film with good warming effects is preferred, while in warm regions where winter protection is not required, black film can be used. Black film has excellent weed control effects and provides some warming effects, with more stable soil temperature. Black film is commonly used in foreign countries. The width of the film should be about 20 cm wider than the bed width (ridge width) to facilitate soil compaction and wind protection around the edges.

(2) Mulching periods. There are generally two periods for mulching, before winter and before early spring sprouting. In cold regions, it is advantageous to mulch before winter when the average daily temperature is 3 – 5 ℃. Mulching too early may cause leaf curling and discoloration, while mulching too late may result in freezing damage. Before mulching, the soil should be sealed with freeze-proof water. In Shandong and central Hebei, mulching is usually done in mid to late November. The mulching time can be appropriately delayed in warmer regions. For spring mulching, it should be done when the soil begins to thaw and after removing winter protection materials.

(3) Mulching method. Prepare the ground by leveling it, breaking up clumps of soil, and removing crop residues. Choose a windless day and lay the plastic film smoothly over the strawberry plants. The film should be spread without curling, and the edges should be firmly covered with soil. If the bed is too long, it can be compacted at intervals to prevent the film from being blown up and torn by the wind. Crop straw can be further spread on the film to facilitate insulation and protect the film. When planting strawberry ridges, the ridges should be slightly arched to ensure close contact between the film and the soil.

(4) Removing or uncovering the film. When the soil thaws in spring, remove the coverings on the film and clean the film surface to facilitate the soil temperature increase and early sprouting of strawberries. The film should be removed when the plants have unfolded leaves and exposed flower buds. A small hole should be made on the film directly above the plants, and the strawberry plants should be pulled out from under the film, covering the base of the plants with soil to prevent air from entering and causing the film to bulge. After fruit harvest, remove all the plastic film.

(5) Fertilizer and water management. After strawberry sprouting, apply additional fertilizer combined with irrigation. When applying fertilizer, make holes in the film with a diameter of 2–3 cm and a depth of 5 cm, and place the fertilizer in the holes. Apply 10 kg of urea or 20 kg of compound fertilizer per 667 square meters. Spray 0. 2% potassium dihydrogen phosphate solution twice during the leaf expansion to early flowering stage, spray 0. 3% borax solution once during the peak flowering period, and apply irrigation once during the budding stage, early flowering stage, peak flowering stage, and before fruit ripening, depending on the

soil moisture conditions. Drainage should be considered when there is excessive precipitation.

(6) Removing residual film. After strawberry harvest, remove all residual film pieces and other debris from the soil to avoid soil contamination and affecting the quality of subsequent crops.

045 What are the key points of semi-forcing cultivation using simple plastic tunnels for strawberries?

Simple plastic tunnels for semi-forcing cultivation of strawberries are typically constructed using bamboo or wooden frames. They are about 1.5 meters wide, 50 cm high, and 10 - 20 meters long. Excessive length is not conducive to ventilation. The tunnels are covered with 0.06 mm thick transparent polyethylene film in a north-south orientation. Planting is done on flat beds with a width of 1.3 - 1.4 meters. Each bed is planted with 4 rows, with a plant spacing of 20 cm × 30 cm. Approximately 10 000 plants are planted per 667 square meters. To prevent the plastic film from being blown away by strong winds, multiple ropes should be used to secure it. Other management measures are similar to open-field cultivation. Planting should be done as early as possible after flower bud differentiation, preferably in September. Planting before or immediately after bud differentiation can damage the roots and result in fewer flower buds. Late planting can lead to more deformed fruits and poor fruit quality. The best seedlings for planting have 5 - 6 unfolded leaves, new stems with a diameter of more than 1 cm, and a weight of about 30 grams. Planting large seedlings can result in more axillary buds and flower cluster development, increased small fruit rate, and decreased commercial value.

In terms of fertilizer and water management, 5 000 kg of farmyard manure and 25 kg of calcium superphosphate can be applied per 667 square meters as base fertilizer. Nitrogen fertilizer is mainly used for topdressing, and foliar spraying can be done 1-2 times. Manure or urine can also be applied in dilute form. Before mulching and covering the tunnel, sufficient water should be applied. The plastic film is covered in early to mid-January, the plastic tunnels are added in early to mid-February, and fruit harvesting begins in late to mid-April. The management before winter is the same as open-field cultivation. After covering the tunnels, the cultivation shifts from open-field to protected cultivation in the plastic tunnels. The temperature should be initially controlled at 15-20 ℃. After germination, ventilation should be provided as the temperature increases and the plants grow, with the temperature controlled at 20-25 ℃ and not exceeding 30 ℃. In April, when the natural temperature meets the requirements for strawberry growth, the plastic film can be removed.

The use of plastic tunnels with mulching can increase the temperature by 5-6 ℃ and maintain higher air and soil humidity. It generally results in strawberries ripening about 20 days earlier (harvestable in mid-April), extending the harvest period by about 2 weeks, and significantly increasing yield and economic benefits.

Simple plastic tunnels have low costs, easy availability of materials, simple structures, and convenient operation, making them suitable for promotion and application by farmers. However, the warming effect of simple plastic tunnels is limited, and they cannot significantly increase yield or greatly advance or extend the harvest period. Therefore, in areas with suitable conditions, large greenhouses are recommended.

046 How to select suitable varieties for semi-forcing cultivation in ordinary plastic greenhouses for strawberries?

Semi-forcing cultivation in ordinary plastic greenhouses refers to the cultivation of strawberries in sunlight plastic greenhouses without artificial lighting or heating equipment, similar to vegetable greenhouses. It is currently a commonly used cultivation method for strawberries in China. With this method, the yield per 667 square meters can reach 1 500–2 500 kg, and the strawberries can mature in early March, resulting in an income of over 10 000 yuan.

Selecting suitable varieties for greenhouse cultivation is an important factor for successful cultivation. Varieties such as "Baojiao Early" "Gorella" "All Star" "Mingbao" and "Fengxiang" have shallow dormancy and are suitable for growth and development under low light and relatively high temperature conditions. They have good productivity, high fruit quality, and a competitive advantage in the market.

The quality of seedlings is a key factor affecting survival rate and yield. The criteria for high-quality seedlings include strong root systems, short petioles, 4–7 leaves, new stems with a diameter of 1 cm or more, and seedling weight of 20–40 grams. The seedlings should not be too large, and they should be free from diseases and pests.

The measures to cultivate high-quality seedlings include establishing dedicated propagation beds and conducting false planting. After fruit harvest, transplant the mother plants with soil in a timely manner to establish propagation beds. The row spacing is 100–200 cm, and the plant spacing is 50–100 cm. Before planting, it is preferable to conduct false planting to promote the development of strawberry roots and ensure vigorous seedlings for nitrogen control and flower induction. The false planting bed is generally 70–80 cm wide, with 4–5 rows per bed and a plant spacing of 15 cm × 15 cm. If early planting is possible, direct planting of high-quality seedlings without false planting is also feasible.

047 How to cultivate strawberries in ordinary plastic greenhouses?

Select a location with higher terrain, flat ground, loose and fertile soil, and convenient drainage and irrigation facilities for establishing ordinary plastic greenhouses. Install the greenhouse framework first and cover the plastic film for insulation at the appropriate time.

Plant strawberries in the greenhouse from August to early September. If false planting or the previous crop has not been harvested, planting can be delayed, but it should be completed before the soil freezes.

To ensure rapid growth and development within a short period in plastic greenhouses, it is necessary to apply sufficient base fertilizer before planting. Apply 2 500 – 3 000 kg of chicken manure per 667 square meters, mixed with an appropriate amount of phosphorus and potassium fertilizers. The chicken manure should be well-rotted to avoid urate damage to the plant roots. Apply the base fertilizer by furrow application combined with broadcasting.

Plant strawberries in high ridges in the greenhouse, with a ridge height of 20 cm and a ridge top width of 50 cm. The ridge spacing should be around 80 cm. Plant 2 rows per ridge, with a small row spacing of about 30 cm and a plant spacing of 10 – 15 cm. High ridge planting in the greenhouse allows for efficient space utilization, and the planting density can be appropriately increased compared to open-field cultivation. Approximately 11 000–19 000 plants are planted per 667 square meters, with the most suitable density being 12 000–14 000 plants.

During planting, it is preferable to use seedlings with soil attached. If the seedlings do not have soil, soak the roots in a 0.01% napthyl acetic acid(NAA) solution for 2–4 hours to promote root development and seedling vigor.

In Shandong, to achieve strawberry maturity from late February to early April, it is generally necessary to use three to four layers of transparent plastic film for insulation. The outermost layer of the film and the ground cover should be applied first, and granular carbon dioxide fertilizer should be applied under the ground cover at a rate of 40 kg per 667 square meters. In late December to early January, add a plastic tunnel in the greenhouse. If the temperature is too low, an additional layer of plastic film can be added inside the greenhouse.

The timing of starting insulation is crucial for strawberry growth. Starting insulation too early may result in insufficient low-temperature exposure for strawberry dormancy release, leading to stunted growth. Starting insulation too late may result in prolonged low-temperature exposure, causing excessive vegetative growth and delayed maturity. Therefore, insulation should begin after the required period of low temperatures for strawberries has been met. In Shandong, insulation usually begins in late November, resulting in an earlier harvest period by about one month and a significant increase in yield.

Different growth stages in the greenhouse have different temperature management requirements: 15–25 ℃ during sprouting and leaf expansion, with a minimum temperature above 0 ℃; 20–28 ℃ during flowering and fruit set, with a minimum temperature not lower than 5 ℃ and a maximum temperature not exceeding 30 ℃; 10 – 28 ℃ during fruit development. Low temperatures are beneficial for fruit enlargement, while high temperatures promote fruit ripening. The temperature in the greenhouse can be regulated by uncovering the straw mulch and providing ventilation. After covering the greenhouse, spray 0.008%–0.01% gibberellin solution 1–2 times to break dormancy and promote early flowering, resulting in longer petioles and flower stalks. Apply 0.5% urea or potassium dihydrogen phosphate solution externally and spray 0.04%–0.05% cytokinin solution multiple times to increase yield and improve fruit quality. During flowering, reduce the air humidity appropriately, use bees or manual pollination, supplement carbon dioxide gas if conditions permit, and remove old leaves and thin flower clusters to increase fruit yield and quality. During the fruit enlargement period, ensure timely irrigation to maintain soil moisture. Drip irrigation or pipe irrigation is recommended.

048 How to manage temperature in a winter greenhouse for strawberry cultivation?

For forcing cultivation, it is recommended to use a winter greenhouse (commonly known as a solar greenhouse) with soil walls on the north, east, and west sides. The greenhouse should be covered in late October in areas between 32° and 35° north latitude to facilitate axillary bud formation into flower buds. After covering the greenhouse, maintain a temperature of 25–30 ℃ during the day and 5–10 ℃ at night. When the nighttime temperature drops to 5 ℃, an additional small greenhouse should be installed inside the main greenhouse for further insulation. Before flowering, the daytime temperature should be around 25 ℃, but should not exceed 30 ℃. During the fruit enlargement stage after flowering, maintain a daytime temperature of 20–25 ℃ and a nighttime temperature of 5 ℃. During the harvest period, the daytime temperature should not exceed 25 ℃, while the nighttime temperature should be maintained at 3–5 ℃.

Under the temperature conditions in the greenhouse, powdery mildew is prone to occur. Therefore, it is important to regularly remove the bottom leaves and apply preventive sprays before flowering, especially for varieties that are not resistant to powdery mildew.

In strawberry forcing cultivation, the plants have a long fruiting period and early yields are high, but the plants are prone to weaken and become stunted. Spraying with gibberellin is an important measure for forcing cultivation as it promotes growth, induces flower bud differentiation, breaks dormancy, and promotes early maturity. When flower buds appear on more than 30% of the plants, focus on spraying 5 mL of 0.005%–0.01% gibberellin solution on the crown leaves of each plant. After spraying, slightly increase the greenhouse temperature, which can advance the flowering of the terminal flower cluster. Spraying too early can cause axillary buds to become runners, while spraying too late only promotes petiole elongation without promoting flowering.

7. Disease and Pest Control

049 What are the characteristics of strawberry viral diseases? What are the methods of prevention and control?

There are dozens of types of viral diseases that affect strawberries, including mosaic, vein banding, crinkling, and strawberry crinkle viruses. The main characteristic of strawberry virus diseases is their hidden symptoms or subtle manifestations. Infected plants typically exhibit stunted growth, inadequate leaf expansion, distortion, deformity, yellowing, mottling, small and dull leaves, stunted plant height, decreased yield, reduced quality, and increased occurrence of malformed fruits. The yield reduction is generally around 20% to 30%, and in severe cases, it can reach 50%. Strawberry virus diseases are mainly transmitted by insects and nematodes with piercing-sucking mouthparts.

The technical methods for preventing and controlling strawberry virus diseases include the following.

(1) Implement crop rotation and avoid continuous or intercropping with solanaceous crops.

(2) Eliminate pathogens by removing and destroying infected plants.

(3) Use virus-free tissue culture techniques to propagate disease-free seedlings.

(4) Select resistant varieties, such as All Star, Red Glove and Mingjing.

(5) Establish a system for supplying disease-free planting materials, using virus-free seedlings, and regularly(every 4-5 years) replacing the plants.

(6) Timely and effective control of vector insects such as aphids and nematodes. Pay particular attention to the control of strawberry aphids, peach aphids, and cotton aphids.

(7) Disinfect the soil using chemical or high-temperature methods.

050 What are the symptoms and methods of prevention and control for strawberry gray mold disease?

Gray mold is a major disease of strawberries, which can cause a yield reduction of 10% to 30% and even more than 50% in severe cases. In the early stages of the disease, the fungus can infect petioles, leaves, and flower stalks, but the symptoms are most pronounced during the ripening period of the fruits. Infected leaves initially develop inconspicuous brown, moist spots, which can produce a layer of gray spores under humid conditions. Infected petioles and fruit stalks turn brown and often encircle the affected areas, causing wilting and drying. In the early stages of fruit infection, small brown oily spots appear. Later, the spots rapidly expand to cover the entire fruit, causing softening and the formation of white cottony mycelium on the surface. Gray spore masses develop at the tips of the mycelium, and immature fruits often rot. Infected petals turn yellow-brown.

The fungus overwinters as mycelium or sclerotia in infected plant debris, and the disease spreads through conidia in the air. The disease is most prevalent during the fruit ripening period. High humidity is an important condition for disease development, and it is more prevalent during continuous rainy days.

To prevent and control strawberry gray mold disease, the following measures can be taken.

(1) Select resistant varieties, practice proper planting density, and use plastic mulch to prevent fruit contact with the ground.

(2) Remove old and senescent leaves in early spring, reduce overuse of nitrogen fertilizer, and lower humidity levels.

(3) Before the flowering stage, spray 50% chlorothalonil at 800 times dilution, or 50% pyrimethanil at 500–700 times dilution, or 50% carbendazim at 500 times dilution. Bordeaux mixture, oxytetracycline, iprodione, and other fungicides are also effective.

(4) During the fruit enlargement period, place a small amount of straw under the fruits to reduce surface humidity. Collect and bury infected fruits separately outside the greenhouse.

051 What are the symptoms and methods of prevention and control for strawberry powdery mildew?

Powdery mildew is a common disease of strawberries. It is more prevalent in protected cultivation compared to open-field cultivation and can even cause plant death. Powdery mildew primarily affects the leaves, but it can also infect petioles, runners, calyxes, fruits, and fruit stalks. In the early stages of the disease, infected leaves develop small, inconspicuous brown spots, which gradually enlarge and become circular or elliptical patches. The center of the spots turns grayish-white, resembling powdery mildew. When multiple spots merge, they can cause extensive leaf tissue necrosis. Infected leaves exhibit reduced photosynthesis, decreased cold resistance, and weakened disease resistance. The pathogen is a powdery mildew fungus that survives on petioles during the winter and spreads through the air. The disease occurs continuously throughout the growing season, with optimal development at temperatures between 15 ℃ and 25 ℃. The fungus is a low-temperature pathogen and rarely occurs during hot summer seasons.

To prevent and control strawberry powdery mildew, the following measures can be taken.

(1) Clear the field in winter and spring, and burn infected plant residues.

(2) Increase plant spacing appropriately, remove old leaves that are in contact with the ground in a timely manner, and ensure good ventilation in the field. Pay attention to drainage after rainfall.

(3) Control the application of nitrogen fertilizer and choose resistant varieties.

(4) Before the appearance of flower clusters and before flowering, spray Bordeaux mixture at 2 000 times dilution. In severe cases, repeat the spray every 10 days until harvest.

052 What are the symptoms and methods of prevention and control for strawberry leaf spot and leaf blight diseases?

Strawberry leaf spot primarily affects the leaves but can also infect petioles, runners, sepals, fruits, and fruit stalks. It occurs mildly before flowering and becomes severe after fruit harvest. It occurs to varying degrees in strawberry planting areas in China. The disease initially causes small reddish-brown spots on the leaves, which gradually enlarge and become round or elliptical. The margins are reddish-purple, and the centers become gray-white, resembling snake eyes. When numerous spots occur, they can lead to leaf withering. The disease affects photosynthesis, cold resistance, and disease resistance. The pathogen has sexual and asexual generations and belongs to the class Dothideomycetes. The pathogen overwinters on fallen stems and leaves and spreads through conidia in the air during the following spring.

To prevent and control strawberry leaf spot and leaf blight diseases, the following measures can be taken.

(1) Use the same control methods as for powdery mildew.

(2) Remove small amounts of infected leaves when a small number of diseased leaves are found. For heavily infected areas, remove all leaves after harvest, followed by cultivation, fertilization, and irrigation to promote new leaf growth.

(3) Before the appearance of flower clusters, spray Bordeaux mixture at 2 000 times dilution. In severe cases, repeat the spray every 10 days until harvest.

053 What are the symptoms and methods of prevention and control for strawberry yellow wilt disease?

When strawberries are affected by yellow wilt disease, the young shoots become stunted, exhibiting boat-shaped curling and yellow-green coloration. Out of three small leaves, one or two may become deformed. The leaves lose their luster and turn reddish-purple, starting from the leaf margins and gradually wilting and drying. The whole plant shows poor growth and eventually dies. During high-temperature periods, the symptoms include wilting and death, while during low-temperature periods, yellowing symptoms are observed. The crown and petiole vessels turn brown. The roots turn blackish-brown from the outside or the tips and soon rot. The central column of the root appears normal.

Strawberry yellow wilt disease occurs in Shandong, Liaoning, Shaanxi, and other regions of China. The disease occurs within a temperature range of 8–36 ℃, with optimal development at 28 ℃.

To prevent and control strawberry yellow wilt disease, the following methods can be employed.

(1) Avoid continuous cropping.

(2) Select disease-free planting materials.

(3) Do not introduce seedlings from infected areas.

(4) Remove infected plants in a timely manner and burn them.

(5) Choose resistant varieties.

(6) When preparing seedlings, soak the roots in benomyl (Benlate) or thiophanate-methyl (Topsin) at 500 times dilution or drench the soil. For diseased fields, soil disinfection with chloropicrin can be performed.

(7) Utilize sunlight for soil disinfection.

054 What are the symptoms and methods of prevention and control for strawberry bud rot disease?

When strawberries are affected by bud rot disease, the young buds become wilted and shriveled, and the dead buds turn blackish-brown. The flower buds start browning from the base of the small flower stalks, ultimately resulting in bud withering and death. The leaves become wilted and droopy. Newly emerging leaves are stunted, with red-colored petioles. Browning starts from the base of the petioles, followed by wilting and death. The roots appear normal. The optimal temperature range for the development of the pathogen is 22–25 ℃. Good development can still occur at lower temperatures, while development ceases at temperatures above 33 ℃. The disease is more likely to occur under high humidity and excessive soil moisture. Overcrowding and excessive planting depth also contribute to disease development. The pathogen primarily spreads through infected seedlings, and the nursery and mother plant gardens serve as sources of infection.

To prevent and control strawberry bud rot disease, the following measures can be taken.

(1) Avoid using seedlings from infected areas.

(2) Avoid deep planting and excessive planting density.

(3) Avoid excessive irrigation and ensure proper ventilation in greenhouses.

(4) Control the application of nitrogen fertilizer.

(5) During the visible bud stage, spray oxytetracycline.

055 What are the symptoms and methods of prevention and control for strawberry root rot disease?

When strawberries are affected by root rot disease, the above-ground growth is poor. Starting from the base of the leaves, they turn reddish-brown and gradually wilt and die. In severe cases, the stems and leaves rapidly wilt, secondary roots rot first, and soon the entire root system turns brown. In the early stages of infection, browning starts from the root tips, and the central column of the roots turns red or light reddish-brown. Later, browning spreads from the center, resulting in rotting. The disease is more likely to occur when the soil temperature is below 25 ℃. It is less prevalent in high-raised beds. Excessive flooding can lead to the development of this disease. The disease mainly occurs from March to April in open-field cultivation, while it occurs less frequently in greenhouses and small tunnels.

To prevent and control strawberry root rot disease, the following methods can be employed.

(1) Select healthy mother plants.

(2) Avoid continuous cropping in infected areas.

(3) Pay attention to drainage and adopt high-raised bed planting.

(4) Avoid excessive flooding between rows; if conditions permit, use drip irrigation.

(5) Implement soil disinfection.

056 What are the symptoms and methods of prevention and control for strawberry leaf spot disease?

Strawberry leaf spot is a common disease of strawberries. The pathogen infects the leaves and petioles, causing purple-red circular or elliptical spots with distinct concentric rings. The spots gradually enlarge, and the central area becomes brown and necrotic. Larger spots exhibit clear concentric rings, with purple-brown margins that often crack and die. Black spore masses develop on the dead tissue. The petioles develop long elliptical reddish-purple spots. Severe infections can cause extensive leaf necrosis. The pathogen is the Colletotrichum acutatum fungus, which thrives in warm temperatures around 28 – 30 ℃. The pathogen overwinters on petioles and spreads through the air. The disease is more prevalent during periods of high temperature and rainfall. Distinguishing between leaf spot disease and false leaf spot disease in the field can be challenging, but the latter is a low-temperature disease that rarely occurs at temperatures above 28 ℃.

To prevent and control strawberry leaf spot disease, the following measures can be taken.

(1) Employ the same control methods as for powdery mildew.

(2) Practice good ventilation and control soil humidity in protected cultivation.

(3) Cultivate robust seedlings and apply hormone treatments to promote sugar accumulation in fruits.

(4) In areas with severe infections, after harvest, remove all leaves and spray 70% methyl thiophanate at 1 000 times dilution or 50% thiram at 800 times dilution, or 30% tebuconazole at 5 000 times dilution.

057 What are the symptoms and methods of prevention and control for strawberry brown spot(leaf blight) disease?

Strawberry brown spot(leaf blight) is a major disease of strawberries that primarily affects the leaves, fruit stalks, and petioles. Infected leaves initially develop reddish-brown small spots, which gradually enlarge into round or elliptical patches. The center of the spots becomes brown, while the outer part turns purple-brown, and the outermost edge becomes purple-red. The spots have a diameter of 1-3 mm, with distinct boundaries between healthy and diseased tissues. Later, brown dots (conidiophores) form on the spots, arranged irregularly in concentric circles. When multiple spots merge, they can cause extensive leaf tissue necrosis. The pathogen is the Colletotrichum acutatum fungus. The disease occurs rapidly at temperatures between 20 ℃ and 30 ℃, especially during rainy periods. The disease affects strawberries throughout the growing season, with a peak occurrence from June to August in northern China.

To prevent and control strawberry brown spot(leaf blight) disease, the following methods can be employed:

(1) Select resistant varieties.

(2) Cultivate vigorous strawberry seedlings and control the application of nitrogen fertilizer.

(3) Before planting strawberries, soak the seedling roots in 40% methyl thiophanate at 500 times dilution for 20 minutes to reduce the disease source in the following year.

(4) Remove infected leaves in a timely manner, burn rotten stems and leaves in winter and spring.

(5) Use fungicides such as chlorothalonil at 800 times dilution, or mancozeb, or carbendazim at appropriate intervals during the flowering period and before fruiting.

058 How to prevent and control strawberry slime mold disease?

In recent years, strawberry slime mold disease has been spreading. It is more likely to occur during the peak fruiting period in spring and autumn and affects the stems, leaves, petioles, and fruits. In the early stages of infection, the affected areas are covered with slime. Subsequently, yellow cylindrical particles are produced in an orderly arrangement. Other fungi can grow on the infected areas, causing the stems, leaves, and fruits to turn blackish-brown. Under dry conditions, the infected areas develop a gray-white powdery hard shell, and the plants wither, and the fruits rot. The pathogen is a saprophytic slime mold that survives as spores and can also affect grasses.

To prevent and control strawberry slime mold disease, the following methods can be employed.

(1) Before planting strawberries, clear the field, level the ground, and prevent water accumulation in low-lying areas.

(2) Avoid using seedlings from infected areas and avoid excessive planting density. Overcrowding in greenhouses can exacerbate the disease.

(3) Cultivate robust seedlings and increase phosphorus and potassium fertilization.

(4) During autumn and the early fruiting period, combine the prevention and control of leaf spot disease by spraying 2–3 applications of carbendazim and mancozeb at 600–800 times dilution.

059 What are the symptoms and methods of prevention and control for strawberry fruit whitening disease?

Strawberry fruit whitening disease results in white or pale yellow-white fruits, or a portion of the fruit surface becomes noticeably whitened with a clear boundary. Fruits affected by whitening disease are typically normal in size but lack color, flavor, and firmness. They appear soft, have poor appearance, and are prone to rotting. Almost all strawberry varieties can be affected, with American varieties being more susceptible than Japanese varieties. Varieties with high sugar content are less affected. The disease is more likely to occur during warm weather with cloudy and rainy conditions, especially before the peak fruiting period. The exact cause of the disease is not fully understood, but it is not caused by pathogens. It is primarily attributed to environmental factors and physiological disorders. Excessive nitrogen fertilizer, vigorous plant growth, excessive fruit set, poor leaf growth, cold weather before and after planting strawberries, inadequate light during the fruiting period, and low sugar content are all factors that contribute to the disease.

To prevent and control strawberry fruit whitening disease, the following methods can be employed.

(1) Select strawberry varieties suitable for local climatic conditions.

(2) Avoid excessive application of nitrogen fertilizer and ensure sufficient phosphorus and potassium fertilization.

(3) Practice good field management, including hormone treatments to promote sugar

accumulation in fruits.

(4) Implement protected cultivation.

060 What are the common types of nematodes that affect strawberries? How can they be prevented and controlled?

The common nematodes that affect strawberries include strawberry nematodes, lesion nematodes, and bud nematodes.

When strawberries are infected by nematodes, their vitality decreases, making them more susceptible to bacterial and fungal infections, and some nematodes can also transmit viral diseases. Symptoms of nematode infestation include stunted growth, deformities, discoloration, wilting, and weakened plants. Strawberry plants affected by bud nematodes may exhibit distorted or deformed new leaves, increased leaf coloration, and red or yellow discoloration of buds and petioles, leading to a condition known as "strawberry red bud". Infested flower buds can become deformed or fail to develop, significantly affecting strawberry yield. Bud nematodes primarily spread through infected runners in the soil.

To prevent and control nematodes in strawberries, the following methods can be employed.

(1) Establish a virus-free propagation base.

(2) Practice crop rotation, avoid continuous cropping, and disinfect the soil to eliminate infected plants.

(3) After transplanting, drench the roots with a 500-800 times dilution of dichlorvos. Repeat the drenching every 7-10 days for a total of 3-4 times.

061 What are the main types of spider mites that affect strawberries? How can they be prevented and controlled?

There are several types of spider mites that can affect strawberries, with the two main ones being the two-spotted spider mite (also known as the two-spotted spider mite) and the strawberry spider mite. The two-spotted spider mite has a wide range of host plants (such as cotton, soybeans, eggplants, watermelons, and weeds), and spider mites from different host plants can transfer and infest strawberries. They can produce more than 10 generations in a year, with the overwintering female adults staying in the soil. They lay eggs in spring, and the hatched nymphs become active. The mites reproduce rapidly in hot and dry weather. The strawberry spider mite primarily affects greenhouse-grown strawberries but can also infest field-grown strawberries. Both types of spider mites feed on the undersides of leaves, sucking plant juices. Infested areas initially develop small white spots, which later turn red. Severe infestations can cause leaves to turn rusty brown, resembling burnt leaves, leading to inhibited plant growth and significant yield reduction. Adult spider mites do not have wings and spread through wind, rain, seedlings, tools, and human contact.

To prevent and control spider mites in strawberries, the following methods can be employed.

(1) Before the flowering stage, remove old leaves from the lower part of the strawberry plants, reducing the density of mites. Dispose of leaves with mites and disease residues outside the field to reduce the source of infestation.

(2) Before flowering, spray a 1 000 times dilution of 2.5% chlorpyrifos, 500–750 times dilution of 20% fenpropathrin, or other acaricides effective against eggs and mites. Spray once every 7 days for a total of 2 times. When spraying, direct the nozzle upwards to ensure the undersides of the leaves are covered with the pesticide.

(3) During the fruit enlargement period, spray a 5 000–8 000 times dilution of 20% cypermethrin twice at a 5-day interval. Stop spraying 2 weeks before harvest.

(4) After harvest, spray the plants with an 800 times dilution of 20% trichlorfon, adding 0.2% benomyl to control mites.

062 How do aphids harm strawberries? How can they be prevented and controlled?

There are several species of aphids that can harm strawberries. Aphids not only suck sap from the plants, leading to stunted growth, but they can also transmit strawberry viral diseases. During winter, aphids overwinter in the soil near strawberry, vegetable, and rapeseed crops. Cotton aphids are migratory pests that overwinter as eggs on plants such as Chinese prickly ash, summer cypress, and plantains. They hatch in spring when the temperature rises above 10 ℃ and start feeding on strawberry plants. They reproduce multiple generations in a year, with each adult aphid capable of producing 20–30 nymphs, resulting in high population growth. The peak occurrence of aphids is during the hot season.

To prevent and control aphids in strawberries, the following methods can be employed.

(1) Timely removal of old leaves, cleaning the field, and eliminating weeds.

(2) Before flowering, spray a 1 500 times dilution of dimethoate or a 2 000 times dilution of imidacloprid. Spray once or twice at a 7-day interval.

(3) During the propagation or nursery bed period, pay attention to aphid control using appropriate methods.

063 How do strawberry ground beetles harm strawberries? How can they be prevented and controlled?

The main ground beetle that harms strawberries is the ground beetle. The adult beetles are approximately 5 – 6 mm long and have a bronze color. They use their needle-like mouthparts to suck sap from the seeds at the tops of young fruits, destroying their contents and resulting in empty seeds. The tops of the fruits do not develop properly, and clusters of empty seeds form, leading to deformed fruits and significant reduction in fruit quality.

To prevent and control strawberry ground beetles, the following methods can be employed.

(1) Clear the field of weeds and eliminate sources of infestation.

(2) In small areas heavily infested with ground beetles, manually capture and kill the beetles during the spring and autumn seasons.

(3) During the occurrence of adult beetles, spray a 3 000 times dilution of 2.5% bromophos or an 8 000 times dilution of 20% carbofuran. Repeat the spray if necessary.

Chapter 2 Q&A on Practical Techniques for Grape Cultivation

Manual of Horticultural Techniques
园艺技术手册

1. Selection of Excellent Grape Varieties

001 What are the excellent grape varieties suitable for cultivation in northern regions?

The main excellent grape varieties suitable for cultivation in northern regions include Kyoho, Phoenix 51, Muscat, Fujiminori, Xiyanghong, Pink Yadumi, Red Ruby Seedless, Zuijinxiang, Molixiang, Liaofeng and Lanbaoshi.

002 What are the characteristics of Kyoho grape?

It is a European-American hybrid variety and belongs to a tetraploid variety. The plant has strong growth potential, high bud germination rate, and strong fruit setting ability. It is a hermaphroditic flower. The grape clusters are large, conical in shape, with an average weight of 365 g and a maximum weight of 730 g. The berries are large, nearly round, with an average weight of 9 g and a maximum weight of 13. 5 g. The skin is thick, purplish-black, and has a high amount of bloom. The flesh is soft, with flesh pockets, abundant juice, and a sweet and sour taste. It has a distinctive strawberry aroma and a soluble solid content of 14%. The seeds are 1–2 in number and easily separate from the flesh. The lateral buds and shoots have strong fruit setting ability. It bears fruit early, has high yields, and is greatly influenced by the load at the ripening stage. In the North China region, the fruit begins to color around July 15, and it fully ripens around August 10. The growth period from bud emergence to fruit maturity is 125 days. This variety has large berries, above-average quality, and strong disease resistance. However, when the load is excessive or management is poor, the ripening period may be delayed, making it susceptible to grapevine trunk diseases and causing severe fruit drop. Therefore, it is necessary to pay attention to purification and rejuvenation, control the yield, and implement measures such as shoot thinning and cluster thinning to reduce fruit drop.

003 What are the characteristics of Liaofeng grape?

Liaofeng grape is a medium-ripening variety, with a full maturity period of about 132 days from bud emergence to fruit ripening. In the Liaoyang region, the fruit ripens in the early to mid-September. The vines are strong, with a stem thickness of up to 14 mm at 1m from the main shoot. The fruiting branch rate is 61%, with an average of 1.8 flower clusters per fruiting branch and an average of 1.46 flower clusters per new shoot. The leaves are large and nearly circular, with an average diameter of 27 cm for mature trees and up to 35 cm for larger leaves. The fruit clusters are long and conical, tight, with a length of 27 cm and a width of 18 cm, and an average weight of 600 g per cluster. The individual berries weigh 12 g for young trees and 14 g for mature trees, with a longitudinal diameter of 35 mm and a transverse diameter of 31 mm, appearing round or oval in shape, and having uniform size. The skin is dark purple to black, thick, with a thick bloom. The flesh is green, relatively firm, and sweet in taste. It is suitable for cultivation on trellises, with a vine spacing of 0.7–0.8 m and a row spacing of 3.5–4 m. Pruning should be done to maintain a well-shaped canopy and shorter shoots.

004 What are the characteristics of the Fujiminori grape?

This is a hybrid variety from Europe and America. It has vigorous tree vigor, thick branches, green new shoots, and light purple young leaves with a glossy leaf surface. The leaf back has moderate fuzz, and the mature leaves are large, 5-lobed, with slightly shallow upper lobes and deep lower lobes, and the leaf stalk is arched. It has hermaphroditic flowers. It has good fruit set, with minimal flower and fruit drop. The average weight of each fruit cluster is around 600 g, and the average berry weight is about 15 g, making it the largest berry variety among the Kyoho series. The fruit skin is dark purple. The flesh is relatively dense, with a high juice content. It has a sugar content of 18%, moderate acidity, and a slight off-flavor, resulting in average quality. It matures in mid-August, but the berries are prone to drop after ripening, and it has poor storage and transportation capabilities. It starts bearing fruit early and has a high fruit set rate. It has strong tree vigor, does not grow excessively, and the branches mature well. It has strong disease resistance and is easy to manage. It is a large-berry variety that can be planted near urban areas.

When cultivating Fujiminori grapes, attention should be paid to the following points.

(1) Control tree vigor to promote ventilation, light penetration, and use of secondary shoots for fruiting.

(2) Apply organic fertilizer more and nitrogen fertilizer less, especially be cautious in using nitrogen fertilizer after fruit set, as it can delay ripening.

(3) Strictly control the crop load, pay attention to thinning flowers and fruit, and leave around 30 berries per cluster, avoiding exceeding this number to ensure the formation of large clusters and berries.

(4) Harvest at the right time to prevent berries from dropping due to overripening.

005 What are the characteristics of the Zuijinxiang grape?

The young leaves are green with a slight gloss on the leaf surface, and the underside of the leaves has fine hairs. Mature leaves are exceptionally large, heart-shaped, 3–5 lobed, with deep lobes and a rough, green surface that has blister-like elevations. The leaf stalk is arrow-shaped and long, with a purple color. Mature branches are light brown, long, and stout. The fruit clusters are exceptionally large, with an average weight of 800 g and a maximum weight of up to 1 800 g. They have a conical shape and are tightly packed. The berries are exceptionally large, with an average weight of 13 g and a maximum weight of 19 g. The berries are inverted-ovate in shape, and when fully ripe, the skin turns golden yellow, with consistent ripening, uniform size, prominent navel, abundant bloom on the skin, thick skin, strong fragrance, excellent quality, and a sugar content of 16. 8%. The plant grows vigorously with a bud sprouting rate of 80. 5% and a fruiting branch rate of 55%. Each fruiting branch has an average of 1. 32 flower clusters. It exhibits strong resistance to fungal diseases such as downy mildew and white rot. In early September, the berries fully ripen in the Shenyang area of Liaoning province. From bud sprouting to full fruit maturity, it takes approximately 126 days. This variety is suitable for cultivation on trellises or pergolas with medium to short pruning. During the early tree stage, it is important to maintain strong tree vigor without excessive growth and promote a balance between vegetative and reproductive growth. After fruiting, it is necessary to ensure sufficient nutrients and water, with particular emphasis on autumn application of organic fertilizer. Nitrogen fertilizer should be applied in moderation, while phosphorus and potassium fertilizers should be used more.

006 What are the characteristics of the Muscat grape?

The tender shoots are green with a slight reddish-brown color. Young leaves are yellow-green with a purple-red hue, glossy, and the backside is densely covered with white hairs. Mature branches turn yellowish-brown in one year. Mature leaves are large, nearly circular, 5-lobed, with undulating and wavy surfaces, large serrated margins, and reddish-brown leaf stems. The plant has bisexual flowers. The fruit clusters are large, measuring 17–19 cm in length, 10–14 cm in width, and weighing 292–428 g, with the largest cluster weighing up to 1 200 g. They are conical in shape, loosely or moderately compact, and have shoulders. The fruits are of medium size, with a longitudinal diameter of 19–20 mm, a transverse diameter of 18–19 mm, and a grain weight of 4–5 g. They are oval-shaped, red-purple or black-purple in color, with moderately thick skin and a thick layer of fruit powder. There may be variations in grain size. The flesh is juicy, with a strong rose fragrance, a sugar content of 17%, an acid content of 5–7 g/L, and a juice yield of 76%. Each fruit branch bears an average of

1.75 clusters, and the secondary fruiting can be easily induced through proper summer pruning. It has high yields and is a mid-maturing variety with excellent fruit quality. The Rose Fragrance variety is suitable for cultivation in hilly and mountainous areas, well-drained fertile soils, and sandy loam soils near beaches. However, if not properly managed, it is prone to physiological disorders such as color change disease, also known as "water tank disease", when individual plants bear excessive fruit load. Therefore, it is necessary to strengthen management by removing the growing point before flowering and applying additional phosphorus and potassium fertilizers to promote high-quality and high-yield production.

007 What are the characteristics of the Pink Yadumi grape?

The fruits are purple-red and elongated oval in shape. The average weight of Pink Yadumi grapes is 8–9 g, with a maximum weight of 13 g. The fruit clusters are of medium size, with the largest cluster weighing up to 1 500 g and an average cluster weight of 500 g. The grains are moderately compact. The skin of Pink Yadumi grapes is thin and brittle, with a moderately thick layer of fruit powder. The flesh is crispy, juicy, and sweet, with a soluble solids content of 15%–18% and an acid content of 0.5%–0.6%. The quality is excellent. It is an early-maturing variety with strong plant growth and moderate secondary fruiting ability, resulting in high yields. It has moderate disease resistance and adaptability. It is suitable for cultivation on trellises or pergolas, with a focus on medium and short pruning. This variety has beautiful fruit shape, a sweet and tangy taste, a pleasant fragrance, and is early-maturing, storability and disease resistance.

008 What are the characteristics of the Bronx Seedless grape?

Also known as seedless red. A hybrid variety from Europe and America. Originally from the United States, it was introduced to China in 1973. The young shoots are green with slight dark red stripes and sparse fuzz. The young leaves are yellow-green, glossy, and densely covered with fuzz on the back. The mature leaves are nearly circular, shallowly 3-lobed. The back of the leaves has sparse prickles, the leaf margins curl downward, the saw teeth are large and blunt, and the leaf petiole cavity is completely closed or has small gaps. It has bisexual flowers. The fruit clusters are conical, with closely spaced grains. The average cluster weight is 450 g, with a maximum weight of up to 1 200 g. The grains are oval-shaped, with an average weight of 3.5 g, and a maximum weight of around 4.0 g. The fruit skin is rose red or pink, relatively thin and tough, the flesh is soft and juicy, with a soluble solid content of 16%, acidity of 0.45%, and a sweet and refreshing taste. It has a strawberry fragrance and is of superior quality. It has poor storage and transportation resistance. The tree has a strong vigor and the fruiting branches account for 40% to 55% of the total number of branches. Each fruiting branch has 1.0-1.5 fruit clusters, while the secondary branches have weak fruiting ability. It is cold-resistant and has strong resistance to anthracosis and anthracnose, and is high-yielding. It matures in early August in Xingcheng, Liaoning Province, and in mid-August in Shenyang. The growth period is about 110 days. It is one of the varieties that can be eaten fresh or used for juice.

009 What are the characteristics of the Thompson Seedless grape?

Eurasian variety. It is the main variety in Xinjiang region of China, with limited cultivation in Northeast, North China, and Northwest regions. The fruit clusters are conical in shape, with tightly or moderately compacted grains. The average cluster weight is 380 g, with a maximum weight of 1 000 g. The grains are oval-shaped, with an average weight of 1.95 g. The thin peel is green-yellow in color. The flesh is crisp, with low juice content and no seeds, and has a sweet and sour taste. It contains 22.4% sugar and 0.4% acid, with a drying rate of 20% to 30%. It is the main variety for drying in China, and is also suitable for canning and fresh consumption. The tree has a strong vigor and high fruit set rate, with two fruit clusters per fruit branch and strong secondary branch fruiting ability. The fruits mature uniformly and do not shatter. In the Turpan region of Xinjiang, it sprouts in early April, blooms in mid-May, and the fruits mature in late August. It is suitable for cultivation in small sheds or trellises. Under high temperature and dry conditions with good fertilization and irrigation, 4-year-old trees can yield up to 6.5 kg. The yield is moderate. It has strong disease resistance, but disease prevention and control should be strengthened in rainy areas.

010 What are the characteristics of the Venus Seedless grape?

Euro-American hybrid variety. The young shoots are green with a dense covering of trichomes and glandular dots. The young leaves are green with a light red edge and are densely covered in trichomes. The mature leaves are large, dark green, heart-shaped, moderately thick, 3-5 lobed with shallow lobes, rough on the upper surface, densely covered in trichomes on the lower surface, with blunt serrations along the leaf margin. The petioles are long and either closed or narrowly arched. The plant has perfect flowers. The fruit clusters are conical, with an average weight of 260 g and a maximum weight of 500 g. The grains are closely packed and uniform in size. The grains are nearly round, with an average weight of 4.0 g and a maximum weight of 4.5 g. The skin is blue-black and relatively thick. The flesh is soft, juicy, and has a rich aroma, combining the characteristics of both the Eurasian and American species. It contains 15.2% to 17% soluble solids and has an acid content of 0.97%. The fruit stems are long, and there is no cracking or shattering. The overall quality is above average. In the Shenyang region, it ripens in mid-August, with a growth period of approximately 110 days from sprouting to fruit maturity. In the Chaoyang region of Liaoning Province, it sprouts in early May, blooms in early June, and the fruit matures in early August. In the Nanjing region, it ripens in mid-July. The fruit has good storage and transportability. It is suitable for short pruning and cultivation in small sheds or trellises. The plant exhibits strong cold resistance and disease resistance.

011 What are the characteristics of the Red Ruby Seedless grape?

Eurasian species, performs well in Liaoning, Hebei, Jiangsu, Beijing, and other provinces and cities. The plant exhibits vigorous growth. The fruit clusters are conical, with moderate grain attachment. The average cluster weight is 650 g, with a maximum weight exceeding 1 500 g. The grains are short oval-shaped, with an average weight of 3. 7 g and a maximum weight of 4. 5 g. The skin is purplish-red, thin, and the flesh is crisp with a rich, sweet and refreshing taste. It contains 17. 5% soluble solids and has excellent quality. In Beijing and Shenyang regions, the fruit matures in early September and mid-September respectively. The yield is moderate. It has medium resistance. The young shoots are green with sparse trichomes. The young leaves are yellow-green with orange edges and covered with white trichomes. The mature leaves are large, heart-shaped, 5-lobed with deep lobes, smooth on the upper surface, deeply serrated, with long deep red petioles that close the petiole groove.

012 What are the characteristics of the Mars Seedless grape?

Euro-American hybrid variety. The young shoots are green with dense white trichomes and spherical glands. The young leaves have pink edges and the undersides are covered in white trichomes. One-year-old mature branches are dark brown, prickly, with medium-length internodes. The mature leaves are large, nearly circular, shallowly 3-lobed, rough on the upper surface, and densely covered in trichomes on the lower surface. The leaf margin has blunt serrations. The petioles are relatively long, light red in color, and have a narrow arched petiole groove. The fruit clusters are long conical, with an average cluster weight of 266 g and a maximum weight of 502 g. The grains are closely packed, nearly round, with an average weight of 4. 3 g. The skin is purple-black, relatively thick, quickly colored, and moderately coated with bloom. The flesh is soft, juicy, and contains 16. 2% soluble solids and 0. 62% acid. It has a sweet and sour taste with a strong aroma characteristic of American varieties. The overall quality is above average. The tree has strong vigor, with a fruiting branch rate of 89. 2% and a double cluster rate of 91. 2%. The terminal bud has strong fruiting ability, making it suitable for medium to short pruning. The plant exhibits strong cold resistance and disease resistance. It is highly productive, with 5-year-old trees yielding 24.6 kg. The fruit clusters mature uniformly and do not shatter. In the Chaoyang region of Liaoning Province, it sprouts in early May, blooms in early June, and the fruit matures in early August. In the Shenyang region, it ripens in late August. This variety is suitable for cultivation in colder and rainy regions and is one of the early-maturing seedless varieties.

013 What are the characteristics of the Cabernet Sauvignon grape?

Also known as Cabernet and Sauvignon, Cabernet Sauvignon is a Eurasian variety belonging to the Western European population and originated in France. It is an ancient French variety. Currently, Cabernet Sauvignon grapes are extensively cultivated in Yantai, Longkou, Zibo, Jinan, and other areas in Shandong province. It is also cultivated in Beijing, Liaoning, Henan, and Jiangsu provinces. It is the main raw material variety for producing high-quality dry red wines. This variety has strong vigor, with buds sprouting in late April, flowering in early June, and maturing from late September to early October. The ripening period is consistent, and the growth period is 155 – 165 days. It requires an effective accumulated temperature of 3 150 – 3 300 ℃. The fruiting ability is weak, with a fruiting branch rate of about 70% on trellises and around 40% on pergolas among all the newly sprouted shoots. Each fruit branch usually bears 1 – 2 clusters. It has moderate disease resistance and strong drought tolerance. The clusters of grapes are medium-sized and conical in shape, with a cluster length of about 15.5 cm and width of around 10 cm. The grains are moderately packed, with an average cluster weight of 170 g. The grains are medium-sized, nearly round, with an average weight of 182 g per hundred grains. They have a dark purple-black color and thick bloom. The skin is thick, and the flesh is relatively firm. The juice yield of the fruit is 72% to 78%. The juice has a ruby red color, is clear and transparent, with a soluble solids content of 17% to 21% and an acidity level of 0.8% to 1.0%. It has a sweet and sour taste with a hint of grassiness. The wine produced from this variety has a gem-like red color, is clear, and has a harmonious and mellow flavor. This variety has strong growth potential and is easy to manage, but the vigor should be appropriately controlled to prevent excessive growth. It has strong adaptability and resistance, and is suitable for cultivation in poor and gravelly slopes. Sufficient fertilizer and water supply are needed during the fruiting period to avoid low yields. This variety prefers warm and sunny climates and is suitable for trellis cultivation, preferably with medium to long pruning. To improve grape quality and reduce fruit diseases, 3 to 5 leaves at the base of the leaf canopy that affects sunlight exposure to the clusters should be removed one month before harvest, allowing most of the clusters to be exposed to sunlight.

014 What are the characteristics of the Cabernet Gernischt grape?

This is a Eurasian variety, originally from France and is an ancient French variety. It was introduced to Yantai, China in 1892 and has been widely promoted and developed in the Jiaodong grape-growing region. There is also a small amount of cultivation in Beijing, Tianjin, Henan, and other areas. In recent years, regions such as Changli in Hebei province have also seen development in wine grape production, making it one of the main cultivated varieties. This variety has strong vigor, with buds sprouting in late March to early April, flowering in mid to late May, and fruit ripening in early to mid-September. The growth period is over 150 days, requiring an accumulated temperature of over 3 200 ℃. The bud sprouting rate is relatively high, with fruiting branches accounting for about 70% of the total buds, and an average of 1. 6 flower clusters per fruiting branch. The clusters are medium-sized, measuring around 15–16 cm in length and 11 cm in width, with a weight of over 230 g. They are cylindrical or conical in shape, and the clusters are tightly packed. The grapes are medium-sized, round, with an average weight of about 210 g per hundred grapes. The skin is thick, purplish-red in color, and covered with a thick bloom. The taste is sweet and juicy, with a grassy flavor. The soluble solids content ranges from 17. 8% to 19. 5%, and the acidity level is between 0. 6% and 0. 8%. The juice yield is over 75%, and each fruit usually contains 1–2 seeds. It has strong disease resistance, drought tolerance, and can tolerate poor soil conditions. It is suitable for cultivation in loose soil, sandy loam, and dry or wet areas. It is recommended to use medium to short pruning. Moderate fertilizer and water supply are required.

015 What are the characteristics of the French Blue grape?

This variety, also known as Blue Franconian, is a Eurasian species originally from Austria. It was introduced to China by the Yantai Zhangyu Grape and Wine Company in 1915 and is now cultivated in the North China and Yellow River old course regions. The tree has moderate vigor, with buds sprouting in mid-April, flowering in late May, and ripening in early to mid-September. It is a mid-ripening variety with a growth period of 130–137 days, requiring an accumulated temperature of 2 890–2 980 ℃. The bud sprouting rate is 68%, and the fruiting branch rate is around 55%, with an average of 1. 6 clusters per fruiting branch. The clusters are relatively large, conical in shape, and have secondary clusters, generally weighing 250–300 g. The grapes are closely packed, blue-black in color, with thick bloom and tough skin. The flesh is soft and juicy, with a soluble solids content of 17%–18%, acidity level of about 0. 6%, and juice yield of 68%–70%. This variety has moderate growth vigor, is easy to control, and is manageable. It does not have strict soil requirements and can grow well even in relatively barren mountainous and sandy conditions. It is suitable for cultivation in dryland and sloping areas. The wine made from this variety is ruby red in color, clear and bright, with a soft and refreshing taste.

016 What are the characteristics of the Carignan grape?

This variety is a Eurasian species and is one of the ancient and excellent wine grape varieties in Western Europe. Currently, it is widely cultivated in regions such as Shandong, Hebei, and Henan in China. The plants have strong growth vigor, with a high bud sprouting rate. Each fruiting branch has an average of 1.9 flower clusters, and the secondary branches have strong fruit-setting ability. Young trees enter the fruit-bearing period early and are highly productive. It has strong adaptability and salt-alkali tolerance. It is suitable for cultivation on trellises and small pergolas, with mixed pruning. In Jinan region, the phenological period starts with bud sprouting in early April, flowering in mid-May, and ripening in early September. The growth period is about 150 days, requiring an accumulated temperature of 3 500 ℃. The clusters are medium-sized or large, conical in shape, with divergent shoulders, weighing between 270 g and 650 g. The grapes are tightly packed, medium-sized, with varying ripening periods. They are elongated and oval in shape, purplish-black in color, and weigh between 250 g and 300 g per hundred grapes. The flesh is soft and juicy, with a sweet and sour taste. The soluble solids content ranges from 15.5% to 18.5%, acidity level from 0.71% to 0.91%, and juice yield from 75% to 80%. This variety is one of the ancient red wine grape varieties in the world. The wine made from it is ruby red in color, with a good flavor and aroma. It is suitable for blending with other varieties, and the skin can be used to make white or rose wines.

017 What are the characteristics of the YAN 73 grape?

The Yantai Zhangyu Grape Brewery Company created the variety YAN 73 through the hybridization of Alicante Bouschet as the female parent and Muscat as the male parent in 1966.

It was officially identified as a grape coloration variety cultivated in China in 1981. This variety has vigorous growth and adapts well to various soil types, showing relatively good disease resistance. It is important to pay attention to the application of basal fertilizer and organic fertilizer during cultivation, as well as the appropriate use of micro-fertilizers to promote increased sugar content and full pigment formation. YAN 73 is an excellent variety for producing dry red and sweet red wines with enhanced coloration. Wines made from YAN 73 after coloration exhibit a deep purple-red color and pure aroma, surpassing the world's best original coloration variety, Purpureas Seedling. This variety has strong growth vigor and fruit-setting ability, with a fruiting branch rate of 57. 3% and an average of 1 - 2 clusters per fruiting branch. It bears fruit early and has high yield. The fruit matures in early September. It shows relatively good disease resistance and strong adaptability. The clusters are conical in shape, medium-sized, with an average cluster weight of around 250 g. The berries are elliptical, red-black in color, closely spaced, medium-sized, with consistent ripening, and weigh approximately 226 g per hundred berries. The juice is deep purple-red in color. It has a soluble solids content of 16. 5%, acidity level of 0. 71%, and juice yield of 68%.

2. Requirements for Safe and High-Yield Cultivation Techniques for Grapes

018 What are the effects of air pollution on grapes?

Air pollution, including sulfur dioxide, fluorides, nitrogen oxides, chlorine gas, as well as gases, solids, and liquid particles such as dust and smoke, can cause damage to the flowers, leaves, and fruits of grapevines. The accumulation of pollutants in the plant can lead to acute and chronic poisoning when consumed by humans.

019 What are the effects of soil pollution on grapes?

Soil pollution refers to the entry of harmful substances discharged from human production and life into the soil, which affects the growth and development of grapevines and subsequently impacts human and animal health. Common pollutants include heavy metals, toxic substances. The main sources of soil pollution are industrial waste, pesticide and fertilizer pollution, and pollution caused by sewage irrigation. Soil pollution directly deteriorates the biochemical characteristics of the soil, affects the growth and development of grapevines, and impacts human and animal health.

020 What are the effects of water pollution on grapes?

Water pollution sources mainly include domestic sewage, industrial wastewater, agricultural wastewater, and livestock wastewater. Water pollution increases the harmful salt content in the soil, which inhibits the growth of grapevines.

021 What are the effects of pesticide pollution on grapes?

Pesticides play an important role in the prevention and control of grape diseases and pests but also bring serious negative effects to grape production. Pesticide pollution manifests in two aspects: contamination of the vineyard and direct contamination of the grape produce.

022 What are the effects of fertilizer pollution on grapes?

Excessive and imbalanced use of fertilizers can lead to the accumulation of nitrosamines (carcinogenic substances) and nitrogen dioxide, causing harm to the grapevines and human and animal health.

023 What are the requirements for grapevine site selection?

Grapes have strong adaptability, and in general, they can be cultivated in mountainous areas, flatlands, or riverbanks to achieve good yields. However, the growth, yield, and quality of grapes vary in different soil, terrain, and slope conditions. Mountainous vineyards have ample sunlight, good air circulation, large diurnal temperature variations, and good grape quality with minimal pest and disease damage. However, mountainous areas are prone to soil erosion and are more affected by drought, so it is important to pay attention to soil and water conservation and apply organic fertilizers to create a favorable growth environment for grape roots. Flatland vineyards have the advantage of fertile soil, sufficient water supply, vigorous plant growth, and high yields. However, they have inferior conditions in terms of sunlight, ventilation, and drainage, resulting in poorer grape quality and storage ability, as well as more severe pest and disease damage. Therefore, when selecting a vineyard site, it is necessary to conduct a detailed investigation of the local topography, soil, water sources, transportation, market sales, and other factors, taking into account the advantages and disadvantages and selecting a suitable site to lay a good foundation for high grape yields. Additionally, it is important to avoid competition with staple crops and cotton when selecting vineyard sites and make use of available rural courtyard spaces and scattered land. Areas with poor sunlight conditions, excessively heavy or sticky soils, and poor ventilation and light transmission are not suitable for grape cultivation.

024 What are the temperature requirements for grapes?

Temperature conditions are a key factor to consider when developing grape production and selecting grape varieties in a specific region. Generally, the absolute minimum temperature for grapevines during winter is around −17 ℃, which is the boundary between burying the vines for winter protection and leaving them exposed. Grapevines have poor cold tolerance, and European grape varieties can withstand temperatures as low as −15 ℃ during dormancy, but freezing damage can occur at around −17 ℃−−16 ℃. Fully mature one-year-old shoots can tolerate short-term low temperatures of −20 ℃, while older canes experience freezing damage at temperatures between −26 ℃ and −20 ℃, and tender shoots can be damaged at −1 ℃. Grapevine roots have the poorest cold resistance and can be damaged at temperatures of −7 ℃−−5 ℃. Flower buds can be damaged at −6 ℃, and low temperatures during the flowering period can cause freezing damage to the floral organs.

Among grape varieties, the Beida and muscadine types have the strongest cold resistance, with their roots able to withstand temperatures of −16 ℃−−14 ℃. Therefore, these varieties are commonly used as cold-resistant rootstocks in cold regions.

025 What are the rainfall requirements for grapes?

Grapes have strong drought tolerance, and they can be cultivated in areas with annual rainfall ranging from 350−1 200 mm. The seasonal distribution of rainfall has a significant impact on grape growth, fruit quality, and yield. During the bud break and shoot growth period in spring, sufficient rainfall is beneficial for the differentiation of flower primordia and new shoot growth. Grapevines require clear and warm weather with relatively dry conditions during the flowering period. If the weather is humid or continuously rainy and cool, it can hinder normal flowering, pollination, and fertilization, leading to fruit drop. Excessive rainfall or continuous rain during the grape ripening period(July to September) can lower the sugar content, promote disease development, and cause fruit cracking, severely affecting grape quality. In the late growth period(September to October), excessive rainfall can lead to poor maturity of new shoots and susceptibility to frost damage. Therefore, overall, grapes are most suitable for cultivation in areas with abundant sunshine, loose soil, good drainage, and dry climates with irrigation conditions.

026 What are the light requirements for grapes?

Grapes are typically sun-loving crops, and under sufficient sunlight, the leaves are thick and dark, the plants grow vigorously, and the flower bud differentiation and yield are high, resulting in good fruit quality. Insufficient light can have the opposite effect. Different grape varieties have different requirements for light intensity, with European and Eurasian varieties requiring higher light conditions compared to American varieties. For example, varieties like Cabernet Sauvignon can achieve good coloration under diffused light, while varieties like Muscat and Cabernet Franc require direct sunlight for proper color development. Seedless white varieties have even higher light requirements. Areas with poor light conditions, excessively heavy soils, excessive moisture, and poor ventilation and light transmission are not suitable for grape cultivation.

027 What are the soil requirements for grapes?

Grapes are not very demanding on soil conditions. They can be grown in various types of soil, except for heavily saline-alkaline soils, marshlands, areas with groundwater levels less than 1 meter, and soils that are excessively heavy and poorly ventilated. However, the most suitable soils for grapes are well-drained gravelly loam and sandy loam, especially for high-quality wine grape varieties that have strict requirements for soil texture and structure. Grapes have a wide adaptability to soil pH, generally being able to be cultivated in soils with a pH range of 5. 8-8. 2. The optimal pH range for grape growth is typically between 6. 5 and 7. 5. Different grape varieties have varying abilities to tolerate salt and alkali, as well as iron deficiency-induced chlorosis. European and Eurasian varieties generally have better salt and alkali tolerance, while European-American hybrid varieties have poorer tolerance and are prone to leaf yellowing symptoms in saline-alkali soils. However, overall, areas with excessively wet and heavy soils and severe soil salinization are not suitable for grape cultivation.

028 What are the conditions for producing pollution-free grapes?

The requirements for air quality in pollution-free grape production mainly involve the content of sulfur dioxide, nitrogen dioxide, fluorides, and total suspended particulate matter (Table 2-1).

Table 2-1 Environmental Air Quality Indicators

Item	Concentration Limit	
	Daily Average①	1 h Average②
Total Suspended Particles(Standard State)/(mg/m^3)	≤0. 30	-
Sulfur Dioxide(Standard State) /(mg/m^3)	≤0. 15	≤0. 50
Nitrogen Dioxide(Standard State) /(mg/m^3)	≤0. 12	≤0. 24
Fluorides(Standard State) /($\mu g/m^3$)	≤7	≤20

Note: ① Daily Average refers to the average concentration of any 1 day.

② 1 h Average refers to the average concentration of any 1 hour.

The quality requirements for irrigation water in farmland mainly involve pH, chemical oxygen demand, chloride, and the quantity of coliform bacteria. The concentration requirements are shown in Table 2-2.

Table 2-2 Irrigation Water Quality Indicators

Item	Concentration
pH	5. 5-8. 5
Chemical Oxygen Demand /(mg/L)	≤150
Total Mercury/(mg/L)	≤0. 001
Total Cadmium/(mg/L)	≤0. 005
Total Arsenic/(mg/L)	≤0. 05
Total Lead/(mg/L)	≤0. 10
Chromium(Hexavalent)/(mg/L)	≤0. 10
Fluorides/(mg/L)	≤2. 0
Cyanides/(mg/L)	≤0. 50
Petroleum/(mg/L)	≤1. 0
Fecal Coliforms/(individuals/L)	≤1 0000

Soil environmental quality mainly involves the content of hexachlorocyclohexane(HCH), dichloro diphenyl trichloroethane (DDT), arsenic, mercury, lead, copper and other heavy metals(Table 2-3).

Table 2-3 Soil Environmental Quality Indicators

Item	Content Limit		
	pH<6. 5	pH 6. 5-7. 5	pH>7. 5
Cadmium/(mg/kg) ≤	0. 30	0. 30	0. 60
Mercury/(mg/kg) ≤	0. 30	0. 50	1. 0
Arsenic/(mg/kg) ≤	40	30	25
Lead/(mg/kg) ≤	250	300	350
Chromium/(mg/kg) ≤	150	200	250
Copper/(mg/kg) ≤	50	100	100

Note: The items listed above are average values based on elemental quantities and are applicable to soils with a cation exchange capacity > 5 cmol(+)/kg. If the cation exchange capacity is ≤5 cmol(+)/kg, the standard values are half of the values shown in the table.

It is necessary to establish necessary roads, drainage systems, shelterbelts, and auxiliary buildings in pollution-free fruit production orchards. Slopes below 6° should be oriented north-south, while slopes between 6° and 25° should be planted horizontally. Vineyards should not be established on slopes steeper than 25°. When selecting a vineyard site, it is necessary to consider the local climate conditions, soil conditions, irrigation conditions, environmental conditions, topography, and terrain. Adhering to the principle of suitable site selection, it is important to stay away from cities, transportation arteries, and industrial areas with pollution from "three wastes". To achieve scale and facilitate mechanized operations and the application of high-tech, vineyards should be concentrated and contiguous.

029 How can pollution be controlled in different stages of grape production?

Developing pollution-free grape production is an inevitable requirement for China's economic and social development. As a developing country, China is facing the dual tasks of development and environmental protection. To limit the use of fertilizers and pesticides, control production process pollution, and change traditional agricultural production methods, it is necessary to rely on a complete standard system to regulate pollution-free grape production through pre-production, mid-production, and post-production stages. Protecting and improving the ecological environment of agriculture, promoting the application of pollution-free agricultural production technologies, and adopting a rational approach to the use of pesticides and fertilizers are essential requirements for the development of pollution-free grape production.

The over-standard residue of pesticides, nitrite, heavy metals and hormone-like substances in grapes has greatly impacted China's grape exports. Meanwhile, the influx of foreign grapes into the Chinese market necessitates the development of pollution-free, harmless, and nutritionally safe grape production in order to ensure China's share in both domestic and international markets. There are significant price differences between ordinary, harmless, green food, and organic food in the market. The price of harmless grapes is 20% to 30% higher than ordinary grapes, and some are more than 50% higher. Therefore, vigorously developing harmless grape production and developing harmless grapes with local characteristics can not only improve agricultural economic benefits and increase farmers' income, but also protect the ecological environment and promote sustainable agricultural development.

Adhere to the policy of "prevention first, integrated control", take agricultural prevention as the foundation, combine manual, physical and biological control, scientifically use chemical pesticides to effectively control pests and diseases. Chemical control plays an important role in the prevention and control of grape diseases and pests. However, it is also the aspect with the most issues. Due to a lack of knowledge about disease and pest control and pesticide use, improper pesticide applications, excessive concentrations, and the use of highly toxic pesticides without following regulations can lead to pollution and the development of pesticide resistance, creating a vicious cycle. It is necessary to strictly follow the

"Pesticide Use Guidelines for Green Food" and prohibit the use of highly toxic, high residue, carcinogenic, teratogenic, and mutagenic pesticides. The use of broad-spectrum and highly resistant pesticides should be limited, and the use of hormone-based pesticides should be strictly controlled. Whenever possible, non-toxic, residue-free, or low-toxicity, low-residue pesticides should be used. Plant-based pesticides, animal-based pesticides, microbial pesticides, and mineral-based pesticides, such as sulfur and copper compounds, should be actively used. Proper understanding of the scope, target pests, dosage, and frequency of pesticide use is necessary, and increasing the dosage or concentration without authorization should be avoided. Long-term use of the same pesticide should be avoided, and attention should be paid to alternating and rotating pesticide use to delay the development of resistance and control other diseases, saving labor and reducing pesticide use.

030 What are the characteristics of grape root growth and development?

Grapes have a well-developed root system, strong regenerative capacity, and excellent nutrient absorption ability, making them resistant to drought, infertile soil, and saline-alkali conditions. They are not easily damaged by cultivation. Both the above-ground stems and underground root system of grapes have strong growth potential, large growth volume, and long lifespan. Throughout the year, as long as the conditions are suitable, grapes can continuously grow new roots. Under open-field cultivation conditions, there are generally two growth peaks in a year.

031 What are the characteristics of grape shoot growth and development?

Grapevines are trailing plants and cannot grow upright, requiring support structures. The branches are called shoots, which grow rapidly with long internodes. They climb and grow upwards with tendrils. The annual growth can reach 1 to 10 meters. New shoots grow from single axes and compound axes, and the tendrils exhibit a regular distribution pattern. Grapevines do not form terminal buds, and as long as the temperature is suitable, new shoots can continue to grow until leaf fall. Under natural growth conditions, nutrient accumulation is not easy, and upper branches and vines often mature poorly. Pruning should be done regularly to limit excessive growth. Shoots and vines exhibit clear apical dominance and vertical dominance.

032 What are the characteristics of grape bud growth and development?

The bud of grapes is a compound bud, also known as a "bud eye". Each leaf axil contains two buds, namely the summer bud and the winter bud. The summer bud is a naked bud, which sprouts as a lateral shoot during the growth of new shoots. Some varieties can produce flower clusters on lateral shoots. The winter bud is covered by scales and generally sprouts in the spring of the second year. In addition to the primary bud, it may also contain 3-8 dormant buds, among which only 2-3 develop well. In spring, double or triple buds often occur on leaf buds. When the primary bud is damaged, the dormant buds can sprout and replace it. The flower bud of grapes is a compound bud, but the differentiation of the inflorescence primordia before winter is shallow, making it difficult to distinguish from leaf buds based on appearance. Generally, from the first to second node at the base of the branch (fruiting cane), and up to the 20th node or more, each node can form a flower bud. Weaker varieties tend to have lower positions of flower bud initiation, while stronger varieties have higher positions. Additionally, this is also influenced by climate and cultivation techniques. The best quality flower buds are often found in the lower to middle sections of the branch(4th to 5th node, 10th to 15th node). Similar to the primary bud, the dormant buds of grapes can also undergo differentiation into flower clusters, but the degree of differentiation is generally lower and varies depending on the variety.

033 What are the characteristics of grape flower growth and development?

Grape flowers are compound inflorescences(panicles), typically borne on the 3rd to 8th nodes of the current year's fruiting shoots. Each inflorescence can have 2-4 flower clusters. Each cluster can contain 200-1 500 small flowers, varying depending on the variety. Nodes without flower clusters will develop tendrils instead. There are various transitional types between inflorescences and tendrils. Most grape varieties are hermaphroditic and can self-pollinate to set fruit. Grape flowers and fruits experience significant drop. Fruit drop begins 3-7 days after flowering, with a peak around 9 days after flowering, lasting for about 2 weeks. After this period, fruit drop is minimal. Apart from varietal factors, fruit drop can be caused by low-temperature and rainy conditions during the flowering period, excessive shoot growth, weakened tree vigor, and boron deficiency in the soil. Some grape varieties exhibit monoecious fruiting or partial seed abortion, resulting in seedless grapes, such as seedless white varieties. Occasionally, "bean-shaped" fruits can develop within normal grape clusters, causing variations in berry size, maturity, and leading to uneven ripening, which should be avoided.

034 What is a trellis? What are the different types?

A trellis is a structure that is erected vertically, resembling a fence, hence the name "trellis". It is also referred to as an upright trellis due to its vertical orientation. Trellises can be categorized into single-arm trellises and double-arm trellises.

035 What are the characteristics of a single-arm trellis?

A single-arm trellis facilitates ventilation and light penetration, improves grape quality, and allows for easy field management. It also enables high-density planting, leading to early and abundant yields, making it suitable for large-scale winemaking operations. Additionally, it facilitates mechanized operations such as cultivation, spraying, shoot thinning, harvesting, and soil covering for winter protection, saving labor. However, its disadvantages include being influenced by the polar growth of the plants, resulting in excessive vegetative growth, dense foliage, upward relocation of the fruiting zone, and difficulty in control. The lower fruit clusters are closer to the ground, making them more susceptible to contamination and pest and disease infestation.

036 What are the different types of trellises?

Trellises include small trellises, large trellises, funnel-shaped trellises, horizontal trellises and ridge trellises.

037 What are the characteristics of a small trellis?

The main characteristics of a small trellis are as follows: Suitable for most grape varieties' growth requirements, promoting early and abundant yields. The vine shoots are only 5–6 meters long, making it easy to manage and manipulate the trellis. The main vine is relatively short, allowing for easy adjustment of tree vigor, resulting in higher and more stable yields. Additionally, it has a quick recovery after renewal, with minimal impact on production. The selection of trellis materials is relatively easy.

3. Grapevine Spring Management (March to May)

038 What are the effects of early or late uncovering and trellising on grapes? What should be considered during uncovering and trellising?

When the spring temperature reaches 10 ℃, the grapevines covered with soil for winter protection should be uncovered and trellised in a timely manner. Uncovering too early can expose the vines to late frosts and freezing temperatures, while uncovering too late can result in the sprouting of young buds in the soil and potential damage to the tender shoots. The correct timing of uncovering should be determined based on the local climate and the phenological stage of the grape variety. Generally, it should be done before bud swelling. In the central and northern regions of Liaoning Province, uncovering and trellising usually take place in mid-April.

During the process of uncovering and trellising grapevines, it is important to be careful and cautious to prevent damaging the shoots and buds, which could lead to bleeding. After trellising, it is recommended to spray a mixture of 3-5 Bordeaux mixture and 0.3% sodium pentachlorophenolate to eliminate any overwintering pests and diseases remaining on the shoots.

039 How should pesticide application and irrigation be managed after trellising grapevines?

After trellising grapevines, it is crucial to immediately irrigate them to ensure sufficient water penetration. Adequate irrigation should be provided without excessive watering to avoid lowering the soil temperature and delaying bud sprouting. At the same time, it is recommended to spray a mixture of 3-5 Bordeaux mixture and 0.3% sodium pentachlorophenolate to eliminate any overwintering pests and diseases on the shoots. When spraying, pay attention to thoroughly cover the shoots and the planting furrow.

040 How should grapevines be trellised after uncovering?

Since grapevines have vigorous growth, trellising is a regular task during the summer management period. In the spring, as soon as the grapevines are uncovered, trellising should begin. It is important to ensure the even distribution of shoots on the trellis, with approximately 50 cm spacing between each shoot. As the new shoots grow to a certain length, adjust the binding angle based on their growth vigor to achieve balanced growth.

041 How can blind shoots on grapevines be controlled?

Blind shoots, also known as "blind eyes", refer to the dormant buds on mature grapevines that fail to produce new shoots. This phenomenon can lead to a disorderly vine structure, weakened vigor, and hindered rejuvenation.

To control blind shoots on grapevines, it is necessary to understand the causes and take corresponding measures based on the actual situation. During the early stages of vine growth, organic fertilizers should be primarily used for fertilization, and excessive application of quick-acting nitrogen fertilizers should be avoided. Organic fertilizers can be applied in the autumn after harvest, while additional fertilization should mainly occur in the first half of the year, especially during the rainy season, when the application of quick-acting nitrogen fertilizers should be avoided. Pay attention to vineyard drainage during the rainy season, and control irrigation during the later stages of grape growth. From April to June, during the shoot growth and development period, it is recommended to spray a 0.3% urea solution and a 0.2% potassium dihydrogen phosphate solution every 10 days to meet the needs of bud differentiation and fruit enlargement. Thinning flowers and fruits at the appropriate time and adjusting the grape's crop load are also important. Prune an appropriate number of fruiting branches, ensuring they are not excessively long or numerous. Properly manage shoot tipping, removing lateral shoots and handling secondary shoots to ensure good light conditions and bud differentiation. Timely remove or retract densely growing shoots, overlapping shoots, and excessively long shoots. For weak shoots emerging from blind spots, leave 2-3 short nodes to allow for the development of new shoots in the following year and rejuvenation of the old vine. If the shoots are vigorous, long shoots can be left for the current year and tipped, cultivating them as fruiting canes or mother vines.

042 How to select a grapevine orchard site and varieties?

(1) Selecting the orchard site based on local natural conditions. Grape cultivation is relatively easy, and grapes have strong adaptability. However, not every location is suitable for grape cultivation. Before establishing an orchard, a detailed investigation of the local climate and soil conditions should be conducted to determine if they align with the ecological requirements of the cultivated varieties. Additionally, it is important to utilize the characteristics of small areas and microclimates, overcoming and mitigating unfavorable

factors. For example, in areas with high rainfall and humidity in summer and autumn, it is advisable to choose open slopes for orchard establishment to facilitate ventilation and drainage. In hot climate regions, selecting sites at higher elevations is recommended. Specific requirements include elevated and dry terrain, sufficient sunlight, convenient irrigation and drainage, groundwater level below 1 meter in flat areas, and access to ample water sources. Deep and fertile soil with good structure and a pH value of 6.5–7.5 is desirable.

(2) Determining the orchard scale and variety development based on market demand. One of the prominent features of modern agriculture is industrialized production. Without a certain scale, it is difficult to establish a production area, cultivate a market, and build a brand, which hinders participation in market competition and achieving greater profitability. To achieve better economic benefits and avoid significant economic losses caused by blind development, it is necessary to conduct market research and forecasting before establishing an orchard. The development scale and cultivated varieties should be determined based on market demand and economic benefits, ensuring that the chosen varieties are suitable and in line with supply and demand coordination.

043 What elements are included in grapevine orchard planning and design?

Grapevine orchard planning and design mainly include plot layout, trellis systems, road networks, and irrigation and drainage systems.

044 When is the best time to plant grapes in cold regions during the winter in the north?

In cold regions during the winter in the north, spring planting is commonly practiced. In the central and northern parts of Liaoning Province, planting is usually done in mid to late April, as long as the seedlings haven't sprouted. It is recommended to plant slightly later for better survival.

045 How should the planting spacing be determined for grapevine orchard establishment?

In different regions of China, grape planting systems mainly include trellises and pergolas, which result in significant differences in plant density per unit area. Currently, commonly used planting spacing is approximately 0.5–1 meter between vines and 2–3 meters between rows for trellises, and 0.5–1 meter between vines and 4–6 meters between rows for small pergolas.

046 How to improve the survival rate of planted grapevines?

To improve the survival rate of planted grapevines, it is essential to select strong and healthy seedlings, ensure sufficient water supply, use rooting agents, apply plastic mulch, and properly control the planting depth.

047 How to perform bud rubbing and shoot tipping for grapevines?

After the spring budburst, dormant buds that sprout on old vines should be removed, except for those that need to fill space or are planned for renewal. Generally, vigorous shoots with flower clusters should be retained, which are usually sprouted from the primary bud, while shoots sprouted from secondary buds should be removed. During the vigorous growth period of young vines, some shoots with flower clusters can also be left, but this is not necessary for mature vines. For pergola systems, it is recommended to retain approximately 15 new shoots per square meter. When there are excessive and overcrowded new shoots on the trellis, weak shoots and undeveloped shoots without fruit clusters should be removed. After shoot tipping is completed, the trellis should have full foliage, even distribution, good ventilation, and light penetration. It is also important to retain a certain proportion of fruitless shoots at the base of the plants to increase the photosynthetic area, achieve an appropriate leaf-to-fruit ratio, ensure normal cluster development, and provide robust fruiting canes for the following year.

048 How to perform shoot tipping for grapevines?

Shoot tipping involves removing the growing tip of the new shoots, along with several young leaves, which helps temporarily suppress apical growth, channel more nutrients into the flower clusters, promote cluster development, increase fruit set rate, and reduce flower and fruit drop. The timing of shoot tipping varies depending on the variety, growth vigor, and cultivation conditions. For varieties with lower fruit set rates, such as Muscat and Kyoho, shoot tipping can be performed around one week before flowering until the fruit is fully developed, resulting in better effects. Generally, 4–8 leaves should be left on the cluster. For varieties with higher fruit set rates, shoot tipping before flowering is not meaningful. Shoot tipping can be performed during the early fruiting stage after flowering.

For non-flowering developmental shoots, the timing of shoot tipping should be determined based on the planned application. Generally, it should be done when the new shoots have reached the desired length. For example, if the developmental shoots are planned to be used as main canes, shoot tipping can be performed after they have grown to 8–9 leaves. In the central and northern regions of Liaoning Province, shoot tipping for extended canes is generally done in early August, and additional shoot tipping should be performed appropriately in the later stages to promote berry enlargement, sugar accumulation, and cane maturation. Pay attention to the timing and severity of shoot tipping. To promote shoot vigor, continuous shoot tipping can be performed on secondary shoots based on their growth vigor and planned applications, promoting maturation.

049 How to manage grape flower clusters?

Thinning flower clusters should be done based on the position of fruiting branches and their growth vigor. Generally, for vigorously growing fruiting branches in the upper part of the vine, two flower clusters can be retained. For moderately growing fruiting branches in the middle and lower parts, one flower cluster can be retained. For weakly growing branches that are planned as renewal canes, no flower clusters should be retained.

To ensure the uniformity of grape clusters, cluster shaping should be performed. Usually, within one week before flowering, remove lateral shoots, shoulders, and larger small clusters. At the same time, pinch the cluster tip, typically removing 1/5 to 1/4 of the cluster.

To promote the even development of clusters and improve their market value, it is necessary to thin the berries. Remove poorly developing or densely clustered berries and retain a certain number of berries based on the characteristics of the variety. For example, Kyoho variety requires around 30 berries per cluster, while Red Globe requires around 50 berries.

050 How to tie grapevines?

When the new shoots reach a length of 30–40 cm, they should be tied to the trellis to prevent them from breaking due to wind or being damaged by fruit weight. Young shoots need to be tied 2–4 times as they grow, while fruiting canes need to be tied 1–2 times. When tying the vines, remove tendril growth because tendrils consume nutrients and interfere with the desired vine shape. When tying the vines, it is important to secure them to the wire without excessive tension, allowing the vines to thicken and grow without hindrance. For this purpose, plastic ropes can be fixed to the wire using "pig's foot buckles", and then the ropes can be twisted twice before tying the vines.

051 How to perform flower and fruit thinning for grapes?

Flower cluster thinning should be carried out one week before flowering, and it should be completed within 2–3 days. Thinning too early makes it difficult to distinguish the clusters, while thinning too late affects the effectiveness. For large clusters, remove lateral clusters first, then remove 1–8 secondary axes at the base of the cluster axis. For small clusters, fewer or no secondary axes should be removed. If the secondary axes at the base of the cluster axis are not removed, the base of the cluster becomes large, affecting the cluster's appearance. Additionally, flowers in the middle of the cluster tend to open earlier and have better fruit development compared to the flowers at the base.

Cluster tipping should be done just before flowering, aiming to complete the process within 2–3 days. Tipping too early can result in lateral growth of the clusters, affecting their appearance. Remove approximately 1/4 of the cluster tip, and for hybrid grape varieties, it is recommended to retain around 12–14 small clusters. European grape varieties have larger clusters, so more clusters can be retained.

Fruit thinning should be done when the young fruits reach the size of a soybean, about two weeks after flowering, generally around May 20th. Retain well-developed fruits with thick and long pedicels, uniform size, and fresh green color. Remove poorly fertilized fruits that protrude outward, have short or elongated pedicels, or are small, deformed, diseased, or infested by insects. To prevent unexpected risks such as diseased or cracked fruits, it is also necessary to increase the number of retained fruits by 20%–30%. When thinning the fruits, be careful to avoid damaging the remaining fruits or clusters by using pointed scissors.

052 Why do grapes need bagging?

Bagging the fruits effectively prevents infections from diseases such as black spot, white rot, anthracnose and sunburn. It also helps protect the clusters from various pests, including birds, animals, bees, ants, flies, mealybugs, thrips, ladybugs and noctuid moths. Bagging prevents fruit contamination and the accumulation of toxic residues. It also reduces the occurrence of fruit cracking, resulting in smooth and tender skin, abundant bloom, attractive appearance, thick flesh and delicious taste, enhancing the marketability of the fruits.

053 What are the characteristics of grape nutrient requirements?

Grapes have high nutrient requirements due to their vigorous growth and high yield. They are often referred to as potassium-demanding fruit trees because their demand for potassium during growth and development significantly exceeds that of other fruit trees. Under normal production conditions, the ratio of nitrogen, phosphorus, and potassium requirements is approximately 1: 0. 5: 1. 2. If higher yields and improved quality are desired, the demand for phosphorus and potassium fertilizers will increase. It is essential to pay attention to the potassium supply throughout grape production. In addition to potassium, grapes also have higher demands for calcium, iron, zinc, manganese, and other elements compared to other fruit trees.

054 How to apply additional fertilizers to grapes?

Additional fertilizers should be applied during the grape growing season, with 2 - 3 applications recommended for high-yielding vineyards. The first application should be done when the buds start to swell in early spring. At this stage, the flower buds are still differentiating, and the new shoots are about to grow vigorously, requiring a large amount of nitrogen nutrients. It is advisable to apply a mixture of well-decomposed human manure, urine, ammonium nitrate, or urea, with the application amount accounting for 10%-15% of the annual fertilizer requirements. The second application should be done during the early stage of fruit enlargement after flowering, focusing on nitrogen fertilizers and supplemented with phosphorus and potassium fertilizers. This application not only promotes fruit enlargement but also benefits flower bud differentiation. This stage is the vigorous growth period of grapes and a critical period for determining the yield of the following year. It is also known as the "water and fertilizer critical period". Proper water and fertilizer management in the vineyard is crucial during this period. For this application, well-decomposed human manure, urea, and wood ash should be the main fertilizers, with the application amount accounting for 20% - 30% of the total annual fertilization. The third application should be done during the early stage of fruit coloring, focusing on phosphorus and potassium fertilizers, with the application amount accounting for approximately 10% of the annual fertilizer requirements. The additional fertilizers can be applied in conjunction with irrigation or directly to the soil around the plant's roots during rainy days.

055 How to determine the concentration of foliar fertilization for grapevines?

Foliar fertilization is a widely used method in grapevine management, providing rapid nutrient supply to support vine growth. The nutrient requirements of grapevines vary at different stages of growth. Generally, during the new shoot growth period, spraying 0. 2% - 0. 3% urea or 0. 3%-0. 4% ammonium nitrate solution promotes new shoot growth. Spraying a 0. 1% - 0. 3% borax solution before flowering and during the peak flowering period can improve fruit set rates. Spraying 0. 5% - 1% monopotassium phosphate, 1% - 3% calcium superphosphate solution, or a 3% wood ash extract solution 2–3 times before berry ripening significantly increases yield and improves quality. When symptoms of iron or zinc deficiency appear in the vines, spraying 0.3% ferrous sulfate or 0.3% zinc sulfate can be beneficial. However, when using foliar fertilization, it is important to pay attention to the fertilizer concentration to avoid phytotoxicity.

056 What are the main watering times for grapes throughout the year? How to adapt watering based on local conditions?

In general, established grapevines require 5–7 watering sessions during the bud break period, before and after flowering, berry enlargement period, and after harvest. The number of watering sessions may vary depending on the amount of rainfall in a given year.

(1) Bud break watering. Also known as "thawing water" in northern regions, it should be applied from the emergence of grape shoots to the bud burst stage. It is recommended to apply this watering after applying bud-breaking fertilizers for better results. In areas with dry springs and little rainfall, an additional thorough watering should be given to meet the water requirements for bud break and shoot growth.

(2) Controlled watering during flowering. Approximately 10 days before flowering, one watering session should be applied. During the flowering period, watering should be controlled. If it rains during this period, proper drainage should be ensured as it significantly affects pollination, fertilization and fruit set rates.

(3) Watering during berry enlargement for fruit growth. When the berries reach the size of a soybean and the new shoots are growing vigorously, the temperature is increasing, and leaf transpiration is high, requiring additional nutrients and water. Therefore, it is necessary to combine fertilization with watering for fruit growth. If rainfall is scarce during this stage, watering should be done every 10–15 days to meet the water requirements of new shoots and berries.

(4) Watering after autumn fertilization. After fruit harvest, preparations for autumn fertilization should be made. Watering should be combined with the application of base fertilizers to promote nutrient accumulation in the plant, which plays a vital role in the following year's growth and yield. Additionally, in dry winter and spring regions, another watering session should be given before frost protection to reduce frost and early damage. However, in regions with soil mound protection, the vines should be aired for 2–3 days after watering, allowing the soil surface to dry before applying mound protection to prevent excess soil moisture and bud rot.

To adapt watering to local conditions, there are five key points to consider.

(1) Watering for bud break. Apply watering from the emergence of grape shoots to the bud burst stage. In areas with dry springs and little rainfall, an additional thorough watering

should be given to meet the water requirements for bud break and shoot growth.

(2) Watering for flower induction. Apply one thorough watering approximately 10 days before and after flowering. Combining flower induction fertilizers with watering allows for quick absorption of nutrients by the roots, promoting rapid growth of inflorescences, pistils and pollen, leading to improved pollination and fruit set rates.

(3) Watering for fruit growth. In late June, when the berries reach the size of a soybean and the new shoots are growing vigorously, additional watering should be provided to meet the increasing water demands of the leaves and berries. If rainfall is scarce, watering should be done every 7-8 days to meet the water requirements of new shoots and berries.

(4) Pre-harvest watering. Before the grapes reach full maturity, one watering session should be applied to supplement the water deficiency in the later stages of grape growth. This helps increase yield and improve fruit quality, promoting the maturation of grape clusters.

(5) Watering for vine development. After fruit harvest and before frost protection, 2-3 watering sessions should be applied to improve soil moisture conditions, promote root growth and development, and prevent dehydration. In dry regions, after the frost protection watering, the vines should be aired for 2-3 days before applying mound protection to prevent excessive soil moisture and bud rot.

057 How to collect and store grape cuttings for propagation?

Grape cuttings should be collected during winter pruning. Select well-developed one-year-old shoots with short internodes, normal color, plump buds, and free from pests and diseases. Cut the shoots into 7-8 node sections (approximately 50 cm long). Bundle 50-100 cuttings together, label them with the variety name and collection location, and store them in a trench for preservation. The storage trench should be located in a high and dry area with shade. It should be 60-80 cm deep, and the length and width should be determined based on the number of cuttings to be stored. Before storage, a layer of moist sand about 10-15 cm thick should be spread at the bottom of the trench. The cuttings can be laid flat or upright, but each layer of cuttings should be covered with a layer of sand to reduce respiration heat. If sand is not available, soil with a moisture content of around 10% can be used, and fine soil should be filled between the bundles. It is recommended to arrange the cuttings in three layers. Every 2 meters, place a vertical bundle of straw to facilitate air circulation. After arranging the cuttings, a layer of straw can be placed on top, followed by a 20-30 cm layer of soil. In cold regions such as Northeast China and North China, the soil cover should be appropriately increased. During the storage period, regular inspections should be conducted to maintain the temperature in the storage trench at around 1 ℃. It should generally not exceed 5 ℃ or fall below -3 ℃. Higher temperatures can increase respiration and fungal growth, while lower temperatures can cause freezing damage to the buds. The humidity in the storage trench should also be appropriate. If excessive humidity or mold is found on the cuttings, they should be turned and ventilated in a timely manner.

058 What are the methods to promote rooting in grape cuttings? What should be considered for grape rooting?

The main methods to promote rooting in grape cuttings are bottom heat rooting, mound rooting, electric heating rooting, and chemical agent rooting.

Regardless of the rooting method used, the following points should be considered to ensure good rooting results: The degree of rooting should be appropriate for planting. The ideal stage for planting is when the root primordia break through the bark by approximately 0.5 cm. If the rooting is too long, it may be easily damaged during planting, affecting the survival rate. The timing of rooting should be flexible. If the cuttings will be directly planted in open ground, the rooting process should be slightly delayed to allow for immediate planting. If the cuttings will be planted in a protected environment, the rooting process can be initiated earlier. When using heating methods for rooting, there should be a gradual cooling process during the later stages of rooting. This allows the new roots to adapt to the external environmental conditions before planting. When handling the rooted cuttings, it is important to avoid damaging the root primordia. After planting in the nursery bed, the cuttings should be thoroughly watered to ensure good contact between the new roots and the soil.

059 What are the different methods of grafting grapevines?

There are three methods of grafting grapevines: horizontal grafting, wave-like grafting, and aerial grafting.

060 How to determine the timing for green branch grafting in grapes?

Green branch grafting can begin when both the rootstock and scion have reached the semi-lignified stage. It can continue until the new shoots of the grafted seedlings can mature in autumn. In northern regions such as Heilongjiang and Jilin, grafting can start from mid-May to mid-June, with a duration of about one month. In southern parts of Liaoning, Shandong, and Hebei provinces, grafting can last for over two months. If combined with protected cultivation, the grafting period can be even longer.

061 How to select grape hardwood grafting rootstocks?

The rootstock should be selected based on the suitability for the local region. To improve cold resistance, dormant hardwood branches of cold-resistant varieties such as Vitis amurensis and Vitis bellula can be used as rootstocks for grafting.

062 What should be considered when transplanting grape seedlings from the nursery?

It is important to strictly determine the timing for transplanting grape seedlings and conduct timely seedling planting, grading, quarantine, and seedling disinfection.

063 How to identify grape water berry disease?

Water berry disease mainly affects the grape berries and usually manifests after the berries start to color. In colored varieties, the symptoms include abnormal coloring and a pale appearance. In white varieties, the berries become blister-like, the sugar content decreases, the taste becomes sour, the flesh becomes soft, and the flesh easily separates from the skin, resulting in a watery acid-filled package. When lightly squeezed, the diseased berries release droplets, hence the name "water berry disease". After the disease occurs, separation layers can easily form between the stem and the berry, leading to easy detachment.

064 What are the symptoms of nitrogen deficiency in grapes?

The symptoms of nitrogen deficiency in grapes include yellowing of the upper leaves of new shoots, thinning and shrinking of new leaves, yellow-green or reddish-purple coloration of older leaves, shortened internodes in new shoots, slender flower clusters, poor flower differentiation, significant flower and fruit drop, and early cessation of growth. In severe cases, the lower leaves of new shoots may also turn yellow and even shed prematurely.

065 How to prevent and treat phosphorus deficiency in grapes?

To prevent and treat phosphorus deficiency in grapes: Foliar spray phosphorus fertilizers such as ammonium phosphate, calcium superphosphate, potassium phosphate, and monopotassium phosphate. Among them, ammonium phosphate and monopotassium phosphate have the best effect. The recommended spray concentration is 0.3%-0.5%. Spray once every 7-10 days during the fruit enlargement period, for a total of 3-4 sprays. Before flowering, apply 20–40 kg of phosphorus fertilizer per 666.7 m^2 to promote flower development and fruit set. During fruit coloring and shoot maturation stages, to enhance fruit coloring, increase sugar content in berries, and promote shoot maturation, apply 20–40 kg of phosphorus fertilizer per 666.7 m^2, or perform 2–3 foliar applications after fruit enlargement. After harvest, during basal fertilization, it is recommended to apply 0.5–1 kg of calcium superphosphate per mature tree, deep into the tree basin or fertilization trench along with other organic fertilizers.

066 How to prevent and treat potassium deficiency in grapes?

Prevention and treatment methods for potassium deficiency in grapes: Mainly increase the application of organic fertilizers to improve soil structure and increase soil fertility and potassium content. From July to August, foliar spray potassium fertilizer every 10 days, using 0.3% monopotassium phosphate or 0.2%–0.3% potassium chloride, until mid-August, for a total of 3–4 sprays. Apply wood ash to the roots or foliar spray 3% wood ash leachate to alleviate potassium deficiency symptoms.

067 What should be done for zinc deficiency in grapes?

Improve soil structure and increase the application of organic fertilizers. Sandy soils have low zinc salt content and are prone to zinc loss. Alkaline soils can convert zinc salts into an unavailable state, which is unfavorable for grape absorption and utilization. Therefore, improving soil structure, enhancing soil management, increasing the application of organic fertilizers, and balancing various elements can improve zinc supply. During the flowering period or after flowering, foliar spray 0.1%–0.3% zinc sulfate every half month. This not only promotes normal berry growth, increases yield and sugar content but also accelerates fruit maturation.

068 What are the symptoms of iron deficiency in grapes?

The main symptom of iron deficiency in grapes is chlorosis of young leaves. The leaves of affected plants turn yellow, while the leaf veins remain green. This is a characteristic symptom of iron deficiency. In severe cases, the leaf surface becomes ivory-colored or even brownish, leading to leaf necrosis. At the same time, the flower clusters become pale yellow, flower buds drop, and fruit set decreases significantly. Iron deficiency in grapes is closely related to soil pH and aeration conditions. When pH is too high (alkaline soil), iron in the soil often converts into insoluble forms that roots cannot absorb and utilize (although soil analysis may not indicate iron deficiency, iron in the soil is not available to plants). Similar symptoms can also occur in soils that are excessively heavy or irrigated with alkaline water. Therefore, the prevention and treatment of iron deficiency involve both iron supplementation and soil improvement to promote the conversion of iron in the soil into an available state.

4. Grapevine Summer Management (June to August)

069 How to prevent flower and fruit drop in grape production?

Improve tree nutrition by applying fast-acting fertilizers rich in nitrogen and phosphorus from bud break to early flowering, followed by timely irrigation. Prune shoots at the appropriate time to adjust nutrition according to standard requirements, ensuring the nutritional needs during flowering. Spray boron by applying a 0.3% borax solution half a month before flowering, which has a significant effect on improving fruit set. Around 7 days before flowering, spray a 0.3%–0.5% Gibberellin solution to control shoot growth. Thin out lateral shoots and trim the tops of clusters. For large-berried varieties with tightly clustered fruit, promptly remove the upper lateral shoots and 2–3 branches from the clusters. For overcrowded small clusters or excessive length of cluster tips, thinning and avoiding excessive growth should be carried out to ensure compact clusters and uniform and attractive berry size. Girdling can be performed at the beginning of flowering to improve fruit set.

070 What should be considered when inter-row cultivation and weed control in grapevines?

Inter-row cultivation is a soil cultivation practice carried out during the growing season of grapevines when the root activity is vigorous. To prevent root damage, inter-row cultivation should be shallow, generally around 3–4 cm. After irrigation or rainfall, timely inter-row cultivation should be done to loosen the soil, prevent soil compaction, and reduce water evaporation. Clearing weeds from the vineyard during the growing season is an important management task, and inter-row cultivation should be combined with weed control.

071 How to select herbicides for weed control in grapevines?

There are various types of herbicides available, and the selection should be based on the weed species present in the vineyard while ensuring no harm to grapevine roots. The commonly used herbicides include glyphosate, which can kill various grass weeds with relatively low damage to dicot weeds and grapevines. Another option is glufosinate, which has strong underground tissue destruction effects on both annual and perennial weeds, and foliar application can lead to complete plant death. However, care should be taken not to spray it on grapevine foliage to avoid damage. Other herbicides such as simazine and paraquat are also available and should be selected according to the instructions and specific weed species.

072 How to select cover crops for young grapevines?

For young vines and vineyards with wide row spacing, it is possible to plant short-stemmed crops such as soybeans, peanuts, peas, medicinal herbs, vegetables, and seedlings between rows. This can increase economic income and facilitate short-term cultivation for long-term vine growth. Green manure cover crops, especially leguminous cover crops, are important measures for achieving high-quality, high-yield, stable production, vigorous vines, and low costs.

073 How to perform deep plowing in a grapevine vineyard?

There are two main forms of deep plowing in grapevine vineyards: full-field deep plowing and alternate-row deep plowing. It is generally carried out in late autumn or early winter in conjunction with the application of base fertilizer or in spring when green manure is incorporated. The depth should be around 25–30 cm, with slightly shallower plowing near the main trunk and slightly deeper in the inter-row spaces.

074 How to effectively prevent and control grape anthracnose?

Thoroughly remove diseased clusters, shoots, and leaves to reduce the source of infection. In regions south of the Yangtze River, bagging can be done immediately after flowering. Strengthen cultivation management, timely pruning, trellising, and shoot tipping to ensure good ventilation. Increase the application of phosphorus and potassium fertilizers while controlling nitrogen fertilizer dosage. Spray a 0.3% sodium pentachlorophenate plus 4% sulfur mixture or 100 times the concentration of thiophanate-methyl as an eradication agent during bud break and woolly bud stage. In the southern regions, spraying should start from late April, and in the northern regions, from late May. Generally, spray once every 10–15 days. Recommended fungicides include 80% carbendazim at 700–800 times dilution, 50% captan at 600–700 times dilution, 50% thiophanate-methyl, or 800 times dilution of mancozeb.

075 What are the characteristic symptoms of grape white rot?

The disease first appears as pale brown, water-soaked, nearly circular lesions on the pedicels and rachis of the clusters. The infected areas rot and turn brown, quickly spreading to the berries, which become soft and brown. Later, small, grayish-white granular conidiomata develop on the surface of infected berries and rachis. When humidity is high, grayish-white masses of conidia are released from the conidiomata. The infected berries are prone to falling off, and when they dry, they become brown or grayish-white mummies. On the shoots and canes, the disease initially manifests as water-soaked, pale brown lesions, which gradually expand into large, brown, sunken spots with grayish-white conidiomata on the surface. As the disease progresses, the epidermis of the infected area cracks and separates from the woody tissue, leading to the shedding of the epidermis. The vascular bundles become a tangled mass of brown color. When the lesion extends around the circumference of a shoot or cane, the upper portion of the shoot or cane dies.

076 How to effectively prevent and control grape powdery mildew?

Remove fallen leaves and infected branches by burying them deeply or burning them. Prune, trellis, improve drainage, and control weeds in a timely manner. Increase the application of phosphorus and potassium fertilizers. Start spraying fungicides at the early stage of disease development. In northern regions, spraying should generally begin in mid to late June, with a frequency of once every 15 days. The best results are obtained with 25% dodine or domestically produced 600-fold dilution of mancozeb. 300-fold dilution of aluminum phosphide, 500-fold dilution of aluminum phosphide, or Bordeaux mixture also have good control effects. Soil application of dodine can also be effective. In regions south of the Yangtze River, spraying can be combined with black spot disease control before early autumn. In autumn, the focus should be on controlling powdery mildew. The experience in Suzhou, Wuxi, and Changzhou areas suggests that Bordeaux mixture is most effective before the disease occurs. If powdery mildew occurs, timely spraying of 700-fold dilution of mancozeb or 500-fold dilution of aluminum phosphide is recommended. Soil application of mancozeb can also be effective.

077 What are the characteristic symptoms of grape black spot disease?

On young fruit, small brown circular lesions appear on the surface before gradually enlarging. The center of the lesion becomes grayish-white and slightly sunken, with black dot-like structures resembling bird's eyes. New shoots, tendrils, petioles, and pedicels are affected, initially showing small brown circular or irregular spots, which later enlarge into nearly elliptical, gray-black lesions with dark brown edges. The central part of the lesion becomes significantly concave and cracks. Ulcerative lesions form on the tendrils, sometimes extending downwards to form layers. Affected shoots stop growing, leading to withering, drying, and turning black. Tender leaves are affected, initially showing small brown or black spots the size of a needle tip. When there are many spots, the tender leaves become wrinkled and eventually die.

078 How to prevent and control grape powdery mildew?

Strengthen cultivation management, apply organic fertilizers, enhance tree vigor, and improve disease resistance. Prune in a timely manner, thin out dense foliage, trellis the vines, and maintain good ventilation and light penetration to reduce disease incidence. Pay attention to orchard hygiene, remove diseased residues, and either burn or bury them to reduce sources of infection. Before bud break, spray a 3–5-degree sulfur mixture, and after bud break, spray a 0. 2 – 0. 5-degree sulfur mixture or 500-fold dilution of thiophanate-methyl, or 1 000-fold dilution of methyl thiophanate, or 1 000-fold dilution of 25% triadimefon wettable powder.

079 What are the characteristic symptoms of grape brown spot disease?

There are two types of brown spot diseases: brown spot and small brown spot. Brown spot disease is caused by Phomopsis viticola and mainly affects the leaves. The initial lesions are light brown, irregularly shaped angular spots. The lesions gradually expand and can reach a diameter of 1 cm. The lesions change from light brown to reddish-brown, with a yellow-green margin. In severe cases, multiple lesions merge into larger spots with clear edges. The surrounding area on the underside of the leaves appears fuzzy. In later stages, the infected areas become necrotic, and under rainy or humid conditions, gray-brown powdery structures may appear. Some varieties may exhibit indistinct rings within the lesions. Small brown spot disease, caused by Guignardia bidwellii, presents as small yellow-green circular spots that gradually enlarge into round lesions of 2–3 mm. The lesions progress from browning to a tea-brown color, and black mold may develop on the underside of the lesions.

080 How to implement integrated management of grape diseases?

(1) Prevention through cultivation management techniques. In the grapevine management process, a series of targeted cultivation techniques can be adopted to enhance plant resistance, prevent pathogen reproduction, spread, and infection. Measures such as selecting disease-resistant varieties, planting virus-free seedlings, raising the height of grape clusters appropriately, timely removal of diseased leaves and fruits, pruning infected branches and shoots, scraping off lesions and peels, improving pruning and soil, fertilizer, and water management, and timely harvesting can not only prevent grape diseases but also improve fruit quality.

(2) Chemical control. Chemical control has multiple functions, including protection, eradication, and treatment, and can effectively control disease occurrence and epidemics in a short period. According to the Good Agricultural Practices for Grape Production, the use of low-toxicity pesticides is allowed to control diseases, while the use of highly toxic and highly residual pesticides is prohibited. For fungal diseases, chemical control measures can be as follows: Before winter dormancy and after emergence, spray sulfur mixture to prevent diseases such as black spot, gray mold, anthracnose, white rot, and powdery mildew. During the growth of new shoots, use 50% carbendazim at 800-1 000-fold dilution or 70% mancozeb zinc at 600-800-fold dilution in the early stage, and use half-dose Bordeaux mixture to control black spot disease in the later stage. During flowering and fruit setting, use 50% carbendazim wettable powder at 800-1 000-fold dilution to control gray mold, and use 80% dodine wettable powder at 600-800-fold dilution to control black spot, powdery mildew, and white rot. Within 20 days after fruit setting, alternate the use of 70% methyl thiophanate wettable powder at 800-1 000-fold dilution and half-dose Bordeaux mixture to control black spot disease, and timely bagging to prevent anthracnose. In July and August, alternate the use of 80% dodine wettable powder at 600-800-fold dilution, 80% iprodione wettable powder at 500-600-fold dilution, and 15% fenarimol emulsion at 1 000-1 500-fold dilution to control anthracnose, powdery mildew, and white rot. After fruit ripening until harvest, alternate the use of 70% methyl thiophanate wettable powder at 800-1 000-fold dilution, 80% dodine wettable powder at 600-800-fold dilution, and 80% iprodione wettable powder at 500-600-fold dilution to control powdery mildew and brown spot.

081 What are the six misconceptions in the prevention and control of grape diseases?

(1) Emphasizing treatment over prevention. Many growers tend to apply effective therapeutic agents only after observing obvious disease symptoms. Therapeutic agents(such as

broad-spectrum fungicides like carbendazim and thiophanate-methyl) lose their effectiveness due to the development of varying degrees of resistance in the pathogens from long-term and repeated use. Protective agents(such as Bordeaux mixture) can effectively control the source of infection and protect plants without promoting pathogen resistance, but they are often overlooked in disease prevention.

(2) Improper mixing of chemicals. When multiple diseases occur, some growers mix different pesticides together without considering the specific targeting of each disease. This practice aims to reduce labor and achieve multiple effects with a single application. However, such mixing, especially with commercially available compound pesticides, can lead to multidirectional resistance in pathogens, rendering subsequent pesticide applications ineffective.

(3) Quantity over quality. There is a tendency to focus on the frequency of pesticide applications while neglecting the quality of each application. In the production process, growers may follow a schedule of applying pesticides every 7-15 days or after rainfall, but they often fail to ensure thorough coverage. For example, insufficient spraying on the underside of leaves for controlling powdery mildew or inadequate coverage of the tender parts of plants for black spot control. Incomplete application of protective agents leaves many areas untreated, allowing for the recurrence of diseases.

(4) Neglecting leaves while focusing on fruits. Fruit clusters with diseases easily attract attention, while leaf diseases are often overlooked. It is common to see extensive leaf drop caused by diseases like powdery mildew.

(5) Prioritizing management before harvest but neglecting post-harvest. Fine management practices are often implemented before grape harvest, but after the harvest, the plants are left untreated, leading to a significant increase in diseases during the later stages, early leaf drop, and a severe impact on the next year's yield.

(6) Overemphasizing treatment without prevention. The occurrence of grape diseases is closely related to factors such as ventilation, light penetration inside and outside the canopy, soil fertility, and tree vigor. Relying solely on chemical control measures is insufficient to fundamentally solve the problem.

To address these misconceptions in grape disease prevention and control, it is important to adhere to the principles of prevention as the primary approach, targeted use of pesticides, proper mixing of chemicals, and rotation of pesticides. Additionally, comprehensive cultivation measures such as improving soil fertility, optimizing light conditions, and enhancing tree vigor should be implemented to achieve more efficient disease prevention and control.

5. Grapeyard Autumn Management (September to November)

082 How to prevent and control grape leafhoppers?

Remove fallen leaves and weeds to eliminate overwintering adults. Strengthen field management during the growing season to improve ventilation and light conditions. Avoid intercropping with soybeans and cucurbits in the vineyard as much as possible. Seize the occurrence period of the first generation nymphs and young larvae for pesticide spraying. Generally, it is advisable to spray highly effective contact insecticides such as deltamethrin, cypermethrin, lambda-cyhalothrin, and bifenthrin at a dilution of 2 000-3 000 times after grape leaf emergence in mid to late May. Repeat the spray every 5-7 days to control the infestation throughout the year.

083 What are the characteristic symptoms of grape erineum mites?

The affected areas of the leaves show convexity on the upper surface and concavity on the lower surface. White woolly felt-like growth appears on the concave side of the leaf, hence the name "erineum disease". In the later stages, the woolly growth on the lower surface turns yellow-brown and eventually dries up and becomes brown. Severe infestations can affect tender shoots, tendrils, and young fruits, hindering the normal development of leaves. When leaf damage is severe, the entire leaf becomes wrinkled and unable to expand, eventually leading to withering and death.

084 What are the harmful characteristics of grape moth?

Grape moth mainly damages grape leaves by creating notches or holes, and in severe cases, it can completely consume the leaves, leaving only the petiole and the base of the leaf veins.

085 How to effectively prevent and control greenish-blue tortoise beetles?

The adults of the beetles tend to gather and cause damage on flowers or injured fruits. They can be captured and killed using insect nets. When the pest population is high, spraying can be done using insecticides such as pyrethroids (deltamethrin, cypermethrin, lambda-cyhalothrin) at a dilution of 2 000 times, mixed with organophosphates (dimethoate) at a dilution of 1 000 times.

086 What are the activity characteristics of grape vine moths?

The small larvae of this insect create holes or notches in the leaves, while the larger larvae consume the leaves entirely, leaving only the main veins and petioles. The presence of large insect feces under the grape arbor can be used to detect the larvae (green bean bugs) for manual capture.

087 How to effectively prevent and control grape berry moth?

It is more difficult to control the larvae once they bore into the branches and tendrils. Therefore, it is better to focus on pesticide control before and after the adult moth lays eggs and before the eggs hatch. Timely spray 25% fenoxycarb suspension concentrate at a dilution of 2 000 times or 20% chlorfluazuron suspension concentrate at a dilution of 3 000 times, 1–2 times. The pesticide should be sprayed evenly and thoroughly on the whole plant until the liquid forms droplets without flowing, aiming to eliminate adults and eggs and control the number of borers. Due to the appearance of yellow leaves and swollen branches at the affected sites, careful inspection should be conducted in June and July. Prune off infested branches and burn them. During autumn pruning, if infested branches are found but not pruned, the borer holes can be opened, and the feces can be removed using a wire hook. Stuff a cotton ball soaked in a 100-fold dilution of dichlorvos into the hole and seal it with plastic wrap, which can kill the larvae. Alternatively, insert a quarter piece of aluminum phosphide into the hole, wrap it with plastic wrap to kill the larvae. For severely affected plants, during the sap flow period (after the bleeding stops), drill 2–4 holes with a thick awl into each main cane at a distance of 10–15 cm from the ground, reaching the pith. The holes should be inclined downward at an angle of 40–50° to allow the liquid to flow down and prevent backflow. Then, prepare a 10–20-fold solution of oxydemeton-methyl (or phoxim, dimethoate) and use a medical syringe to inject the solution into the drilled holes, approximately 4–5 mL per plant. It is better to seal the holes immediately with adhesive tape, or they can be sealed with mud. If the pests have already caused damage to the branches and tendrils, the injection method can also be used to directly inject the pesticide into the holes from which the feces are expelled. However, the concentration of the pesticide should be around 100 times, and sealing the holes after injection can enhance the sealing and killing effect. For vineyards with heavy larval infestation, cotton balls dipped in a 5–10-fold dilution of dichlorvos can be clamped with tweezers and directly applied to the affected areas, followed by wrapping with plastic film, which also has insecticidal effects. Avoid using pesticides before fruit harvest.

088 Why is autumn application of base fertilizer important for grapevines in the fruiting stage?

Base fertilizer is the most important part of grapevine fertilization, and it can be applied in autumn from grape harvest to soil freezing. However, practical experience shows that the earlier the autumn application of base fertilizer, the better. Typically, well-rotted organic fertilizers (manure, compost, etc.) are applied immediately after grape harvest, along with some quick-acting fertilizers such as ammonium nitrate, urea, calcium superphosphate, and potassium sulfate. Base fertilizer plays a significant role in restoring tree vigor, promoting root absorption, and facilitating flower bud differentiation.

089 How to determine the amount of base fertilizer for autumn application?

A general rule of thumb is "1 kg of fruit requires 5 kg of fertilizer". The amount of base fertilizer applied should account for 50% to 60% of the total annual fertilizer application. In general, for high-yielding and stable grape vineyards, the application rate of mixed organic and inorganic fertilizer is 5 000 kg per 666.7 m^2 (equivalent to 12.5–15 kg of nitrogen, 10–12.5 kg of phosphorus, and 10–15 kg of potassium, with a ratio of nitrogen to phosphorus to potassium of 1:0.5:1).

090 How to determine the maturity of grapes?

The maturity of grapes can be divided into the onset of maturity, full maturity, and over-maturity stages. For colored varieties, the onset of maturity is marked by the fruit starting to change color. For colorless varieties, it is marked by the fruit becoming soft, elastic, and transitioning from green to semi-transparent. The onset of maturity is not the harvest period for consumption because the sugar content in the fruit is not high at this stage, while the acidity is relatively high, making it unsuitable for consumption. Full maturity for colored varieties is reached when the fruit fully exhibits the characteristic color, flavor, and aroma of the variety. For colorless varieties, it is reached when the fruit becomes soft, nearly translucent, the variety's characteristics are fully expressed, and the seed coat becomes hard and turns completely brown. At full maturity, sugar accumulation in the fruit ceases, and the sugar content reaches its peak. In addition to observing fruit color, texture, and seed characteristics, regular measurements of sugar content (every 2 days) can be used to determine full maturity. Full maturity is the optimal harvesting period for fresh consumption and winemaking varieties. For winemaking varieties, only after reaching full maturity can the fruit's sugar content, pigments, and aromatic substances reach their highest levels, and only by pressing fully ripe fruit can high-quality wines with varietal typicity be produced. If not harvested after full maturity, the berries may either fall off due to over-ripening or start to shrink as moisture is lost through the skin, resulting in a passive increase in sugar concentration in the juice due to water evaporation. In some foreign regions, to obtain higher sugar content for processing purposes, intentional delay of harvest until the over-maturity stage is sometimes practiced, but for fresh consumption varieties, over-ripeness should be avoided during harvesting.

091 How to harvest grapes?

Harvesting methods. Grape harvesting should be done on sunny days. It is not suitable to harvest on cloudy or rainy days, or during times of heavy dew or intense midday sun, as it may affect the quality of the clusters. When harvesting table grape varieties, it is generally recommended to leave a stem of about 3–4 cm attached to the cluster to facilitate handling and storage. However, the stem should not be left too long to prevent it from puncturing other clusters. During harvesting, it is important to handle the clusters gently, taking care to protect the bloom (the natural waxy coating on the grapes). Any damaged or diseased berries should be removed promptly. For grapes intended for transportation to other areas for sale, timely packaging is necessary.

092 What are the main aspects of post-harvest grapevine management?

(1) Protecting the autumn leaves. After harvest, it is important to focus on protecting the leaves and enhancing their photosynthetic capacity, as this plays a crucial role in increasing nutrient storage in the plant. During the late growth stage, strict control of fungicides and sulfur-based pesticides should be implemented to prevent premature leaf aging and reduce photosynthetic efficiency. In general, except for the "Kyoho" variety, measures to remove old leaves should be minimized or avoided after fruit harvest. During fruit harvesting,

care should be taken to minimize leaf damage.

(2) Establishing a proper fertilization regime. Base fertilizer should be applied as early as possible after fruit harvest. In cases of heavy fruit set or weak tree vigor, a portion of quick-release nitrogen fertilizer can be applied to supplement tree nutrition, enhance late-season leaf photosynthesis, and promote tree recovery. Organic fertilizers used as base fertilizers should be well-rotted and adequately stacked in advance.

Spraying 1–2 times of 0.5% urea or potassium dihydrogen phosphate after harvest can effectively enhance leaf photosynthetic capacity and improve nutrient storage in the plant.

(3) Strengthening disease and pest control. Post-harvest disease and pest control are crucial, especially for diseases such as downy mildew, powdery mildew, and pests like the grape leafhopper. These can often cause premature leaf drop, leading to late autumn bud sprouting and severely affecting nutrient accumulation and the following year's growth and yield. Therefore, post-harvest disease and pest control should be given significant attention. Generally, spraying 300 times the concentration of ethyl phosphorothioate or 1 000 times the concentration of carbendazim every 10 days before leaf drop can effectively prevent the occurrence of downy mildew. For grape leafhoppers, spraying with 1 000 times the concentration of oxidized organophosphate can be used for control.

In summary, late-stage management centered around leaf protection and enhancing photosynthesis plays a crucial role in ensuring high and quality grape yields year after year. Sufficient attention must be given to these practices in production.

093 What are the specific methods for grape cellar management?

After grape harvest, pre-cool the grapes in a cool place for 2 days. The pre-cooling temperature should be controlled below 10 ℃ to dissipate field heat. Then carefully place the grapes on racks in the cellar. Control the temperature and humidity in the cellar. Initially, due to the higher external temperatures, ventilation measures can be employed to maintain the temperature below 10 ℃. After entering winter, as the temperature drops, the "daytime ventilation and nighttime closure" method can be used to maintain the cellar temperature at 0–1 ℃. The relative humidity should be around 80% – 90%, and if the humidity is insufficient, water can be sprayed on the floor to maintain humidity. When the external temperature drops below 10 ℃, attention should be given to sealing the cellar door. Strengthen inspections and promptly remove diseased clusters and rotten berries.

094 What are the differences between long, medium and short pruning of grapevines?

For mature branches, pruning 2–4 buds is referred to as short pruning, 5–7 buds as medium pruning, and 8-11 buds as long pruning. Pruning to 1 bud or only retaining latent buds is known as very short pruning and is suitable for use as renewal spurs.

095 What methods are generally used to update fruiting branches in grapevines?

To maintain compact fruiting branches and prevent outward growth of fruiting positions, regular updates of the fruiting branches are necessary. The methods commonly used for updating fruiting branches are single cane renewal and double cane renewal.

096 How is cane renewal performed in grapevines?

To prevent the upward movement of fruiting positions, it is important to utilize newly sprouted shoots from dormant buds. Select vigorous lower and middle parts of the vine to replace the upper parts, and prune the older sections of the vine. Adjust the growth vigor, spacing, and quantity of the main canes during winter pruning.

097 How to determine the timing for soil mound protection in grapevine production?

Generally, soil mound protection should begin about 15 days before the local soil freezes. If soil mound protection is done too early, the high soil temperature and humidity may cause bud rot. If it is done too late, the soil may freeze, making it difficult to mound the soil properly. In the central and northern regions of Liaoning, early November is generally a suitable time to start soil mound protection.

098 What are the methods for soil mound protection in grapevine cultivation?

(1) Above-ground mound method. This method involves not digging trenches on the ground. After pruning, the canes are laid down in the same direction, with one vine pressed against another to keep the main cane straight. The canes are tied in sections with straw ropes, and cushions(soil or straw) are placed under the main cane to prevent breakage. Straw or other materials for insulation are placed on both sides of the canes, tightly packed, and then bundles of straw are placed on top of the canes, followed by a layer of fine soil for proper coverage.

(2) Underground mound method. Trenches of about 50 cm deep and wide are dug between the grapevine rows. The canes are pressed into the trenches and covered with soil. In

extremely cold areas, to enhance the insulation effect, a layer of plastic film, dry straw, or leaves can be first placed on the plants before mounding the soil.

099 What are the key points to consider during soil mound protection?

At the bending point below each grape stem, prepare a cushion of soil or straw to prevent the main vine from being broken when burying it in the plant. At the lower part of the vine, dig a shallow trench about 35 cm deep to accommodate the placement of the vine. When burying the vine, loosely bundle it and place it in the trench, compacting the soil on both sides. Then cover the vine with soil, tamping it down while adding soil to prevent air pockets in the soil mound.

100 What aspects are included in grapevine cold protection cultivation techniques?

(1) Use cold-tolerant varieties. This is crucial for cold protection cultivation.

(2) Utilize cold-tolerant rootstocks. Using rootstocks like Beta or American grape can significantly reduce the required soil mound thickness.

(3) Deep trench planting. Digging 60–80 cm deep trenches during planting, applying sufficient bottom fertilizer, and gradually increasing the soil layer thickness each year to encourage deep root growth and enhance the plant's cold resistance.

(4) Opt for trellis shaping. Using wider trellis spacing allows for wider soil mounding areas without damaging the root system. Therefore, in colder regions, smaller trellises should be preferred for grapevine cultivation.

(5) Proper cane management. In shorter growing seasons of northern regions, some varieties may have poor cane maturity in early cooling years. Therefore, early topping and additional phosphorus and potassium fertilization are recommended to promote cane maturity and bud development at the base. Short pruning during winter should be done to retain the best buds and well-matured canes.

(6) Enhance nutrient and water management. Early-stage cultivation requires increased fertilization and irrigation, timely topping, while the later stage requires phosphorus and potassium foliar spraying. Control nitrogen fertilization and water supply, and ensure proper drainage during rainy autumns to promote cane maturity and improve winter cold resistance.

6. Application of New Technologies in High-Yield and High-Quality Cultivation of Fresh Grapes

101 How should grape clusters be managed before bagging?

(1) Thinning and cluster management. After the flower clusters appear, thinning should be done based on the overall load of each vine. It is recommended to leave 2-3 clusters on elongated shoots and 1 cluster on other new shoots. Weak and small clusters should be removed, and each shoot should not have more than one cluster. Remove the shoulder clusters to ensure compact and aesthetically pleasing cluster shape. In trellis cultivation, clusters that are pointing upwards or growing on top of the trellis or between branches should be carefully arranged below the trellis to allow natural drooping. Trim the excessive length of the cluster tip, generally removing around 1/5 to 1/4 of the cluster. After flower shedding, gently shake each cluster to remove underdeveloped or deformed berries, damaged berries, or excessively compacted berries to save nutrients. For each cluster, the number of berries should be around 40-50 for small clusters, 50-80 for medium clusters, and 80-100 for large clusters. Thinning the berries ensures uniform distribution.

(2) Hormone application and pest and disease control. When the grape berries reach the size of soybeans, applying hormones such as gibberellin, brassinolide, or Shuofeng 481 can increase berry size and reduce deformities, resulting in neat and attractive clusters. Before bagging, spraying with 800-fold dilution of ethephon aluminum, or 1 000-fold dilution of mancozeb and zinc, or 1 000-fold dilution of carbendazim can prevent cluster diseases.

102 What aspects are involved in the management of grape clusters after bagging?

(1) Pest and disease control. Strengthen the prevention and control of leaf diseases and diseases that affect both leaves and clusters. Focus on diseases such as powdery mildew, black spot, white rot, cluster stem necrosis, and anthracnose. Regularly check the clusters inside the bags for pest and disease occurrence to enable early prevention and control. In general, cluster-specific treatments are not commonly applied during production.

(2) Sunburn prevention. For vineyard sections prone to sunburn, besides using double-layer bags made of plastic film or newspaper, ground irrigation should be carried out before high temperatures occur to increase air humidity and alleviate sunburn.

(3) Phosphorus and potassium fertilizer supplementation. Under shaded conditions, grape flavor may decrease. Therefore, in the later stages of grape growth, additional phosphorus, potassium fertilizers, and multi-element compound micronutrients should be applied to improve grape quality.

(4) Bag removal. Bags made of special paper or homemade newspaper bags should be removed two weeks before harvest to allow clusters to fully color. Bags made of plastic film do not need to be removed before harvest and can be harvested with the bags. For bags made of plastic film and newspaper, the newspaper layer should be removed one week before harvest, while the grape umbrellas are generally removed one week before harvest.

103 What should be considered in grape bagging?

(1) Selection of bag types. The choice of grape bags should be based on the vigor of the vines, production goals, and economic capacity. As grape production in China is mainly for domestic consumption, it is advisable to choose reliable domestically produced double-layer bags and minimize or avoid the use of plastic film bags and homemade newspaper bags.

(2) Bagging timing. Grape bagging is generally done after physiological fruit drop (15–20 days after fruit set), when the berries are about the size of soybeans and after thinning and cluster management. Bagging should be done on sunny days between 9–11 am and 2–6 pm.

(3) Bagging techniques. Before bagging, place the entire cluster bag in a moist area to regain moisture and flexibility. Carefully remove any debris attached to the young cluster. Hold the paper bag with the left hand, open the bag mouth with the right hand, or blow open the bag mouth with the mouth to expand the bag and open the ventilation and drainage holes at the bottom corners. Hold the bag mouth about 2–3 cm below and insert it over the cluster, ensuring the stem is positioned at the base of the bag opening (avoid including leaves and branches inside the bag). Then, fold the bag mouth from both sides in a fan-like manner, secure it with a tying wire at the folding point, and tear the tying wire from the connection point above the bag opening to rotate it 90 degrees. Tighten the bag mouth by rotating it once around the bag opening to center the young cluster within the bag, allowing it to hang freely inside the bag to prevent friction with the bag surface. Avoid entangling the tying wire around the stem, and exert gentle force while bagging, avoiding pulling off the young cluster. The force should be applied upwards, and the bag mouth should be tightly secured to prevent pests from entering the bag and damaging the cluster, as well as to prevent the paper bag from being blown off by the wind.

104 What should be considered when using ethrel for grape ripening?

When using ethrel for grape ripening, the following points should be considered: The concentration should be appropriate. If the concentration is too low, the effect may not be significant, while concentrations higher than 500 mg/kg can lead to fruit drop. The treatment timing should coincide with the onset of fruit ripening, which is when colored varieties start to show color and white varieties have a slight yellowing of the berries. This period yields the best results. Different grape varieties have different optimal concentrations for ripening. It is necessary to conduct trials to determine the best treatment concentration and method. Ethrel can promote abscission layer formation, which often leads to fruit drop when used alone. To mitigate this effect, adding 10–20 mg/kg of naphthylacetic acid to the ethrel treatment can effectively prevent fruit drop. In recent years, the use of abscisic acid (ABA) for ripening "Kyoho" grapes and enhancing berry coloration has shown significant results. ABA is sprayed at a concentration of 100–200 mg/kg when the fruit begins to ripen. ABA treatment promotes coloration and ripening of "Red Globe" "Xianfeng" and "Jingchuan" red grape varieties and also increases sweetness. However, ABA is not widely produced in China currently, so ethrel remains the primary choice for ripening in large-scale production.

105 What should be considered when seedless grapes are treated? Why do grapes crack?

When treating seedless grapes, it is important to consider the timing and concentration of the seedless agent used.

Grapes can experience cracking before and during the veraison stage, resulting in reduced yield and lower quality. The main cause is improper use of ripening agents or seedless agents. Excessive concentration or improper timing of spraying ripening agents like ethrel or seedless agents can lead to fruit cracking. Therefore, when using ethrel, it is important to dilute it strictly according to the regulations and spray it when the grapes have a slightly sweet taste and tolerable acidity.

106 What should be considered when girdling grapes?

When girdling grapes, the following points should be considered: the girdling period should be appropriate. The timing may vary depending on the purpose of girdling, but in the North China region, girdling should not be delayed beyond mid-July to ensure proper healing of the girdling wound before winter and to avoid affecting the maturity of the branches. The girdling width should be suitable. The width of girdling depends on the thickness of the branches and should generally be within the range that can completely heal within the same year. Girdling should be performed on fruiting branches and not on the main stem. For varieties that use single-cane renewal, a tail branch should be left below the girdling ring. For varieties that use two-cane renewal, girdling should only be performed on fruiting branches and not on renewal branches. Girdling is only suitable for vigorously growing plants. Weak and slender plants or branches should not be girdled. Girdling only affects the direction of nutrient transport. High yield and high-quality grapes cannot be achieved solely through girdling. Girdling must be combined with other agricultural techniques to achieve the desired results.

107 What should be noted when using paclobutrazol to control lateral shoot growth?

When using paclobutrazol, the following points should be noted. Foliar application should be carried out during the early stage of grapevine shoot growth, generally around the end of flowering. Late application may not effectively inhibit shoot growth. Paclobutrazol is mainly used on vigorously growing young and middle-aged grapevines and should not be applied to weakly growing trees. Paclobutrazol has a significant inhibitory effect on shoot growth but may also cause delayed budbreak, excessively compact clusters, and the formation of small green berries. Late application can lead to delayed ripening and reduced sugar content. Different varieties and regions may have different responses to paclobutrazol, so it is necessary to conduct trials to determine the appropriate dosage and application method before use.

Chapter 3 Q&A on Practical Techniques for Fresh Tomato Production

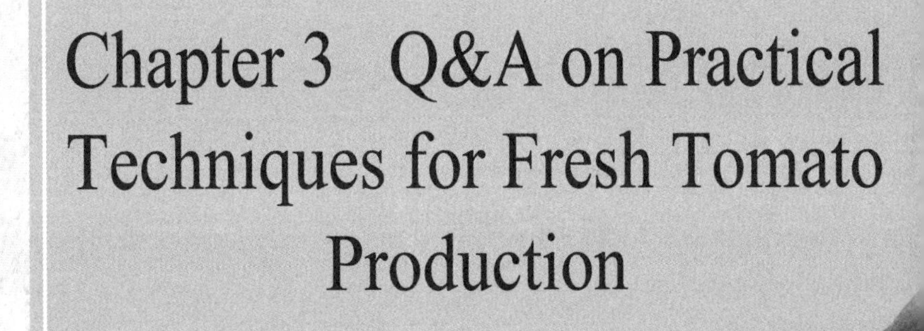

Manual of Horticultural Techniques

园艺技术手册

1. Tomato Production Basics

001 What types of tomatoes can be classified based on their branching habit?

Tomatoes can be classified into two types based on their branching habit: determinate and indeterminate.

(1) Determinate type. The main stem of the plant stops growing after reaching a certain point, and the number of fruit clusters on the main stem is limited. Determinate varieties are usually shorter in height, with a more concentrated fruiting period. They are often early-maturing varieties and exhibit characteristics such as high fruit set, fast maturity, rapid reproductive organ development, and high photosynthetic intensity, resulting in a shorter growth period.

(2) Indeterminate type. After the inflorescence forms at the top of the main stem, lateral shoots continue to grow and bear fruit, without reaching a terminal point. Indeterminate varieties have a longer growth period, taller plant height, larger fruit size, and are often mid-to-late maturing varieties. They generally have higher yields and better fruit quality.

002 What are the characteristics of tomato roots?

Tomato roots are well-developed and have a wide and deep distribution. During the fruiting period, the main roots can penetrate the soil to a depth of 150 cm or more, and the lateral spread of the root system can reach approximately 250 cm. However, under seedling conditions, the main roots are cut during transplantation, resulting in increased lateral root branching and horizontal development. Most of the root system is distributed in the soil layer of 30–50 cm, with minimal distribution in the soil layer below one meter.

Tomatoes easily develop lateral branches not only on the main roots but also on the root collar or stem nodes. Adventitious roots can quickly extend, reaching approximately 1 meter in length within 4–5 weeks under favorable growth conditions. Therefore, tomato propagation through cuttings is relatively easy to establish.

The development capacity, depth, and range of tomato roots are influenced by soil structure, fertility, soil temperature, cultivation practices, as well as transplanting, pruning, and topping measures. They are also correlated with the growth of above-ground stems, leaves and fruits.

003 What are the characteristics of tomato stems?

Tomato stems exhibit sympodial branching(pseudomonopodial branching), where flower buds form at the stem tip. Therefore, observable differences in growth and development stages can be seen based on the morphology of the growing point. In indeterminate tomatoes, after the first inflorescence differentiates at the stem tip, a vigorous lateral shoot grows from the axil below the inflorescence, forming a sympodium(pseudomonopodium) with the main stem. The same pattern occurs with the subsequent inflorescences, resulting in the continuous growth of the pseudomonopodium. In determinate plants, after 3-5 inflorescences, the lateral shoots below the inflorescences transform into flower buds and no longer grow into lateral branches, causing the pseudomonopodium to stop elongating.

Tomato stems are semi-erect or semi-trailing, with some varieties being erect. The base of the stem becomes woody. The stems have strong branching ability, with lateral branches developing in each leaf axil, and the first lateral branch below the inflorescence grows the fastest. Without pruning, tomatoes can form dense bushy plants.

Productive stem morphology of tomatoes: the internodes are relatively short, and the thickness of the stem is similar in the upper and lower parts. In excessively vegetative plants, the internodes become elongated, gradually thickening from the bottom to the top. In aging plants, the opposite occurs, with shortened internodes gradually thinning from the bottom to the top.

004 What are the characteristics of tomato leaves?

Tomato leaves are simple, deeply or fully lobed, with 5-9 pairs of leaflets. The size and shape of the leaflets vary depending on their position on the leaf. The first and second leaflets are smaller in size and number, while the number of lobes increases as the leaf position ascends(Figure 1). Leaf size, shape, and color vary depending on the variety and environmental conditions. Leaf characteristics can be used to identify different varieties and provide ecological indications for cultivation practices. For example, early-maturing varieties generally have smaller leaves, late-maturing varieties have larger leaves, and tomatoes grown in open fields have darker foliage compared to those grown in greenhouses or plastic tunnels. Under low temperatures, leaves may turn purple, while under high temperatures, leaflets may curl inward. Tomato leaves and stems have hairs and secretory glands that can secrete a distinctive odor. This characteristic can deter pests such as the tomato hornworm, resulting in fewer insect infestations.

Productive leaf morphology of tomatoes: the leaves resemble a palm shape, with a flat midrib and leaf color ranging from green to large in size. In excessively vegetative plants, the

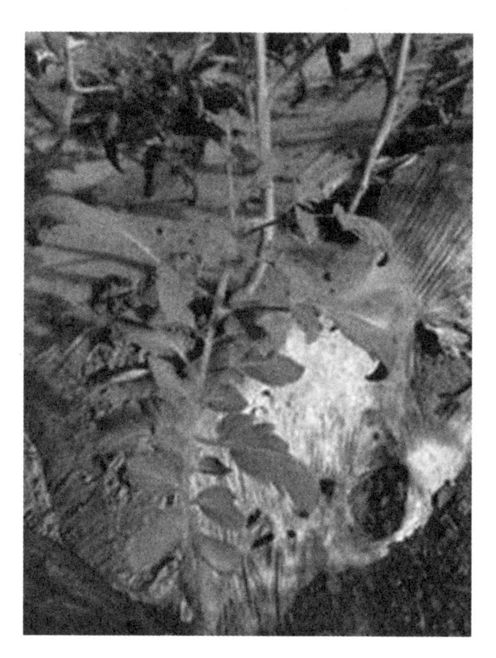

Figure 1 Feather-like Leaf Blade

leaves are triangular in shape, with prominent midribs and a deep green color. In aging plants, the leaves become small, dark green, or pale green, and the top leaves become reduced in size.

005 What are the characteristics of tomato flower organs?

Tomatoes are complete flowers with umbel inflorescences. Small-fruited varieties often have compound inflorescences, and the inflorescences are borne at the nodes. The flowers are yellow in color. The number of flowers per inflorescence varies greatly among different varieties, ranging from five or six to more than ten flowers. Even within the same plant, different inflorescences can show variations under different environmental conditions. The stamens usually range from 5-9 or more and are fused into a cone, surrounding the pistil. The anther splits longitudinally and releases pollen when mature, facilitating self-pollination. Some individual or certain varieties, under certain conditions, may have the stigma extending beyond the stamens, allowing for cross-pollination. The natural hybridization rate ranges from 4%-10%. The ovary is superior and located on the central axis.

Tomato flowering and fruiting habits can be classified into two types based on the pattern of inflorescence development: Determinate growth type. Generally, the main stem starts to produce the first inflorescence when it has grown to 6-7 true leaves. Subsequently, an inflorescence forms every 1-2 leaves. Typically, after 2-4 layers of inflorescences have developed on the main stem, the lateral buds below the inflorescence cease development, no longer branching, and no new inflorescences are formed. Indeterminate growth type. In this type, the first inflorescence appears when the main stem has grown to 8-10 leaves, or even up to 11-13 leaves in late-maturing varieties. Subsequently, an inflorescence forms every 2-3 leaves. Under favorable conditions, the plant can continuously produce inflorescences indefinitely, continuously branching and bearing flowers and fruits.

Each flower of the tomato plant has a distinct "collar" in the middle of its pedicel, which is formed by several layers of differentiated cells during the process of floral bud formation. During the floral differentiation stage, a layer of secondary meristematic tissue appears around 20 layers beneath the apical meristem in the epidermis. As the floral bud develops, the number of layers gradually increases, reaching 10-12 layers. At this point, the number of differentiated cells stops increasing, but the cells in the upper and lower parts of the collar continue to enlarge and grow. This results in the formation of a concave ring-shaped "collar". When environmental conditions are unfavorable for floral organ development, the cells in the collar region separate, leading to flower drop.

The size and uniformity of tomato fruits of the same variety are related to floral development. Generally, larger flowers produce larger fruits, and malformed flowers tend to develop into malformed fruits. For example, flowers formed under low temperatures often have more petals and a thick, flattened stigma, resulting in the development of malformed fruits.

Productive morphology of tomato flowers: In a single inflorescence, the flowers bloom uniformly, with medium-sized floral organs, yellow petals, and moderately sized ovaries. In elongated inflorescences, the flowers do not bloom uniformly, often with enlarged floral organs and ovaries, and intense yellow petals. In aging plants, flowering is delayed, the floral organs are small, the petals are pale yellow, and the ovaries are small.

006 What are the characteristics of tomato fruits and seeds?

The shape, size, color, and number of locules (seed cavities) of tomato fruits vary among different varieties. Tomato fruits are juicy and composed of pericarp (mesocarp) and placental tissue, with good varieties having thick flesh and small seed cavities. Cultivated varieties generally have multiple locules, and the number of locules is correlated with the number of sepals and fruit shape. Fruits with 3-4 locules have a smaller diameter and poor fruit enlargement, while fruits with 5-7 locules develop well and are nearly spherical. Increasing the number of locules results in larger and flatter fruit shape. The number of locules is determined by genetic factors as well as environmental conditions.

The color of tomato fruits is determined by the color of the fruit skin and flesh. If both the skin and flesh are yellow, the fruit appears deep yellow. If the skin is colorless and the flesh is red, the fruit appears pink. If the skin is yellow and the flesh is red, the fruit appears orange-red. The red color of tomato fruits is due to the presence of lycopene ($C_{40}H_{50}$), while the yellow color is attributed to the presence of carotenes and xanthophylls. The formation of tomato carotenes and xanthophylls is influenced by light exposure, while lycopene formation is primarily governed by temperature.

Tomato seeds mature earlier than the fruits. Generally, the seeds start to acquire germination ability around 35 days after pollination, but the embryo development is completed around 40 days after pollination. Therefore, seeds from 40-50 days after pollination have fully developed germination ability, and the seeds reach complete maturity between 50 and 60 days after pollination. Tomato seeds are surrounded by a layer of gelatinous material within the fruit. Due to the presence of germination inhibitors and the osmotic pressure of the fruit juice, the seeds do not germinate inside the fruit. The thousand-seed weight of tomato seeds is around 3.0-3.3 grams.

007 What are the characteristics of the germination period of tomatoes?

The germination period of tomatoes is the time from seed germination to the emergence of the first true leaves (cotyledon breaking). Under normal temperature conditions, this period lasts for 7-9 days.

Tomato seeds store nutrients in the endosperm. The entire embryo (radicle, plumule and cotyledons) is surrounded by the endosperm. During germination, the radicle starts to grow first, while the cotyledons remain in the seed and absorb stored nutrients from the endosperm. Subsequently, the curved hypocotyl begins to grow, penetrating through the soil layer and bringing the cotyledons above the ground.

The successful completion of the germination period depends mainly on temperature, humidity, ventilation, and soil covering thickness. Tomato seeds undergo two stages of water absorption: the first stage involves rapid water absorption, where seeds can absorb around 35% of their dry weight in water within about 0.5 hours at a water temperature of 20-30 ℃ and 60%-65% of water within 2 hours. The second stage involves slower water absorption, where seeds can absorb approximately 25% of their dry weight in water, close to saturation, after 5-6 hours. The fastest germination occurs at a temperature of 25 ℃ and oxygen content above 10%. Root emergence begins around 36 hours, followed by the appearance of cotyledons after 2-3 days.

Under the same conditions, the difference in germination speed among individuals is mainly related to seed quality. Larger and uniformly filled seeds can produce early, uniform seedlings, which often result in better production outcomes. This characteristic should be taken into account when sowing tomato seedlings.

Tomato seeds are small and contain limited nutrients, which are quickly utilized by the developing seedlings. Therefore, ensuring timely and necessary nutrient supply to the seedlings plays a crucial role in their growth and development, especially in the early formation of reproductive organs.

Similar to other crops, young tomato seedlings during the germination period exhibit significant physiological plasticity. Treating germinating tomato seeds with low temperatures (-2℃-0 ℃) or temperature variations(20 ℃ for 8-12 hours, 0 ℃ for 12-16 hours) often leads to consistent seedling growth under lower temperature conditions and promotes early maturity.

008 What are the characteristics of the seedling stage of tomatoes?

The period from the emergence of the first true leaves to the appearance of large flower buds is known as the seedling stage of tomatoes. Although tomato seeds are small and contain limited stored nutrients, they undergo a relatively rapid transition from heterotrophic to autotrophic growth, providing a favorable foundation for the rapid growth of tomato seedling roots. After seed germination, the initial growth of the root system is dominant, particularly pronounced in early-maturing varieties. After 20-30 days of germination, the primary root of the seedling can reach 40-50 cm in length and develop numerous lateral roots. Under direct sowing conditions, the root system can penetrate the soil to a depth of more than 80 cm after approximately 60 days of germination. When transplanting seedlings, the root system may experience damage, but tomato roots have strong regenerative capacity, and the occurrence of abundant lateral roots promotes rapid growth of the aboveground parts of the seedling.

The seedling stage of tomatoes can be divided into two distinct phases. With 2-3 true leaves before the differentiation of flower buds, it is the stage of basic vegetative growth. This stage lays the foundation for flower bud differentiation and further vegetative growth. Additionally, the hormones produced by cotyledons and expanded true leaves have a significant promoting effect on the differentiation of tomato flower buds. Therefore, the size of the cotyledons directly influences the timing of the first inflorescence differentiation, and the leaf area of the true leaves affects the number and quality of flower bud differentiation. Thus, cultivating thick, dark green cotyledons and larger first or second true leaves is an essential foundation for cultivating vigorous seedlings. Around 25 - 30 days after sowing, when the seedling has 2-3 leaves, flower bud differentiation begins, entering the second phase of the seedling stage, which is the stage of flower bud differentiation and development. From this point onwards, vegetative growth and flower bud development occur simultaneously. Approximately 35 - 40 days after sowing, the second inflorescence starts to differentiate, followed by the third inflorescence after 10 days.

The node position where the flower bud begins to differentiate is governed by the variety and seedling conditions. Early maturing varieties can have flower buds appearing as early as after 6 true leaves, but if the seedling conditions are poor, the differentiation node position will correspondingly increase. After the flower bud begins to differentiate, a flower bud is generally differentiated every 2–3 days. At the same time, the upper lateral bud adjacent to the flower bud begins to differentiate and grow, continuing to differentiate the leaf blade. When the differentiation of the first inflorescence flower bud is about to end, the next inflorescence has already begun to differentiate the initial flowers. This development continues upwards. When the first inflorescence shows large buds, the third inflorescence flower buds have completely differentiated. From this, it can be seen that the early and rapid differentiation of flower buds and the continuity of flower bud differentiation are the main features of tomato flower bud differentiation.

Looking at a single flower bud, its development starts from the outer organs gradually moving inward, that is, from the sepals to the petals, then the stamens, and finally the pistils. From the initial stage of differentiation to the formation of the sepals, the growth speed of the flower bud is relatively slow. After entering the sepal formation stage, the development speed accelerates, and during the formation of the pollen mother cell and the formation of the pollen tetrad, the growth speed of the flower bud becomes even faster. The process from flower bud differentiation to blooming takes about 30 days, meaning it takes about 55–60 days from sowing to blooming.

The initiation time and development process of flower bud differentiation vary among different varieties and seedling conditions. Early-maturing varieties exhibit earlier flower bud initiation, and the development of flower buds in each inflorescence is faster compared to medium and late-maturing varieties, which is consistent with the characteristics of root development in early-maturing varieties mentioned earlier.

During the seedling stage, the differentiation and development of flower buds occur concurrently with the vegetative growth of the seedlings, with the latter serving as the foundation for the former. The development of the root system and leaf area of the seedlings is closely related to the differentiation and development of flower buds. However, stem thickness is more closely related to flower bud differentiation. The standard stem thickness for the initiation of the first inflorescence is approximately 2 mm, while for the second and third inflorescences, it is around 4–5 mm and 7–8 mm, respectively. Although the criteria for stem thickness in flower bud differentiation may vary among different varieties and seedling conditions, it remains a relatively reliable indicator as it reflects the synthesis and accumulation of nutrients.

Creating favorable conditions, preventing excessive elongation and aging of seedlings, and ensuring robust seedling growth and normal flower bud differentiation and development are the primary tasks of cultivation management during this stage.

009 What are the characteristics of the fruiting period of tomatoes?

The tomato is a crop that continuously blooms and bears fruit. The flowering and fruiting period referred to here only includes the stage from the appearance of large buds in the first inflorescence to fruiting. When the seedlings are transplanted, this period is in the early stage after planting. Although this stage is not long, it is a turning point for the tomato to transition from predominantly vegetative growth to equal development of reproductive and vegetative growth, directly related to the formation of product organs and yield, especially early yield.

The timing of flowering directly affects the early maturity of tomatoes, and flowering time is determined by the variety, seedling age, and temperature conditions after planting. Under normal conditions, it takes about 30 days from the start of bud differentiation to flowering, but in production, it often takes longer than 30 days. This is because the exercise in the late stage of seedling cultivation and the seedling recovery after planting will all delay the growth and development of the seedlings.

In this stage, the contradiction between vegetative growth and reproductive growth is more prominent. From the perspective of above-ground growth, this stage is a critical period for determining the balance of vegetative growth and reproductive growth. Excessive vegetative growth or even rapid growth will inevitably cause a delay in flowering and fruiting or falling flowers and fruits. Vigorous growth of mid-to-late maturing varieties and excessive application of nitrogen fertilizer, poor sunlight, excessive soil moisture, high night temperature, especially when planting young seedlings under these conditions, are most likely to cause this phenomenon. On the other hand, for early maturing varieties, poor management after planting, especially improper squatting, can easily lead to fruit drop and seedling fall. Due to the small volume of plant nutrition, slow fruit development will result in low yield. Promoting early root growth and paying attention to flower and fruit preservation are the main tasks of cultivation at this stage.

010 What are the characteristics of the fruiting period of tomatoes?

The fruiting period of tomatoes encompasses the period from the fruit set of the first inflorescence to the end of fruit development(crop harvest). During this period, both fruit and vegetative growth occur simultaneously, and there is an ongoing conflict between vegetative and reproductive growth. Peaks of vegetative growth and fruit growth occur periodically, but the intensity of this conflict depends on cultivation management techniques. Generally, high-yielding tomatoes exhibit a more balanced conflict between vegetative and reproductive growth. On the other hand, when the conflict is pronounced, yield distribution becomes uneven, and overall yield decreases. By adjusting the balance between vegetative and reproductive growth during flowering and fruiting, and with appropriate nutrient and water management, fruit drop can be minimized. Conversely, improper pruning, excessive shoot growth, and inadequate nutrient and water management can lead to excessive vegetative growth, which must be controlled.

Tomatoes are characterized by continuous flowering and continuous fruiting. While the first inflorescence is undergoing fruit development, the second, third, fourth, and fifth inflorescences are also developing to varying degrees. During fruit development, especially within the first 20 days after flowering, a large amount of carbohydrates is transported to the developing fruit, and there is noticeable competition for nutrients among different flower clusters. Generally, nutrients produced by lower leaves are primarily supplied to the root system and other nutritional organs, mainly meeting the needs of fruit development in the first inflorescence. Middle leaves primarily transport nutrients to the developing fruit, while upper leaves supply nutrients to both upper fruit clusters and the growing shoot tip.

Due to nutrient allocation, excessive nutrient consumption by lower fruit clusters can cause the stem axis to become thinner, resulting in poor development of upper flower clusters with smaller flower organs and inferior fruit set. This not only reflects the nutrient distribution conflicts among different flower clusters but also highlights the conflict between vegetative and reproductive growth. If the plant's vegetative growth is well-balanced, with uniform stem growth from bottom to top, even if the lower flower clusters have a higher fruit load, the upper flower clusters can still develop, flower, and set fruit normally.

During this stage, there is a prominent contradiction between vegetative growth and reproductive growth. In terms of aboveground growth, this stage is crucial for balancing vegetative and reproductive growth. Excessive vegetative growth, or even excessive elongation, can delay flowering and fruit set. Vigorous growth in medium- and late-maturing varieties, excessive nitrogen application, poor sunlight exposure, excessive soil moisture, and high night temperatures, especially when planting young seedlings under such conditions, can lead to delayed flowering and fruit set. Conversely, in early-maturing varieties, fruit drop may occur due to poor post-transplant management, particularly improper pruning. Due to the small nutrient reserves in the plant, slow fruit development can result in low yields. Promoting early root development andensuring proper flower and fruit retention are the main tasks during this stage of cultivation.

In tomato production, there are various patterns of yield formation, such as high early yields with high later yields, high early yields with low later yields, low early yields with high later yields, and low yields throughout the season. The formation of yield depends on the process of yield formation and cultivation management techniques. Generally, there is a trade-off between early and total yield. However, under certain conditions, this trade-off can be reconciled. During the fruiting period, creating favorable conditions to promote both fruit and vegetative growth, maintaining a balanced cycle, and ensuring continuous fruit set are crucial for achieving early and high yields.

011 What are the temperature requirements for tomatoes?

Tomatoes are warm-season vegetables, and the optimum temperature for photosynthesis is around 20 - 25 ℃. Temperatures below 15 ℃ can inhibit flowering or result in poor pollination, leading to reproductive growth disorders such as flower drop. When temperatures drop to 10 ℃, plant growth ceases, and prolonged exposure to temperatures below 5 ℃ can cause chilling injury. The lethal temperature range is typically - 2 to - 1 ℃. When temperatures rise above 30 ℃, photosynthesis significantly decreases, and temperatures above 35 ℃ can disrupt and damage reproductive growth, even short periods of temperatures above 45 ℃ can cause physiological disturbances, leading to flower and fruit drop or poor fruit development.

The temperature requirements and responses of tomatoes vary during different growth stages. The optimal temperature for seed germination is around 28-30 ℃, with a minimum germination temperature of approximately 12 ℃. During the seedling stage, the ideal daytime temperature ranges from 20 - 25 ℃, while nighttime temperatures should be around 10 - 15 ℃. Tomato seedlings have strong adaptability to temperature, and they can tolerate temperatures as low as 6-7 ℃ for an extended period or even short periods of -3 ℃ or 0 ℃ under certain conditions. The flowering period is particularly sensitive to temperature, especially 5 - 9 days before and 2 - 3 days after flowering. The ideal daytime temperature during this period is 20-30 ℃, while nighttime temperatures should be around 15-20 ℃. Temperatures below 15 ℃ or above 35 ℃ are unfavorable for the normal development of flower organs and flowering. During the fruiting stage, the ideal daytime temperature is 25-28 ℃, and nighttime temperatures should be around 16-20 ℃. Low temperatures slow down fruit growth, while temperatures above 30 - 35 ℃ promote faster fruit growth but result in fewer fruit set. High nighttime temperatures are detrimental to nutrient accumulation and can lead to poor fruit development. Temperatures above 26 - 28 ℃ can inhibit the formation of lycopene and other pigments, affecting normal fruit ripening.

The optimum soil temperature for tomato root growth is around 20-22 ℃. Increasing soil temperature not only promotes root development but also significantly increases the content of nitrate nitrogen in the soil, accelerating plant growth and increasing yield. Therefore, as long as the nighttime temperature is not high, maintaining soil temperatures around 20 ℃ throughout the day and night can prevent excessive elongation and is beneficial for protected tomato production. Root nutrient and water absorption are hindered at 5 ℃, and root hair growth ceases at 9-10 ℃.

The optimal temperature range is closely related to other environmental conditions, especially light intensity, nutrient availability, and CO_2 concentration. Under low light conditions, the optimum temperature for photosynthesis decreases significantly. Increasing CO_2 concentration under high light conditions raises the optimum temperature for photosynthesis. When CO_2 concentration reaches 0.12%, the optimum temperature for photosynthesis can be increased to 35 ℃. The interaction between temperature, particularly nighttime temperature, and nitrogen nutrition significantly affects tomato growth and fruiting. Generally, as long as the nighttime temperature is suitable, tomatoes can fruit normally with high or slightly low nitrogen concentrations. However, under high nighttime temperatures, low nitrogen concentrations can inhibit fruiting, even with normal nitrogen fertilization, resulting in nitrogen deficiency symptoms.

012 What are the light requirements for tomatoes?

Tomatoes are photophilic crops, and their light saturation point is around 70 000 lux. As the light intensity decreases, photosynthesis significantly decreases. Therefore, it is necessary to ensure good lighting conditions during cultivation. Generally, a light intensity of 30 000–35 000 lux or higher is required to maintain normal growth and development. In winter greenhouse cultivation, poor lighting conditions can lead to reduced plant nutrition, causing a significant amount of flower drop and affecting normal fruit development, resulting in reduced yield. Generally, strong light does not cause harm. However, if accompanied by high temperatures and dry conditions, it can cause leaf curling or sunburn on the fruit surface, affecting yield and product quality.

Tomatoes are short-day plants and require short-day conditions during the transition from vegetative growth to reproductive growth, specifically during the process of flower bud differentiation. However, the requirement for short-day conditions is not strict. Some varieties can initiate flower buds and bloom earlier under short-day conditions, while most varieties flower earlier under 11 – 13 hours of daylight, resulting in robust plant growth. Many experiments have shown that tomatoes grow best under 16 hours of daylight. Therefore, extending the photoperiod significantly increases dry matter production.

Tomatoes require a complete spectrum of sunlight for normal growth and development. Tomato seedlings grown under covered facilities sometimes tend to become elongated, mainly due to a lack of short-wavelength light such as ultraviolet light. The lower vitamin C content in tomatoes produced during winter or in greenhouses is also related to this.

013 What are the water requirements for tomatoes?

Tomato plants have lush aboveground stems and leaves, and they have a relatively high transpiration rate, with a transpiration coefficient of around 800. However, tomatoes have well-developed root systems and strong water absorption capacity. Therefore, their water requirements are considered semi-drought tolerant. They require a sufficient amount of water but do not require frequent heavy irrigation. They also do not require high air humidity and generally prefer an air relative humidity of 45%-50%. High air humidity not only hinders normal pollination but also leads to severe diseases under high-temperature and high-humidity conditions.

Water requirements vary during different growth stages. During the seedling stage, to avoid excessive elongation and disease occurrence, soil moisture should not be too high, and irrigation should be controlled appropriately. Before the fruit set of the first flower cluster, excessive soil moisture can lead to excessive stem elongation and poor root development, resulting in flower drop. After the fruit set and rapid fruit enlargement of the first flower cluster, there is a rapid growth of branches and leaves, requiring increased water supply. During the fruiting period, when the fruit is rapidly enlarging, a large amount of water is needed. At this time, the temperature is high, and the transpiration rate of the plants is high. Failure to replenish water regularly can affect the normal development of the fruit. It has been reported that tomatoes in the rapid fruit enlargement stage can absorb 1-2 liters of water per plant per day, excluding the water lost through soil evaporation. For an area of 667 square meters, 5 - 10 cubic meters of water should be supplied daily. Improper soil moisture management during the rapid fruit enlargement stage can contribute to blossom end rot. Excessive soil moisture and poor drainage during the fruiting period can hinder normal root respiration and, in severe cases, lead to root rot and seedling death. The soil moisture range should be maintained at 60%-80% of the maximum water-holding capacity. Additionally, irregular drying and wetting of the soil during the fruiting period can lead to fruit cracking, so regular and even irrigation should be observed.

014 What are the soil and nutrient requirements for tomatoes?

Tomatoes are not very demanding in terms of soil conditions. However, to achieve high yields and establish a good foundation for root development, it is advisable to select deep, well-drained, and fertile soil rich in organic matter. Tomatoes have high requirements for soil aeration, and when the oxygen content in soil air drops to 2%, the plants will wither and die. Therefore, low-lying and poorly structured soils prone to waterlogging are not suitable. Sandy soils have good permeability and quick soil temperature rise, which can promote early ripening. Clay loam soils with good water retention and fertility, especially those rich in organic matter and well-drained, can improve yield. The pH of the soil should be around 6-7, and acidic or alkaline soils should be amended. In slightly alkaline soils, seedling growth may be slow, but once the plants grow larger, they exhibitgood growth and better quality.

Tomatoes require a significant amount of nutrients to be absorbed from the soil during their growth process. It has been reported that to produce 5 000 kg of fruit, tomatoes need to absorb approximately 33 kg of potassium oxide, 10 kg of nitrogen, and 5 kg of phosphoric acid from the soil. Around 73% of these elements are stored in the fruit, while the remaining 27% are distributed among the stems, leaves, roots, and other nutritional organs. Nitrogen fertilizer plays a crucial role in the growth of stems and leaves and the development of fruit and has the closest relationship with yield. Before the rapid enlargement of the first fruit cluster, the plant's nitrogen uptake gradually increases. Throughout the entire growth process, nitrogen is absorbed at a relatively constant rate, reaching its peak during the fruiting stage. Therefore, sufficient nitrogen supply is essential for high yields, as long as adequate light is provided, night temperatures are lowered, and other nutrients are properly applied. Excessive nitrogen fertilization will not cause excessive elongation but is an indispensable condition for high yields. Phosphoric acid uptake is not as significant in quantity, but it plays a significant role in the development of tomato roots and fruit. Approximately 94% of the absorbed phosphoric acid is stored in the fruit and seeds. Additional phosphorus fertilizer during the seedling stage has a positive effect on flower bud differentiation and development. Potassium uptake is relatively high, especially during the rapid fruit enlargement stage. Potassium is essential for sugar synthesis and transport, as well as for increasing cell fluid concentration and enhancing water absorption by cells.

2. Practical Techniques for High-Quality and Efficient Production of Sunlight Greenhouse Cherry Tomatoes

015 What are the comprehensive quality indicators for cherry tomatoes?

The comprehensive quality indicators for cherry tomatoes can be summarized as bright color, sweet taste, crisp texture, attractive shape, nutritional value, and greenness. Through years of practice in sunlight greenhouse cherry tomato cultivation, Liaoning Vocational College (Tieling) has achieved the production of small tomatoes with excellent appearance, taste, and flavor. The yield per acre in both spring and autumn seasons reaches 3 000 - 4 000 kg, resulting in high economic benefits. It has also served as a good demonstration in facility vegetable production in the northern region of Liaoning. The summarized production techniques are suitable for agricultural ecological parks and pick-your-own gardens.

016 What are the characteristics of some excellent cherry tomato varieties currently available?

Excellent cherry tomato varieties include Chuntao, Qianxi, Bijiao, Fengzhu (409), Jinzhu and Jinmijia. These varieties have their own unique features in terms of fruit color, flavor, and texture, and they also have significant differences in biological characteristics. With proper management, these varieties can be cultivated into delicious fruits.

(1) Chuntao. Developed by Taiwan Nongyou Seed Company, it has peach-shaped fruits with a pink color. The average fruit weight is 30-50 g, and each cluster has around 10 small flowers. It has an attractive shape, good quality, and high sugar content. The plants have strong growth vigor, large leaves, and good cold tolerance (Figure 2).

(2) Qianxi. A Santa-type small tomato developed by Taiwan Nongyou Seed Company. The fruits are pink-red, oval-shaped, with an average fruit weight of about 15 g and a high sugar content of 9. 6%. It has excellent flavor, is less prone to fruit cracking, and each cluster has 14-31 small flowers. It is early maturing, vigorous in growth, high-yielding, and resistant to wilt and storage (Figure 3).

Figure 2 Chuntao

Figure 3 Qianxi

(3) Bijiao. An early-maturing small tomato developed by Taiwan Nongyou Seed Company. It has strong disease resistance, extremely small leaves, self-topping, and strong branching ability. The fruits are elongated, orange-red, with a crisp and tender flesh, high sugar content, and thin skin. The average fruit weight is about 13 g(Figure 4).

Figure 4 Bijiao

(4) Fengzhu(409). A small tomato developed by Taiwan Nongyou Seed Company. The fruits are elongated, with 2-4 small dimples(depressions) below the shoulder. The shape is not fully rounded. The ripe fruits are red, with an average fruit weight of about 16 g and a sugar content of up to 9.6%. It has a sweet flavor and fine texture. The plants have strong

growth vigor and resistance to wilt(Figure 5).

Figure 5 Fengzhu

(5) Jinzhu. A golden-yellow cherry tomato variety developed by Taiwan Nongyou Seed Company. It has an indeterminate growth habit, early maturity, dark green leaves with slight curling, and strong fruiting ability. The fruits are round to slightly oblate, bright orange-yellow in color, with good quality, intense aroma, and a sugar content of up to 10%. The average fruit weight is about 15 g(Figure 6).

Figure 6 Jinzhu

(6) Jinmijia. A high-quality small tomato variety bred by Liaoning Yinong Technology Co., Ltd. It has a good taste, with a sugar content of 11%–13%. It is easy to set fruit, with an average of about 40 fruits per cluster. The average fruit weight is 10–15 g. It has moderate growth vigor and good disease resistance (Figure 7).

Figure 7 Jinmijia

017 How should cherry tomato production be arranged?

Cherry tomatoes can be produced in sunlight greenhouses throughout the year. To fill the market gap during the off-season for fresh fruits, the highest prices are achieved when harvested and sold from December to July of the following year. Based on the conditions of sunlight greenhouses, it is generally recommended to arrange winter-spring crops (planting in September, harvesting from December to May of the following year), spring crops (planting in February, harvesting from April to July), and autumn-winter crops (planting in July, harvesting from October to December).

018 How should basic fertilization and bed preparation be carried out in sunlight greenhouse cherry tomato production?

During land preparation, mix well-rotted farmyard manure (5 000 kg per acre) and potassium sulfate compound fertilizer (N-P-K ratio of 14–16–15, 50 kg). Farmyard manure can include chicken manure, pig manure, cow manure, crop straw, etc., and a combination of two or more types can also be used.

Create raised beds that are 10 cm high and 70–80 cm wide. Use one plastic mulch per two rows, and lay two drip irrigation tapes beneath the plastic mulch with the drippers facing up. The aisle width should be around 80 cm, which helps increase light exposure and air circulation and facilitates visitor picking(Figure 8).

Figure 8 Raised bed specifications

019 What are the requirements for cherry tomato planting?

Plant the seedlings in two rows on the plastic mulch, with the planting holes positioned right next to the drip irrigation tapes. Water the seedlings during planting using the drip irrigation system. For double-stem pruning, it is advisable to plant around 1 800 plants per acre in the greenhouse (1 200 plants for Chuntao). Three days after planting, water the seedlings with a slow irrigation to ensure proper water penetration and seal the planting holes to prevent stem base rot in tomatoes.

020 What are the key points for fruit set management in cherry tomatoes?

(1) Double-vine pruning. Retain the main vine and one side branch just below the first flower cluster on the main vine, and train them to grow upward. Remove other side branches and pre-flower branches. Keep 10–12 fruit clusters per plant, and remove the growing tip of the last fruit cluster after leaving two leaves at the front end.

Ensure that the plants grow upward and are evenly distributed, making full use of the greenhouse space. Bijiao cherry tomato variety is self-topping, so multiple topping and vine training on upper lateral branches are needed to achieve high trellis cultivation.

(2) Pollination and fruit set. Preferably use bumblebees for pollination(Figure 9). The advantages of bumblebee pollination are that it increases the number of seeds and juice content in tomato fruits, resulting in better flavor and taste. When bumblebees are present in the greenhouse, avoid using insecticides that are harmful to them. Additionally, the application of "fruit set hormone" can be used by spraying a 30-50 mg/L solution on the flower peduncles when the tomato buds are opening. Flowers on the same inflorescence open at different times, so they need to be treated in stages.

Figure 9 Bumblebee pollination

As cherry tomatoes are intended for fruit consumption, it is recommended to control the average fruit weight at around 15 g(25-50 g for Chuntao). Therefore, it is advisable to have a higher number of fruits per cluster, typically 15-20 fruits per cluster(10 for Chuntao). Remove deformed fruits and excess small flowers to increase the commercial yield of the fruits.

021 What are the key points for fruit expansion management in cherry tomatoes?

(1) Moderate watering. After fruit set, water the plants evenly to prevent fruit cracking. Proper water control helps increase the sweetness of tomatoes. Overwatering can lead to excessive vegetative growth, reducing yield and quality, while severe water shortage can result in thickened and hardened fruit skin, affecting the texture. Additionally, water should be applied during each fertilizer application.

(2) Balanced fertilization. After the first fruit cluster sets, apply 50 g of well-rotted soybean cake fertilizer per plant at a depth of 10 cm from the root zone, and apply 5 kg of American Potash per acre with water. After the second fruit cluster sets, apply 20 kg of

seaweed chitosan biofertilizer per acre with water. Subsequently, alternate the application of American Potash and seaweed chitosan biofertilizer to ensure a balanced supply of nutrients. During the fruiting period, fertilize every 10 days. The application of chitosan biofertilizer should be done at least twice to achieve noticeable effects on seedling vigor and sweetness enhancement.

During the fruiting period, spray 0. 1% calcium humate or amino acid calcium 3-4 times as foliar supplementation. The foliar application of calcium fertilizer can prevent blossom-end rot, increase fruit firmness, reduce acidity, and enhance sweetness.

(3) Suitable temperature. Based on the growth and development characteristics of tomatoes, the recommended temperature during the fruiting period is 25-30 ℃ during the day and 10-15 ℃ at night. Adequate daytime high temperature promotes photosynthesis and increases yield, while low nighttime temperature reduces respiration and prevents excessive vegetative growth, thereby improving quality.

If the nighttime temperature is too low, around 0-3 ℃, the plants may suffer from cold damage, resulting in slow growth and decreased yield. Severe cold damage can cause purple leaf curling and a decline in quality.

(4) Increase light exposure. Tomatoes are high-light-demanding vegetables. Increasing light intensity and duration helps accumulate photosynthates, thereby improving yield and quality. Proper planting density, adjusting light exposure of the plants, controlling excessive vegetative growth, and removing aging lower leaves at the base are all measures that promote light exposure.

(5) Enhance ventilation. Good ventilation reduces the relative humidity inside the greenhouse, which not only reduces the occurrence of cherry tomato diseases and pests but also improves the smooth and glossy appearance and flavor of the fruits, enhancing both external and internal quality. This measure has been validated during tomato bagging treatments. When plastic bags are used to cover tomato clusters, the lack of air circulation increases humidity, resulting in rough and dull fruit skin and sour taste. On the other hand, when paper bags are used, the ventilation is good, resulting in bright-colored fruits with a sweet taste(Figure 10).

Figure 10 Paper bagging

022 Why is it important to focus on leaf protection in the later stages?

Leaves are important organs for photosynthesis and nutrient production. Insufficient leaf quantity, aging, yellowing (reduced chlorophyll content), disease and pest damage, dust pollution, chemical damage, freezing damage, and other conditions can lead to reduced yield and decreased quality. During the fruiting stage, functional leaves are more susceptible to diseases. Tomato leaf mold, powdery mildew, and other diseases often start from functional leaves, while older leaves and young leaves are less affected.

During the growing season, it is recommended to spray a large amount of the water-soluble fertilizer "Green Field God" ($N + P_2O_5 + K_2O \geqslant 500$ g/L, B: 3 - 30 g/L) from Changchun Hongfeng Fertilizer Technology Co., Ltd. on the leaves, diluted at a ratio of 200 times, 2 - 3 times. This helps maintain the healthy growth of tomatoes by supplementing nutrients, alleviating chemical damage, fertilizer burn, and cold damage, increasing chlorophyll content in the leaves, and reducing the harm of chlorosis virus disease.

023 How to develop cherry tomato picking and sales models?

Cherry tomatoes are usually harvested with small stems and packed in boxes for sale, with moist and low-temperature storage for freshness. In 2016, under the guidance of Liaoning Vocational College, Shenyang Zhicheng Duoduo Children's Farm (Tieling) produced greenhouse fruits and vegetables in accordance with the pollution-free agricultural product regulations. They named their delicious small golden tomatoes "Jinguo Guo" and the tender and sweet green cherry tomatoes "Hongguoguo". These products were sold in Shenyang supermarkets at a retail price of 29.6 yuan per kilogram. They also developed a special processed product called "Cherry Tomato Rock Candy Skewers", which was highly popular among children.

3. Key Techniques for High-Yield Cultivation of Late-Extended Autumn Cherry Tomatoes in Sunlight Greenhouses

024 How is the production of late-extended autumn cherry tomatoes in the northern region of Liaoning (Tieling) in sunlight greenhouses?

The high-raised cultivation mode of late-extended autumn cherry tomatoes with one spring crop of cucumbers (or beans) is an important facility vegetable production mode in the cold winter of Tieling, Liaoning Province, China. In this production mode, the growth and development of tomatoes undergoes the process of hot summer to severe cold winter. Although tomatoes are one of the most adaptable vegetable crops, they are still affected by adverse factors such as high temperature, high humidity, excessive plant growth, and increased susceptibility to diseases and pests. By summarizing successful management experience and implementing a series of key techniques, the balance between vegetative growth and reproductive growth can be adjusted, achieving high-yield and high-quality tomato production, as well as obtaining significant economic and ecological benefits.

025 How is the production arrangement for late-extended autumn cherry tomatoes in sunlight greenhouses implemented?

Generally, seeds are sown in rain shelters in early to mid-June, transplanted in July, and harvesting begins from early September to early October. The plants are pruned and replanted around the New Year.

026 What are the main excellent varieties suitable for local production of late-extended autumn cherry tomatoes in sunlight greenhouses?

Select mid-to-late maturing varieties that are heat-tolerant, cold-tolerant, vigorous in growth, disease-resistant, high-yielding, and suitable for storage and transportation. The main excellent varieties include "Baili" "Aoni" "Kaile" "Tianci No. 1" and "Tianmei No. 1".

(1) Baili. A hybrid generation of tomatoes produced by Rijk Zwaan, Netherlands. It is an indeterminate growth type with a large frame and vigorous growth. It has a high fruit-setting rate, early maturity, and good productivity. The fruits are large, round, medium-sized, with an individual fruit weight of 180–200 g, and can reach a maximum weight of around 240 g. They have a bright color, good taste, no cracking, no green shoulders, firm texture, good transportability, and storage resistance. It is suitable for export and long-distance transportation. It is resistant to tobacco mosaic virus disease, blossom-end rot, and wilt. It has good heat tolerance and can set fruit normally under high-temperature and high-humidity conditions, making it suitable for cultivation in sunlight greenhouses in spring and autumn.

(2) Aoni. A hybrid generation of tomatoes introduced from the Netherlands by Liaoning Shenyang Deyi Agricultural Development Co., Ltd. It is a long-frame, storage and transportation type tomato variety with excellent taste. It belongs to the indeterminate growth type. The plants are robust in growth, medium-early maturing, with bright red fruits that are uniform, smooth, and have good firmness. The average individual fruit weight is 220 g (Figure11). It has a high fruit-setting rate, outstanding productivity, and high early yields, with a concentrated ripening period. The fruits have no green shoulders or green rot, thick flesh, and excellent storage and transportation resistance. It has outstanding disease resistance, including resistance to viral diseases, powdery mildew and late blight. The yield per acre in high-raised cultivation is generally above 15 000 kg.

Figure 11 Aoni

(3) Kaile. A large-sized fruit tomato variety with high hardness and resistance to tomato yellow leaf curl virus(TYLCV), developed by Shenyang Ailvshi Seed Industry Co., Ltd.

It is an indeterminate growth type with an average individual fruit weight of 250–280 g, and a maximum weight of up to 360 g. The fruits are tall and round(apple-shaped), with a smooth surface and thick and beautiful calyx. The fruit shape is uniform, with a high percentage of premium fruits. The flesh is very firm, and it has excellent storage and transportation resistance. It is suitable for cultivation in sunlight greenhouses in spring and autumn.

(4) Tianci No. 1. A hybrid generation produced by Shenyang Guyu Seed Industry Co., Ltd. It is an indeterminate growth type with strong vigor, high yield, pink fruits, tall and round fruit shape, and an average individual fruit weight of 220–250 g. It has high firmness, a long shelf life, and a high percentage of premium fruits. It is resistant to TYLCV, tobacco mosaic virus(TMV), and powdery mildew, which are common tomato diseases. It is suitable for autumn tomato facility production.

(5) Tianmei No. 1. A hybrid generation produced by Liaoning Haicheng Sanxing Ecological Agriculture Co., Ltd. It is an indeterminate growth type with strong vigor, pink fruits, good commercial quality, strong fruit-setting ability, an average individual fruit weight of 220–250 g, excellent firmness, a long shelf life, and a high percentage of premium fruits. It is resistant to tomato yellow leaf curl virus(TYLCV), powdery mildew, gray leaf spot, and has strong adaptability. It is easy to cultivate and manage, suitable for late-extended and summer-over tomato facility cultivation.

027 How to cultivate strong seedlings using traditional seedling raising methods?

Currently, seedling companies mostly use substrate-based seedling raising methods(using a mixture of peat, perlite, vermiculite, and various nutrient elements). The traditional nutrient bowl-based seedling raising method is gradually decreasing, but it is still beneficial for cultivating strong seedlings and can be cost-effective by using local materials.

(1) Seed treatment. Sun-dry the seeds for 1–2 days, then soak them. Soak the seeds in water for about 10 minutes, then soak them in warm water or a disinfectant solution for 6 hours at room temperature.

Disinfectant soaking. Soak the pre-soaked seeds in a 100-fold dilution of formalin solution for 10–15 minutes, or soak them in a 10% sodium phosphate solution for 20 minutes, or soak them in a 0.1% potassium permanganate solution for 10–15 minutes. Seeds treated with disinfectant should be rinsed with clean water and germinated in a culture box at 28–30 ℃. When the seeds "crack" and show white radicles, they are ready for sowing. Seeds treated with sodium phosphate or potassium permanganate can be sown directly after washing and drying.

(2) Nutrient soil preparation. Select fertile field soil and well-fermented composted manure, mix them in a 7: 3 ratio after sieving. For 1 m^3 of nutrient soil, add 2 kg of calcium superphosphate, 5–10 kg of wood ash, and 0. 2 kg of potassium dihydrogen phosphate. Mix in 400 g of carbendazim and 1 000 g of seedling mother liquor, and mix well. This mixture is then ready for use.

(3) Sowing. Sow approximately 3 g of seeds per 1 m^2 of seedbed. Water the seedbed thoroughly before sowing. Mix the pre-soaked seeds with a small amount of fine furnace slag and evenly spread them. Cover the seeds with a layer of soil with a thickness of 0. 8–1 cm, and cover with grass on top to maintain moisture. Arch 2/3 of the seeds with soil, and remove the cover promptly to prevent seedlings from elongating.

(4) Seedling stage management. After sowing, set up an arch shed on the seedbed, cover it with insect-proof nets and shade nets to provide shade and cooling, and prevent pest infestation. When the seedlings reach the 2-leaf-1-heart stage, transplant them into nutrient bowls with a size of 10 cm × 10 cm. Deeply plant seedlings that have elongated excessively and water them thoroughly during transplantation. Water every 2–3 days to keep the nutrient soil moist but not waterlogged. Due to the addition of a certain amount of seedling mother liquor in the nutrient soil, excessive watering can lead to aging of the seedlings. Spray insecticides and fungicides every 5–6 days to control pests and diseases, mainly targeting the South American leafminer, whitefly, and seedling diseases. Spray 0. 2% potassium dihydrogen phosphate on the leaves of seedlings 1–2 times during the seedling stage, which can help strengthen the seedlings and prevent diseases. When the seedlings have grown to 4–5 leaves, they are ready for transplanting.

028 What are the requirements for tomato planting during high-temperature periods?

(1) Pre-planting preparation. Retain the old plastic film used during spring production and do not replace it during the summer. Open the lower ventilation openings to increase ventilation, which can prevent rainwater and reduce indoor temperature to prevent the occurrence of certain diseases. Before planting, conduct a high-temperature closed greenhouse treatment to kill various pathogens.

(2) Land preparation and fertilization. Generally, apply 5 000 kg of well-rotted organic fertilizer, 40 kg of potassium sulfate, and 30 kg of ammonium phosphate per acre. After deep plowing, level the soil with a rake. After leveling, open planting furrows with a depth of 10 cm according to a certain row spacing.

(3) Planting. First, place the seedlings in the planting furrows, remove the nutrient bowls, and lightly bury the soil around the rootballs. Then, water the furrows thoroughly, and

2-3 days later, water the seedlings with diluted slow-release nutrient solution. After 1 day of water infiltration, form ridges with a height of about 10 cm. This operation aims to promote wound healing and prevent stem base rot in young seedlings injured during planting. Apply 1 000 g of seedling mother liquor per acre in the planting holes to effectively prevent excessive elongation of the plants after planting. The recommended plant spacing for high-raised cultivation is(40-45) cm ×(80-90) cm, with approximately 2 000 plants per acre.

029 How to enhance management and improve yield and quality after tomato planting?

(1) Timely plant adjustment. When the seedlings reach a height of 30 cm, it is necessary to suspend the vines. Install an 8# iron wire or thin steel wire above the plants in a north-south direction and tie nylon ropes. The lower end of the nylon ropes should be tied below the first cluster of fruits on the stem base. Adopt the method of multiple clusters and single stem pruning, and promptly remove side branches when they are 2-3 cm long to prevent nutrient consumption. Timely guide the vines along the ropes. Pruning should be done in the afternoon. Leave 6-8 clusters of fruits per plant, and remove the growing point 2 leaves before the last cluster of flowers to leave 3-5 fruits per cluster. The plant height can reach about 2 meters, hence the term "high-raised cultivation".

(2) Pay attention to flower and fruit protection. When 2/3 of the flowers in each cluster are open, spray a concentration of 25-50 mg/L of yield-increasing agent No. 2 (growth regulator) on the flower trusses. The spraying is usually done between 8-10 am, with an interval of 5-6 days. To prevent repeated spraying, a small amount of red dye can be added. After fruit set, thinning of flowers and fruits should be done in a timely manner. During summer, tomato plants grow vigorously, and axillary branches tend to grow before flowering clusters. They should be promptly removed to avoid nutrient competition with the fruits.

(3) Temperature management. After planting, increase the ventilation to lower the indoor temperature due to the high external temperature. In mid-September, replace the plastic film with a new one, and in mid-October, cover the greenhouse with straw mats(or quilts) at night to provide insulation and cold protection.

(4) Lighting. Tomatoes thrive in strong light conditions. Perform timely operations such as suspending vines, guiding vines, and removing lower leaves to facilitate ventilation, light penetration, and fruit enlargement.

(5) Strict management of fertilizer and water to regulate growth balance. Before fruit set, generally, no fertilizer is applied, and water is controlled appropriately to prevent excessive vegetative growth. Also, prevent rainwater leakage from the greenhouse and waterlogging to prevent disease occurrence. During this period, the main task is to balance nutrient and reproductive growth to prevent excessive vegetative growth and promote flowering and fruiting.

After fruit set, start fertilization and irrigation. If fertilization and irrigation are done too early, the plants will have vigorous vegetative growth, resulting in smaller and late-maturing fruits in the lower clusters. If fertilization and irrigation are done too late, the plants' nutrient growth will be suppressed, resulting in larger and early-maturing fruits in the lower clusters, but the total yield will decrease. Therefore, it is crucial to grasp the timing of fertilization and irrigation, adjust the balance between nutrient and reproductive growth, and increase tomato yield and efficiency. Apply compound fertilizer(15-15-15) at a rate of 15-20 kg per acre every 10-15 days during the fruiting period. Irrigate with fruit enlargement water every 6-7 days. Spray calcium boron solution on the leaves 2-3 times during the fruiting period to prevent fruit cracking, blossom-end rot, and hollow fruits.

030 What diseases and pests should be prevented in late-extended autumn tomatoes?

In the production of late-extended autumn tomatoes in sunlight greenhouses, the key diseases to prevent and control include late blight, viral diseases, gray mold, powdery mildew, root-knot nematodes, gray leaf spot, and stem base rot. The key pests to control include cotton bollworm, whitefly, and South American leafminer. Physiological disorders such as deformed fruits, fruit abortion, fruit cracking, hollow fruits, and blossom-end rot should also be addressed. It is important to implement a "prevention-first, comprehensive control" approach for tomato diseases and pests and adopt green and pollution-free techniques for tomato production. For specific prevention and control measures, please refer to the section on common diseases and pests and their green control methods in greenhouse tomatoes.

031 What is the yield and profitability of late-extended autumn tomatoes in sunlight greenhouses?

According to market conditions, in the early stage, multiple clusters of fruits are left to inhibit early ripening. After harvesting the first and second clusters, remove the lower leaves and try to increase the management temperature in the greenhouse to promote the ripening of upper fruits. Generally, the yield of tomatoes can reach over 10 000 kg per acre, and planting high-yielding and high-quality varieties can bring remarkable economic benefits.

4. Practical Techniques for Facility Production of Flavorful Tomatoes

032 What are flavorful tomatoes?

With the improvement of people's living standards, leisure agriculture and ecological picking gardens have emerged, but there are disparities in the arrangement of efficient crop varieties and management levels. After years of practice, Liaoning Vocational College has produced flavorful tomatoes by selecting suitable large-fruited tomato varieties and using means to "control" tomato growth. These tomatoes have excellent taste and flavor, surpassing seasonal fruits.

Flavorful tomatoes are different from the hard tomatoes with thick flesh, high disease resistance, long storage period, and poor taste that are currently available in the market. They are also different from cherry tomatoes. By using tomato varieties such as "True Beauty" and "Pink Tarou" distributed by Taiwan's Nongyou Seed Company, which have the characteristics of abundant juice, thick flesh, pink skin, and thin skin, and through careful control of various production processes such as water control, temperature control, and fertilizer application, the quality of tomato fruits can be improved to meet the taste standards comparable to fruits(Figure12). The production of flavorful tomatoes fills the gap in seasonal fruits in the winter and spring in northern regions, especially meeting the demand for fresh fruits and vegetables among individuals with high blood sugar. Developing flavorful tomatoes has high economic value and social benefits. The production techniques for flavorful tomatoes in facilities are suitable for leisure agriculture and ecological picking gardens.

Figure 12 Taste-type Tomato

033 What are the requirements for the production of flavorful tomatoes in facilities?

For the production of flavorful tomatoes, it is advisable to use well-constructed sunlight greenhouses, plastic greenhouses, and other facilities. Rain avoidance cultivation should be adopted to facilitate ventilation, cooling, and moisture drainage, in order to reduce the occurrence of diseases and pests. Select plots with sandy loam soil, high elevation, good drainage, adequate nutrients, and high organic matter content.

034 What are the main varieties suitable for the production of flavorful tomatoes?

The production mainly uses pink-fruited large tomato varieties such as "True Beauty" "Pink Tarou" and "Jingcai No. 6". These varieties have thin skin, sandy flesh, and rich flavor, making them suitable for fresh consumption and meeting the preferences of consumers in different regions, thus helping achieve the desired production goals.

True Beauty. Introduced from Japan, it is an indeterminate type with deep pink fruits. It has strong cold resistance and disease resistance, but it is not resistant to root-knot nematodes. This variety accounts for more than 50% of the tomato market in Tieling City, Liaoning Province throughout the year. It is locally known as "strawberry persimmon". However, its quality varies, with well-managed ones being sweet and delicious, while poorly managed ones have average taste(Figure 13).

Figure 13 True Beauty

Pink Tarou. Introduced from Japan, it is divided into two types: cold-tolerant and heat-tolerant. It has thin skin and a sweet taste, and it is popular among residents of Shenyang City(Figure14).

Figure 14　Pink Tarou

Jingcai No. 6. Imported by Beijing Modern Farmer Seed Technology Co. , Ltd. , it is a new variety of "strawberry persimmon" carefully bred from high-quality foreign resources. It is an indeterminate type, early-maturing, with a single fruit weighing about 130−200 g, round shape, obvious green shoulders on unripe fruits, and pink color when ripe. It has a delicate texture, sweet and flavorful, with a strong tomato taste and a sugar content of up to 7 degrees Brix. It is suitable for fresh consumption as a fruit(Figure15). This variety has vigorous growth, strong overall resistance, high resistance to Tomato Yellow Leaf Curl Virus, and strong resistance to root-knot nematodes and leaf mold.

Figure 15　Jingcai No. 6

035 When is the best time for the marketability and profitability of flavorful tomatoes?

The harvest period should be arranged from November to the following July to avoid the peak season for various fruits in summer and autumn, fill the gap in seasonal fruits, and increase their commercial value. According to statistics from 2008 to 2017, the retail price of "True Beauty" flavorful tomatoes produced in the training garden area of Liaoning Vocational College was around 10 yuan/kg from January to April, around 8 yuan/kg in May, and around 6 yuan/kg in June, July, November, and December. Additionally, during the hot and rainy season, tomato plants are prone to excessive vegetative growth and physiological disorders such as fruit cracking, making it difficult to control their quality.

Therefore, in Liaoning, for the spring crop in greenhouses, sowing is generally done in late November, transplanting in early February, and harvesting and marketing starting in early April. For the autumn crop in greenhouses, sowing is done in early June, transplanting in early August, and harvesting and marketing starting in early October.

036 How to cultivate strong tomato seedlings using substrates?

Seed disinfection treatment must be carried out before sowing. The commonly used method is warm water soaking, which involves continuously stirring and soaking the seeds in water at 55-60 ℃ for 15 minutes. You can choose seedling substrates distributed by Jinan Fengyuan Agricultural Technology Co., Ltd. This substrate is mainly composed of peat, perlite, and vermiculite, with organic matter, humic acid, and plant fibers accounting for more than 60%. It is rich in the necessary nutrients for seedlings and easy to use. Before filling the trays, the substrate should be moistened evenly with an appropriate amount of water. The standard is that when squeezed by hand, it forms a clump that does not drip water and disperses when released. After filling the trays, level and compact the substrate, and create small holes about 0.5-1 cm deep. Sow the disinfected seeds in the trays, then cover them with small bags of vermiculite, and finally spray water to moisten the substrate. Pay attention not to use excessive water during sowing and subsequent management processes to prevent nutrient loss. The suitable temperature during the germination period is 25-30 ℃. After 4-5 days of germination, remove the covering material promptly.

Since a prolonged seedling period is prone to tomato viral diseases, the seedling age should not be too long. In the hot season, it is about 30 days, and in the cold season, it is about 45-60 days. As the growth of flavorful tomatoes is controlled throughout the entire growth period, the plants are subjected to controlled dry conditions after transplanting. Additionally, viruses can be spread by pests such as thrips and whiteflies in the greenhouse, which increases the probability of tomato viral diseases. Seedlings can be transplanted when they have 5–7 true leaves.

037 How to apply organic fertilizer?

Organic fertilizer, such as chicken manure, pig manure, and cow manure, should be fully decomposed and matured. It is recommended to apply 5 000 kg per 667 m^2, using a combination of two or more types of organic fertilizer. For better results, the use of straw bio-reactors is recommended. This involves laying corn straw below the soil cultivation layer and applying decomposed microbial bacteria (with an effective live bacteria count of $\geq 2 \times 10^{10}$ cfu/g) on the straw. This allows the straw to decompose under aerobic conditions, generating heat, CO_2, and releasing readily available nutrients.

038 Why is high bed planting recommended?

Creating high beds with a height of around 20 cm and a bed width of 70–80 cm enhances indoor air circulation and facilitates water control in the field. Planting two rows on the bed and covering with plastic mulch, with an aisle width of around 80 cm, allows for easy picking by visitors and promotes ventilation and light transmission.

039 Why is trellis cultivation preferred?

Trellis cultivation with single stem pruning is recommended, with plant height exceeding 2 meters and 6–8 clusters of fruits per plant. It has been proven that the fruits closer to the upper part of the plant have better taste. For example, the flavor, appearance, and texture of fruits from the top three layers of "True Beauty" tomatoes are superior to those from the lower layers (Figure 16). It is advisable to plant around 2 000 plants per 667 m^2, avoiding excessive planting density that hinders ventilation and increases the risk of diseases and pests, which can affect tomato quality.

Figure 16 Elevated Cultivation

040 What is the ideal plant type for improving tomato quality?

The success of flavorful tomato production depends on whether the taste meets the expected flavor, while considering yield requirements and pollution-free production. The cultivation of an appropriate plant type will yield ideal results. For example, "True Beauty" tomatoes exhibit desirable characteristics when the plant type is ideal, resulting in fruits with green shoulders, distinct longitudinal radial stripes, a balanced sweet and sour taste, and a unique flavor. The ideal plant type should have a main stem that is not too thick, with a stem diameter of 1 cm at the base, 1.0–1.2 cm at a distance of 50 cm from the ground, and 0.7 cm at a distance of 100 cm from the ground. The stem should be symmetrical, with internode lengths of about 6 cm, weak growth but with continuous growth of the growing point. The leaves should be dark green, thick, and relatively small, giving the plant an elegant appearance. Achieving the ideal plant type requires comprehensive control of environmental conditions throughout the tomato growth stages.

041 How to manage the plants after transplanting?

Three days after transplanting, water the plants with an appropriate amount of water. After controlling the dryness for 1–2 days, seal the transplanting holes tightly to prevent stem base rot. Control the watering thereafter based on the growth of the tomato plants. Water lightly when the tomato leaves show moderate wilting at noon on sunny days. To achieve the ideal plant type, manage the plants under high-temperature and drought conditions, with daytime temperatures controlled at around 28–30 ℃ and nighttime temperatures around 10 ℃.

042 How to protect flowers and fruits during the flowering stage?

For flavorful tomatoes, there are two main methods to prevent flower and fruit drop: Spraying anti-drop hormone solution on the flowers. This involves spraying a 30–50 mg/L anti-drop hormone solution on the peduncle of the tomato flowers when they are in full bloom. The advantage is that the fruits enlarge quickly, but it may result in fewer seeds or even no seeds in the fruit cavity. Using bumblebees for pollination. This involves releasing bumblebees to pollinate the tomatoes during the flowering and fruit setting stage. The advantage is that it increases the number of seeds in the fruit cavity, resulting in juicy and flavorful tomatoes with a unique taste. It also saves labor costs(Figure 17). Bumblebees can be purchased from Koppert Biological Systems in the Netherlands for approximately 400 yuan per box. Within 60–70 days, they can complete the pollination task for 500 m^2 of greenhouse tomatoes. There is no need to feed the bumblebees, and the use of insecticides is prohibited when bumblebees are present in the greenhouse.

Figure 17 Bumblebee Setup

For tomatoes consumed as fruits, it is suitable to control the quality of a single fruit to be between 100–150 g, so the number of remaining fruits per bunch should be more, 4–6 are acceptable. Remove the deformed and excess fruits to improve the commercial rate of the fruits.

043 How to manage temperature during the fruiting stage?

A three-stage temperature management model is recommended. During sunny days, the daytime temperature should be controlled at around 25 ℃, the upper half of the night at around 14 ℃, and the lower half of the night at around 10 ℃. The temperature in the later part of the night should not be too high, as it can cause excessive vegetative growth, slow fruit enlargement, and delayed ripening.

044 What is the basis for controlling irrigation during the fruiting stage?

The overall guideline is to water the tomato plants lightly when the leaves show slight wilting at noon on sunny days, and they recover by evening. The plants should have weak growth but continuous growth of the growing point. Watering should be done accordingly based on these indicators (Figure 18). Additionally, water should be applied during each fertilization. Balanced watering after fruit setting is important to prevent fruit cracking. Adequate water control increases the sweetness of the tomatoes, while excessive water deficiency can result in thicker and harder skin, affecting the taste.

Figure 18 Plant Morphology

045 What is the basis for proper fertilization during the fruiting stage?

For fertilization, it is advisable to choose compound fertilizers with high potassium content, chitosan fertilizers, well-decomposed soybean cake, amino acid calcium, etc., which help improve tomato quality. Excessive use of nitrogen fertilizer should be avoided. Soluble fertilizers can be applied through irrigation, while cake fertilizers should be buried underground. Calcium fertilizer is usually applied as a foliar spray in combination with other foliar fertilizers.

After the first fruit cluster of the tomato plant sets fruit, 5 kg of American Potassium Treasure produced by Alpha Agrochemicals (Qingdao) Co., Ltd. can be applied through irrigation per 667 m^2. After the second fruit cluster sets fruit, 20 kg of seaweed chitosan produced by Weifang Nongbangfu Fertilizer Co., Ltd. can be applied through irrigation per 667 m^2. These fertilizers should be applied in rotation, with an interval of about 10 days between each application. A total of 5-6 fertilizations should be carried out during the entire growth period.

046 How to increase light management?

Tomatoes are light-loving crops. To ensure sufficient light duration and intensity in the greenhouse, the following measures should be taken: appropriate planting density, with no more than 2 000 plants per 667 m^2 in trellis cultivation; single stem pruning, with the removal of lower leaves; covering the greenhouse with long-life polyethylene film; uncovering the insulation cover early in the morning and covering it later in the evening during winter; and avoiding ventilation on hazy days.

047 Which diseases and pests should be focused on for prevention and control?

Main diseases to be prevented in greenhouse tomatoes include viral diseases, gray mold, late blight, leaf mold, powdery mildew, root rot, and root-knot nematodes. Common pests to be prevented include greenhouse whiteflies, thrips, cotton bollworms, and Tuta absoluta. Physiological disorders such as fruit blossom-end rot, deformed fruits, and fruit cracking should also be prevented. Following the principle of "prevention first and comprehensive control", strengthen field management, maintain ventilation and dryness, and keep the indoor environment clean to reduce the occurrence of diseases and pests. Monitor in real-time and apply timely treatment when diseases and pests occur. Accurately diagnose the occurring diseases and pests and choose effective, low-toxicity, and low-residue pesticides for targeted treatment to ensure that tomatoes meet the requirements of pollution-free agricultural products. Refer to the following section on "Green Control of Common Diseases and Pests in Greenhouse Tomatoes" for specific prevention and control measures.

048 How to harvest and grade the tomatoes?

In the Tieling area of Liaoning Province, greenhouse tomatoes are generally produced in two crops: early spring and autumn-winter. With proper management, the yield of flavorful tomatoes can reach 4 000–5 000 kg per 667 m^2 per crop. The fruits can be harvested when the surface turns slightly red, offering a sweet and sour taste. Fully ripe fruits can be stored for a few days to develop a unique flavor. After harvesting, the fruits should be graded and packed based on their size, shape, and the presence of green shoulders. This allows for the sale of high-quality tomatoes at premium prices.

5. High-Quality and Efficient Production Technology Model for Overwintering Tomatoes in the Northern Liaoning Region

049 What are the current production status and existing problems of tomato production in the northern Liaoning region?

In the northern Liaoning region, winter is cold, and the outdoor temperature can reach as low as −27 to −20 ℃ during cold waves, which typically last for 5−15 days. Cold waves occur 1 to 2 times per year, mostly from late December to early February, which affects the production schedule of greenhouse fruits and vegetables. As a result, tomato production in the local greenhouses is divided into two distinct seasons: spring and autumn. The spring season involves planting in January or February and harvesting in June, while the autumn season involves planting in June or July and harvesting in November.

One of the problems in local tomato production is the focus on high yield at the expense of quality. The autumn tomato yield in greenhouses can reach up to 10 000 kg per acre, and the spring tomato yield can reach up to 8 000 kg per acre. However, the products are concentrated during the same period, leading to unstable sales prices. For example, in October 2018, the purchase price reached as high as 5 RMB/kg, while in November 2016, it dropped to as low as 0. 6 RMB/kg. This often results in increased production without a corresponding increase in income.

050 What is the significance of the high-quality and efficient production practice of overwintering tomatoes conducted by Liaoning Vocational College?

Liaoning Vocational College has conducted years of practical experience in greenhouse tomato production, aiming to address the contradiction between tomato yield and quality. Instead of solely pursuing high-end and high-priced sales paths, they have developed flavorful tomatoes, which have achieved good economic and social benefits in the local market.

Building on this success, they conducted high-quality and efficient production practice of overwintering tomatoes in greenhouses from 2018 to 2019. The produced flavorful tomatoes (variety: "True Beauty") achieved a total yield of over 5 000 kg per acre, with distinct green shoulders and longitudinal radial stripes, a high sugar content of 8%, and a delightful sweet and sour taste that meets the standards of popular fresh fruits. The wholesale price in January to March reached as high as 16 RMB/kg. During the same period, cherry tomatoes (variety: "Fengzhu") achieved a total yield of over 4 000 kg per acre, with a sugar content of 10%, thin and crispy flesh, and a retail price of 30 RMB/kg. This technology fills the gap in overwintering tomato production in the northern Liaoning region, achieving high-quality and high-priced products. Due to the extended fruiting period, the yield has also increased significantly, resulting in notable economic benefits. This production model is suitable for leisure agriculture and ecological picking gardens.

051 Why is the use of heating equipment recommended?

The greenhouse in the practical training area of Liaoning Vocational College is an improved version of the Anshan III type, with a single-layer cotton cover that provides slightly inferior insulation compared to the traditional high-latitude reinforced insulation greenhouses in Tieling, Liaoning. During the autumn-winter production of flavorful tomatoes in the greenhouse, as the temperature drops in December, the lowest indoor temperature can reach 2–3 ℃, which is only sufficient to prevent freezing damage to the plants. Although the selling price of the tomatoes is high, they cannot flower and fruit normally under such conditions, making continuous production impossible. Therefore, to achieve overwintering tomato production, it is necessary to overcome these low-temperature conditions. In this regard, one automatic intelligent temperature-controlled air heater (with a power of 10 kW or 15 kW) is installed on each side of the greenhouse (length: 70 m). It has been proven that using this auxiliary heating method ensures the safety of overwintering tomato production in the greenhouse while being cost-effective. The advantages of using air heaters are: automatic temp-erature control; rapid heating, with warm air immediately felt after activation; energy-saving and environmentally friendly. The heaters are only used for nighttime auxiliary heating during the period when cold waves occur from late December to early February, maintaining a minimum indoor temperature of 10–12 ℃ (Figure 19).

Figure 19 Air Heater

052 Which excellent varieties are recommended?

It is recommended to choose excellent tomato varieties that are tolerant to low temperatures, have strong disease resistance, vigorous growth, and excellent appearance and flavor. Examples include "True Beauty" and "Fengzhu"(409).

"True Beauty" is a large-fruited tomato variety introduced from Japan. It belongs to the indeterminate growth type and has deep pink fruits. When managed to control excessive vegetative growth, it exhibits distinct radial stripes centered around the fruit pedicle and green shoulders. The individual fruit weight ranges from 50 to 150 g, with a balanced sweet and sour taste. It has strong cold tolerance and disease resistance, being relatively resistant to tomato mosaic virus but not resistant to root-knot nematodes. The plant has an elegant and compact growth habit, with dark green leaves, making it suitable for high-density planting.

"Fengzhu"(409) is an indeterminate growth type cherry tomato developed by Taiwan Nong You Seedling Company. It has elongated fruits with 2-4 depressions below the fruit shoulders. The mature fruits are red, with an average weight of about 16 g. The flesh is fine and the flavor is sweet, with a sugar content of up to 9.6%. The plant has strong growth potential, cold tolerance, and disease resistance, being relatively resistant to tomato mosaic virus, root-knot nematodes, and wilt disease.

053 How to plan the production schedule?

The production schedule should be primarily based on market demand. Seedlings should be raised in early August, transplanted in early September, and harvested from mid-December to mid-June of the following year.

054 What is the role of key cultivation techniques in producing high-quality products?

The current practice of "controlling excessive vegetative growth" has been widely recognized for improving the quality of tomatoes. In May 2008, during the production of flavorful tomatoes in greenhouses at Liaoning Vocational College, accidental tobacco damage occurred during pest control, resulting in the withering of all mature leaves on the tomato plants. However, the tomatoes harvested at that time were unexpectedly delicious, with tender and sweet skin, indicating high quality and market value. Additionally, plants slightly affected by root-knot nematodes in the greenhouse displayed weaker growth, but the fruits they produced were particularly sweet and had good firmness.

The selection of excellent varieties is one of the key aspects in the production of flavorful tomatoes. However, the role of key cultivation techniques in producing high-quality products should be considered as a collective and comprehensive effort, and the impact of individual measures should not be exaggerated. In simple terms, "a good chef can make a good dish even with ordinary ingredients". While having good varieties is important, factors such as cutting techniques, temperature control, secret recipes, types and order of seasonings, and leaf quality can all affect the taste of the final product. Similarly, the taste of tomatoes grown by different individuals can vary, and it is difficult to replicate the same taste consistently. Overall, factors such as variety, fertilizer, water, temperature, light, leaf quality, and pest and disease management can all influence the quality of tomatoes.

Liaoning Vocational College has conducted experiments on pruning and defoliation of tomato plants during the flowering stage but did not achieve the desired effect of "controlling excessive vegetative growth and improving quality". Furthermore, when paper bags were used to cover the fruits during the fruit setting stage, the fruits had a glossy appearance with thin skin and did not crack. However, when plastic bags were used, the fruits had a rough appearance and a sour taste, indirectly verifying that reducing air humidity can improve tomato quality. In the overwintering tomato production in the northern Liaoning region, where there is a large diurnal temperature difference and abundant light, controlling soil and air humidity, applying high-quality organic fertilizers, balanced fertilization, and leaf care are particularly important.

055 What measures should be taken to rejuvenate the plants after overwintering?

(1) Dropping the vines. When the growth point of the tomato plant is close to the iron wire, drop the vines in a timely manner. Open the active buckle that secures the lower end of the plant, lift the plant with the left hand, gradually untangle the nylon rope wrapped around the main stem with the right hand, and adjust the plastic clip accordingly based on the length of the drooping plant. Secure the plant at an appropriate position on the nylon rope, ensuring that it is inclined uniformly towards the south (or north) to reduce plant height. The upper part of the plant should still be wrapped around the nylon rope to guide upward growth. After the fruits gradually turn red, remove the aging leaves at the lower end to improve ventilation and light penetration. The length of the plant can reach over 3 meters, and each plant can produce more than 110 inflorescences (Figure 20).

(2) Strengthening fertilization and water management. After dropping the vines to reduce plant height, increase the supply of fertilizers and water. When the plant resumes vigorous growth, the flowers at the upper part will bloom brightly, and the fruits will have a shiny appearance, resulting in increased tomato yield and improved quality.

Figure 20 Dropping the Vines

056 How to focus on the prevention of diseases and pests throughout the production process?

Overwintering tomatoes have a long growth period, and different periods require different focuses on disease and pest prevention. After transplantation, the key is to prevent stem base rot, viral diseases, thrips, and cotton bollworms. In September and October, the focus is on preventing late blight, powdery mildew, and cotton bollworms. In February and March of the following year, the focus is on preventing tomato leaf mold. Throughout the production process, it is important to prevent gray mold, whiteflies, and Tuta absoluta. Following the principle of "prevention first, comprehensive control", it is essential to strengthen field management, monitor in real-time, and implement measures for "prevention before occurrence, early treatment for existing diseases, and prevention of recurrence after recovery". In case of early detection of diseases and pests, physical methods such as manual removal of infected plants and pests should be employed, followed by the use of biological agents and low-toxicity chemical pesticides for protection to ensure product quality and safety.

6. Sunlight Greenhouse Tomato Cultivation Technology Model with Coconut Coir Substrate Soilless Cultivation

057 What is the current development status of sunlight greenhouse tomato cultivation with coconut coir substrate soilless cultivation in Liaoning region?

With the development of protected vegetable production and the extension of crop planting years, the accumulation of salts and pathogens in the soil has become a key factor restricting vegetable development. Soil secondary salinization, nutrient imbalance, and deterioration of soil microecological environment have become common issues in greenhouse tomato production. The use of coconut coir as a substrate has successfully addressed these problems. Soilless cultivation of tomatoes with coconut coir as the substrate has already started experimental promotion in the Liaoning region, representing a new vegetable cultivation model (Figure 21).

Figure 21 Soilless Cultivation with Coconut Coir Substrate

058 Which varieties are suitable for sunlight greenhouse tomato cultivation with coconut coir substrate soilless cultivation?

For sunlight greenhouse tomato cultivation with coconut coir substrate soilless cultivation, it is advisable to choose varieties with indeterminate growth, strong stress resistance, and excellent quality, such as Jinmijia, Jinzhu, Jinju, and Zhenyoumei.

059 How to prepare planting furrows for sunlight greenhouse tomato cultivation with coconut coir substrate soilless cultivation?

(1) Soil Preparation and Moisture Adjustment. Level the ground with a slope of 0.2% from south to north. If the soil is too dry, water should be applied until it penetrates below 20 cm. If the soil is too wet, it should be dried to a relative humidity of 60%-70%.

(2) Digging Planting Furrows. Start by digging the first planting furrow 1.5 meters wide behind the east wall of the greenhouse. The furrows should be oriented in the north-south direction, with a distance of 10 cm from the northern end to the rear wall passage. The furrows should be 18 cm wide, 15 cm deep, and spaced 150 cm apart, ensuring that the length of each furrow is consistent. Leave a 40 cm distance at the southern end of the greenhouse for drainage (Figure 22).

Figure 22 Planting Furrows

(3) Digging Drainage Channel. Dig a drainage channel 10 cm away from the southern

end of the greenhouse, with a width of 20 cm and an east-west slope of 0.2%. The drainage channel can be made in an inverted trapezoidal shape to prevent collapse.

(4) Laying Isolation Film. Lay the isolation film with the black side facing up and the white side facing down. Start by laying the film for the drainage channel, ensuring that the isolation film is tightly attached to the ground. Use "U" shaped clips to fix it along the southern edge, and cut it with scissors at the intersection with the planting furrows, allowing the drainage from each furrow to flow into the drainage channel. The isolation film for the planting furrows should be laid evenly, avoiding unevenness on either side. The isolation film inside the furrows should be 60 cm longer than the furrows, and both the furrows and the drainage channel should be completely covered. The isolation film should be tightly attached to the ground without wrinkles.

(5) Installing Diversion Plates and Mesh. Calculate the number of diversion plates needed based on the length of the furrows. The diversion plates should be connected with interlocking buckles to increase their compressive strength. The method of laying the mesh is the same as that of the isolation film, ensuring evenness on both sides.

(6) Installing Drip Irrigation System. Calculate the quantity of main pipes, elbows, tees, plugs, and drip tapes required. Select the number of main pipes based on the length of the greenhouse. Straighten the pipes, with the blue line facing up, and fix them with "U" shaped clips. During installation, the drip holes should face up, and the drip tapes should be inserted into the connectors. Finally, seal the ends of the drip tapes by cutting 1.0–1.5 cm of the tubing and folding it over or folding it into three layers to prevent water leakage. The spacing between the drip holes on the drip tapes should be ≤20 cm. The drip tapes should be 30 cm longer than the coconut coir troughs. It is best to install the main pipes of the drip irrigation system on the south side of the greenhouse to avoid installation during excessively high or low temperatures.

(7) Preparation of Coconut Coir Substrate. To facilitate transportation, coconut coir is compressed into blocks (30 cm × 30 cm × 20 cm) with a weight of 5 kg. Before use, the coconut coir blocks should be soaked in water to restore their original loose state. The blocks can be soaked in a spraying pool. If there is no spraying pool, a pit can be dug outside the greenhouse, lined with plastic film to prevent water leakage, and the coconut coir blocks can be evenly placed in the pit, stacked no more than two layers, and soaked with water to allow them to fully expand. There are two types of coconut coir, coarse and fine, which can be soaked together. After soaking, the coarse and fine coconut coir should be mixed evenly, spread loosely, and avoid compression. Filling the planting troughs with coconut coir should not be done too early, preferably 5–6 days before transplanting. Flush the coconut coir with a buffer solution (calcium nitrate and magnesium sulfate in a mass ratio of 2: 1, EC value 2.0 ms/cm) using the drip irrigation system. After soaking for 24 hours, rinse with clean water. Use intermittent flushing, flushing twice per hour for 10 minutes each time. Planting can

proceed when the difference between the inflow EC value and the outflow EC value is less than 0.05 ms/cm. Finally, use a dedicated A+B solution fertilizer for maintenance (EC value 1.5 ms/cm).

060 How to disinfect and transplant for sunlight greenhouse tomato cultivation with coconut coir substrate soilless cultivation?

(1) Pre-planting Treatment. Before transplanting, disinfect the greenhouse with 10% isoprothiolane and 4% carbendazim smoke agents. Disinfect the pipes and drip tapes with a 1%-3% nitric acid solution. The substrate in the tomato seedling trays should be relatively dry to promote root growth.

(2) Preparing for Transplanting. When the seedlings reach the stage of 4 leaves and 1 growing point, select uniformly strong seedlings for afternoon transplanting. Soak the roots in a 30% Mancozeb and Metalaxyl suspension concentrate at a dilution ratio of 1:1 200. After soaking, let the roots dry for about 10 minutes before transplanting. For the first 7 days after transplanting, inspect the seedlings 2-3 times a day and promptly remove any diseased or weak seedlings.

061 How to manage sunlight greenhouse tomato cultivation with coconut coir substrate soilless cultivation after planting?

(1) Early-stage management. Provide shading for the first 3 days after planting, gradually increasing the light intensity as the seedlings establish. During the seedling establishment period, maintain a daytime temperature of 30-32 ℃ and a nighttime temperature of 18 ℃. After the seedlings establish, maintain a daytime temperature of 25-28 ℃ and a nighttime temperature of 13-18 ℃.

(2) Mid-term management. Pollination by bumblebees. Release bumblebees for tomato pollination during the flowering and fruit setting stage. Bumblebees are efficient pollinators and can improve fruit quality. Bumblebees can be purchased from companies like Koppert Biological Systems in the Netherlands. They can complete the pollination task for approximately 500 m^2 of greenhouse tomatoes within 60-70 days. When bumblebees are present, the use of insecticides should be avoided.

Wrap and Prune Branches. Single-stem pruning is implemented when the plant height reaches 30-35 cm. At this point, a specially made plastic clip is used to simultaneously fix the plant and the hanging rope to start hanging the vine. As the vine extends, it is wrapped in an "S" shape. The first two nodes can be wrapped 3-4 times to secure the plant, and then every three nodes are wrapped 2–3 times. When the lateral branches grow to 6–8 cm, they should be removed in time. If the lateral branches are too short, it will be laborious to remove them, and it is easy to damage the growth point. If they are too long, they consume nutrients, and removing them can easily cause large wounds, which are conducive to bacterial infection.

Remove Old and Diseased Leaves. For large-fruited tomatoes, the removal of the lower old leaves begins after the first fruit cluster sets fruit. In winter, this is done once every 15 days on average, and in other seasons, once every 7–10 days, removing 2–3 leaves each time. After the first fruit cluster changes color, all the old leaves below it are removed. In spring, with the rise of external temperature, the plant tends to grow reproductively, and the fruit changes color quickly. At this stage, at least 18 functional leaves should be maintained.

Harvesting. Start harvesting from the first fruit cluster when the sixth or seventh cluster starts flowering. Harvest on sunny mornings, 2–3 times a week for large-fruited tomatoes and daily for small-fruited tomatoes, depending on ripeness and external conditions.

Lowering the Vines. Due to the limited height of the sunlight greenhouse, if long-season cultivation of tomato plants is carried out, the vines need to be continuously inclined and lowered. The first lowering of the vine begins after the first fruit cluster of the tomato is harvested, each time lowering it by 0.5 m is sufficient, too much lowering can affect crop growth. Subsequent lowering begins when the growth point of the plant is at the same height as the hook. However, in the late growth stage of the plant and during the period of high temperature and strong light outside, the distance between the hook and the growth point of the plant can reach 0.8–1.0 m.

Thinning fruit. For large-fruited tomatoes, leave 3–5 fruits per cluster. For medium and small-fruited tomatoes, the number of fruits left per cluster varies from 10 to 20.

(3) Late-stage management. Adjust the fertilization amount to 2/3 of the normal amount and gradually reduce it to 1/2. Stop fertilization 7 days before complete harvest. Minimize or avoid the use of pesticides. After complete harvest and before cleaning the vines, spray a 70% mancozeb and zinc dustable powder at a dilution ratio of 600 times for disease control.

062 How to disinfect the greenhouse during the idle period of sunlight greenhouse tomato cultivation with coconut coir substrate soilless cultivation?

Apply a 0.1% potassium permanganate solution and a sodium hypochlorite solution to the entire greenhouse, with an interval of 3–5 days between the two disinfections. Perform the disinfection after 5 PM, close the greenhouse, and ventilate the next day after 10 AM. Soak the inner wall of the drip irrigation pipes with a 65%–68% concentrated nitric acid solution for disinfection.

063 How to supply nutrient solution for sunlight greenhouse tomato cultivation with coconut coir substrate soilless cultivation?

(1) Irrigation volume management. The irrigation volume per plant depends on the crop size and growth stage. Initially, irrigate 40-50 mL per plant, gradually increasing the volume as the plants grow. When the first fruit cluster ripens, the irrigation volume should reach 100-120 mL per plant. In summer, the drainage should account for 25%-30% of the irrigation volume, while in winter, it should be 20%-25%.

(2) Irrigation timing management. Use the Yamazaki tomato nutrient solution. Generally, irrigate more in the morning, less at noon, moderately in the afternoon, and replenish water in the evening. Start the first irrigation 1-2 hours after sunrise, and the first signs of drainage should appear within 2-3 hours after irrigation. In summer, irrigate every 25 minutes for 3 minutes each time, while in winter, irrigate every 35 minutes for 2 minutes each time. The last irrigation should be done 1-3 hours before sunset. Adjust the timing of the final irrigation based on the moisture content of the substrate, aiming for a moisture decrease of 2%-4%.

(3) EC and pH settings. Adjust the nutrient solution's EC and pH based on the light intensity, temperature, and growth stage of the tomatoes. The initial EC should be 1.6-1.8 ms/cm and gradually increase by 0.2 approximately every 14 days until it stabilizes at 2.6-2.8 ms/cm. The EC can range from 1.6 to 3.0 ms/cm. Adjust the pH of the nutrient solution to be between 5.2 and 6.5 based on the pH of the drainage solution from the coconut coir.

064 How to disinfect the reused coconut coir substrate?

Remove the plant roots from the coconut coir. Use an 800-fold dilution of 32.5% benomyl suspension concentrate to irrigate the coconut coir. After 24 hours, rinse it with clean water 10 times. Add a mixture of 10% coarse and fine new coconut coir. This disinfection method has good efficacy, and the vegetable yield remains unaffected after three years of continuous use. After three years, replace the coconut coir and start with a new cultivation cycle.

7. Diagnosis and Control of Physiological Disorders in Tomatoes

065 Why do tomatoes experience flower and fruit drop?

(1) Reasons for flower and fruit drop. It is mainly caused by physiological disorders resulting from unfavorable environmental conditions. The suitable temperature for tomato flowering is around 20–30 ℃ during the day and 14–16 ℃ at night. Prolonged periods of temperatures above 35 ℃ or below 15 ℃ can affect pollination and fertilization, leading to flower drop. Additionally, when the relative humidity in the greenhouse exceeds 85%, pollen dispersal can be hindered, resulting in flower drop. Improper watering, cloudy and rainy weather, insufficient light, and poor plant nutrition can also cause flower drop. In autumn and winter, when the temperature is low during the fruiting period of greenhouse tomatoes, and the night temperature reaches around 5 ℃, the tomato fruits can easily fall off at the abscission layer position of the fruit stalk(Figure 23).

Figure 23 Tomato Flower Drop

(2) Measures to preserve flowers and fruits. Strengthen field management. Ensure that environmental conditions such as temperature, humidity, and light in the greenhouse are suitable for tomato growth and development. Apply phosphorus and potassium fertilizers and make timely adjustments to balance plant nutrition and reproductive growth.

Application of growth regulators. Spray a 30–50 mg/L concentration of growth regulators on the flower peduncle when tomato flowers are blooming. This can promote fruit enlargement, but it may result in fewer seeds or even seedlessness.

Pollination by bumblebees. Release bumblebees during the flowering and fruit-setting period to facilitate pollination in tomatoes. This method promotes the development of fruits with more seeds, abundant juice, and unique flavors. Bumblebees are adaptable and suitable for temperatures between 8–35 ℃.

066 Why do tomatoes experience fruit cracking, and what are the symptoms?

(1) Symptoms of tomato fruit cracking. Cracking of the fruit refers to the formation of striped cracks on the surface of the fruit. Depending on the location and cause of the cracking, it can be further divided into radial cracking, concentric cracking, and striped cracking.

Radial cracking. Cracks radiate from the pedicel towards the shoulder of the fruit, generally starting with minor cracks during the green mature stage of the fruit. As the fruit changes color, the cracks noticeably deepen and widen, sometimes resulting in severe deep cracks(Figure 24).

Figure 24 Radial Cracking

Concentric cracking. Cracks form in a concentric circle around the pedicel on the nearby fruit surface. In severe cases, the cracks form a ring. This usually occurs when the fruit is about to ripen.

Striped cracking. Vertical, horizontal, or irregular cracks occur at the bottom, top, and sides of the fruit(Figure 25).

Figure 25 Striped Cracking

(2) Reasons for fruit cracking.

Variety. Generally, large-fruited tomatoes with thin skin and a round shape are more prone to cracking.

Improper fertilization. During the fruit growth stage, tomatoes require a large amount of potassium fertilizer, while the demand for nitrogen decreases. Excessive nitrogen application without sufficient potassium can lead to potassium deficiency in the fruits and affect the absorption of calcium and boron, which are also factors contributing to fruit cracking.

Drastic changes in soil moisture. In the early stages of tomato development, water control measures are taken to prevent excessive vegetative growth. However, during the fruiting stage, excessive irrigation and fertilization are often employed to promote fruit development. Under such conditions with abundant nutrients and water, the fruit pulp grows rapidly while the skin's growth rate is relatively slow, leading to fruit cracking.

(3) Prevention measures for fruit cracking. Select varieties with strong crack resistance. Choose medium-sized or small-sized tomatoes with a high-staking growth habit and thin cork layer for cultivation.

Enhance water and fertilizer management. Deep plow the soil, apply organic fertilizers to promote healthy root growth, and buffer drastic changes in soil moisture. Water the plants appropriately, avoiding alternating periods of drought and excessive watering, especially after a prolonged dry period. In the case of rain-fed cultivation, water the plants regularly to prevent rapid fluctuations in soil moisture after sudden rainfall, and ensure proper drainage after rain.

Foliar application of borax. Spray a 0.5% borax solution on the plants during the flowering and fruiting period to supplement boron and prevent fruit cracking. Apply every 7-10 days for 2-3 times.

Timely harvest. During seasons with high temperatures and strong sunlight, tomatoes are more prone to fruit cracking as they approach full maturity. Harvesting slightly earlier can help reduce fruit cracking losses.

067 What are the measures to prevent blossom end rot in tomatoes?

(1) Symptoms. Blossom end rot(BER) is a common physiological disorder in tomatoes. The affected area appears as water-soaked lesions on the blossom end or near the calyx, hence the name "blossom end rot." Initially, the lesions are dark green and usually have a diameter of 1-2 cm, but they can extend to more than half of the fruit. The affected area quickly turns dark brown or black. After fruit infection, healthy portions of the fruit ripen prematurely. The lesion surface becomes leathery, and under humid conditions, pink or black mold-like growth, caused by saprophytic fungi, may appear on the lesions(Figure 26).

Figure 26 Belly Rot Disease

(2) Causes. It is widely believed that BER is caused by calcium deficiency, where the calcium content in affected fruits is significantly lower than in healthy fruits. Insufficient soil calcium content, as well as inadequate calcium uptake due to high temperatures and drought, can lead to the occurrence of blossom end rot. However, research by Japanese scientists (Nunami et al., 1995) has shown that there is no significant difference in calcium content between different tomato varieties in fruits, roots, and leaves, yet the occurrence of blossom end rot varies significantly. This suggests that blossom end rot is not a direct result of calcium deficiency but rather a metabolic disorder controlled by genetic factors.

(3) Prevention measures. Select superior varieties. Choose varieties with thick fruit flesh, strong root absorption capacity, and vigorous growth.

Enhance water and fertilizer management: Increase the application of calcium fertilizers and maintain balanced watering during the fruiting stage. Prevent excessive root burning due to high temperatures and excessive fertilizer application, which can lead to insufficient calcium absorption and the occurrence of blossom end rot.

Calcium foliar application. Spray a 0.1% calcium humate solution on the plants 3-4 times during the fruiting stage to supplement calcium through foliar feeding and prevent blossom end rot.

068 What causes various deformities in tomatoes?

(1) Symptoms. Deformed tomatoes can exhibit fasciation, peach-shaped fruits, eccentric fruits, multi-carpellary fruits, etc. Some fruits may have uneven surfaces or develop wart-like protrusions (Figure 27).

Figure 27 Deformed Fruit

(2) Causes. During the period of tomato seedling bud differentiation and development, prolonged exposure to low temperatures can reduce nutrient consumption for respiration. With sufficient water and light, excessive nutrient accumulation occurs at the growing point, resulting in overactive cell division in the developing flower buds. This leads to an abnormal number of carpels, uneven development of each carpel after flowering, and the formation of multi-carpellary deformed fruits. Improper use of plant growth regulators during flower treatment can also disrupt normal flower and fruit development.

(3) Prevention measures. Select superior varieties. Choose tomato varieties with strong stress resistance, tolerance to low temperatures and weak light, and thick skin.

Cultivate robust seedlings. Maintain suitable temperatures of 20–25 ℃ during the day and 12–16 ℃ at night for seedling growth. Avoid prolonged exposure to temperatures below 10 ℃.

Implement proper flower and fruit management. Avoid using excessive concentrations of growth regulators or repetitive treatments. Spray a suitable concentration (30–50 mg/L) of growth regulators on the flower peduncle during flower opening, ensuring treatment during a period of appropriate temperature and weak light. It is recommended to release bumblebees during the flowering and fruit-setting period, as they are well-adapted and suitable for temperatures between 8–35 ℃.

Thinning flowers and fruits. Conduct timely flower and fruit thinning during production management. Deformed flowers often develop into deformed fruits, particularly the first small flower on each fruit cluster. Careful observation is necessary during the thinning process.

069 Why do tomato plants often produce many parthenocarpic fruits?

(1) Symptoms. Parthenocarpic fruits, also known as "blind fruits" or "bean fruits", are small and hard fruits that fail to develop into normal-sized marketable fruits. They are considered ineffective fruits, leading to reduced tomato yield (Figure 28).

(2) Causes. Poor pollination and fertilization. Insufficient development of floral organs due to high night temperatures and weak light during the seedling stage can result in poor pollination and fertilization, leading to the formation of parthenocarpic fruits. In winter and spring, when tomatoes are exposed to prolonged ultra-low temperatures (night temperatures ranging from 1–6 ℃) and subjected to flower and fruit protection treatments, although fruit setting can occur, the fruit enlargement process is extremely slow, resulting in rigid fruit flesh, fruit arrest, and deformities.

Improper hormone treatment. Treating individual flower buds on tomato inflorescences before they fully open or using low concentrations of hormones can force fruit setting. However, slow fruit enlargement combined with nutrient competition between fruits can

Figure 28 Mummified Fruit

prevent the formation of normal marketable fruits.

Excessive vegetative growth. Excessive nitrogen application, excessive fertilizer usage, and high temperatures can lead to vigorous plant growth. This can result in nutrient competition between the leaves and stems, leading to parthenocarpic fruit formation.

(3) Prevention measures. Cultivate robust seedlings, employ proper flower and fruit management techniques, enhance nutrient and temperature management, increase organic fertilizer application, and avoid excessive nitrogen application. In cases where greenhouse tomato plants exhibit excessive vegetative growth, control measures should be implemented, such as refraining from watering and fertilizing during the first inflorescence flowering and fruit-setting period to promote fruit enlargement.

Chapter 4　Q&A on Practical Techniques for Sweet Melon Production

1. Fundamentals of Protected Cultivation Techniques for Sweet Melon Production

001 What is the relationship between the morphological characteristics of sweet melon and cultivation?

(1) Roots. Sweet melon has a well-developed root system, with the main roots concentrated within the 15-25 cm cultivation layer. The main root can penetrate the soil to a depth of over 1 meter and spread horizontally up to 2-3 meters. The extensive root system allows sweet melon to have strong drought resistance and adaptability to poor soil conditions. Thick-skinned sweet melon has a stronger and deeper root system compared to thin-skinned varieties, resulting in better drought and poor soil tolerance. Thin-skinned sweet melon has better tolerance to low temperatures and wet conditions. The root system requires loose and well-aerated soil, so it is important to apply organic fertilizer. The root system of sweet melon is prone to corking and has limited regenerative capacity, making it difficult to recover after damage. Therefore, it is recommended to use plastic bowls or plug trays for root protection during cultivation.

(2) Stems. During the seedling stage, the stems are short and upright. As the plant grows, the stems become trailing, hollow, with stripes or angles, and covered with prickly hairs. The stem diameter ranges from 0.4 to 1.4 cm, with internode lengths of 5-13 cm. The stems have strong branching ability, and lateral branches can emerge from leaf axils along the main stem. As long as conditions are suitable, the lateral branches can continue to grow indefinitely. In natural growth conditions, the main stem grows weakly and does not exceed 1 meter in length, while the lateral branches grow vigorously and often exceed the main stem. Therefore, timely and proper pruning is necessary. The first lateral branch that emerges from the main stem tends to have weaker growth, so it is generally removed during pruning.

(3) Leaves. Sweet melon has single, alternate leaves that can be blunt pentagonal, heart-shaped, or nearly circular in shape. The leaf petioles and main veins are covered with short stiff hairs, and both the upper and lower surfaces of the leaves are covered with pubescence. The leaf margins can be serrated, wavy, or entire. Leaf axils bear leaf buds, flower buds, and tendrils. The leaf length and width are generally 15-20 cm, with short petioles. The leaves are deep green and not fully expanded.

(4) Flowers. The flowers are borne in leaf axils, and sweet melon plants have separate

male and female flowers on the same plant. Female flowers are solitary, while male flowers are clustered in groups of 3-5. Both male and female flowers have nectar glands and are insect-pollinated, capable of self-pollination and cross-pollination. In protected cultivation, where insects are less abundant, bee pollination or manual pollination using a pollination aid is recommended. Female flowers appear later on the main stem, while they appear earlier on lateral branches and secondary branches. Usually, female flowers appear at the first or second node of lateral branches. Therefore, lateral branches and secondary branches are the main fruit-bearing parts.

During the initial stage of flower bud differentiation, the buds are bisexual and later develop into male or female flowers. Female flowers have both stamens and pistils. Flower bud differentiation occurs early, with the main stem leaf buds differentiating at the 17th node and the flower buds differentiating at the 13th node. Lateral branches start differentiating flower buds from the 1st to 11th nodes, and flower buds on lateral branches from the 1st to 9th nodes have already differentiated. Flower bud differentiation is influenced by temperature and light conditions. Low temperatures and short daylight hours promote the formation of female flowers.

(5) Fruits. The fruit of sweet melon is a pepo, developed from the ovary and receptacle, with 3-4 locules. The young fruit is typically round or oval-shaped and green due to the presence of chlorophyll in the fruit peel. As the fruit matures, chlorophyll gradually breaks down and disappears, and other pigments give rise to various colors such as yellow, white, striped, or green peel. The shape of mature fruits varies depending on the variety, including round, oblate, elongated, oval, elongated-ovate and spindle-shaped. Fruit surface characteristics include smooth skin and netting. The soluble solids content of the fruit pulp varies depending on the cultivar, with thick-skinned sweet melons generally ranging from 12% to 16% and sometimes exceeding 20%, while thin-skinned sweet melons typically have 8% to 12%. Mature fruits are aromatic. Fruit peel color and aroma can serve as indicators of ripeness.

(6) Seeds. Sweet melon seeds are generally milky white or yellow in color, and they can be lanceolate, oval, or sesame-shaped. The thousand-seed weight is 8-12g for thin-skinned sweet melons and 25-80 g for thick-skinned sweet melons. The seed viability is 5-6 years, and stored seeds should be kept under dry and cool conditions.

002 What are the temperature requirements for sweet melon?

Sweet melon is a warm-loving and heat-tolerant crop, extremely sensitive to cold and susceptible to frost damage. The optimal temperature range for its growth is a daytime temperature of 26-32 ℃ and a nighttime temperature of 15-20 ℃. It is sensitive to low temperatures, and when the temperature drops below 18 ℃ during the day and 13 ℃ at night, the plant's growth becomes sluggish. The minimum temperature for its growth is 15 ℃, and below 10 ℃, the growth stops completely and becomes abnormal. On the other hand, sweet melon has a strong adaptability to high temperatures and can still grow and produce fruits within the range of 30–35 ℃.

Different growth stages of sweet melon have varying temperature requirements. The optimal temperature for seed germination is 28–32 ℃. When the temperature is below 25 ℃, the germination time is prolonged and uneven, and lower temperatures result in longer emergence time and potential issues such as damping-off and seedling death. Seeds do not germinate below 15 ℃, so it is essential to ensure a soil temperature above 15 ℃ for early spring seedbeds. The temperature during the seedling stage directly affects the flowering position and fruit setting. Lower temperatures, especially lower nighttime temperatures, promote the formation of female flowers, increasing their quantity and lowering the flowering position. Therefore, it is important to avoid excessively high nighttime temperatures during the seedling stage, as temperatures above 25 ℃ can delay the opening of female flowers and raise the flowering position. The optimal temperature for flowering is 25 ℃, with a nighttime temperature not lower than 15 ℃, as temperatures below 15 ℃ can affect pollination. During fruit development, a daytime temperature of 28–32 ℃ and a nighttime temperature of 15–18 ℃, with a diurnal temperature difference of at least 10 ℃, are beneficial for fruit development and sugar accumulation. High daytime temperatures promote photosynthesis and nutrient production, while lower nighttime temperatures reduce respiration and facilitate the transport of assimilates to storage organs, promoting nutrient accumulation.

Different organs of sweet melon have different temperature requirements. The suitable temperature range for stem and leaf growth is 22–32 ℃, with the optimal daytime temperature of 25–30 ℃ and nighttime temperature of 16–18 ℃. When the temperature drops to 13 ℃, growth becomes stagnant, and below 10 ℃, growth ceases completely. Cold injury occurs at 7 ℃, resulting in loss of leaf greenness. Temperatures above 40 ℃ also cause growth stagnation. The minimum temperature for root elongation is 8 ℃, with the optimal temperature at 34 ℃ and a maximum of 40 ℃. To ensure normal root growth, the soil temperature during the first half of the night should be above 25 ℃ and above 20 ℃ during the second half of the night. Root hair development is affected when the soil temperature drops below 14 ℃, and if the temperature is too low and the soil moisture is excessive, it can lead to damping-off and seedling death.

The accumulated temperature requirements for the entire growth period of sweet melon are 1 800–2 000 ℃ for early-maturing varieties, 2 200–2 500 ℃ for mid-maturing varieties, and over 2 500 ℃ for late-maturing varieties.

003 What are the light requirements for sweet melon?

Sweet melon is a light-demanding crop and requires sufficient light during its growth period. Poor light conditions result in poor growth and development. Normal growth and development require at least 10 - 12 hours of sunlight per day, and the crop has a high requirement for light intensity. The light saturation point for photosynthesis is around $(5.5-6)\times10^4$ lux, and the light compensation point is 4 000 lux. With adequate light, the plant exhibits a compact growth habit, shorter internodes and petioles, thicker vines, larger and thicker leaves, and a deep green color. Under continuous cloudy conditions with insufficient light, the plant shows elongated internodes and petioles, narrow and elongated leaves, pale color, underdeveloped growth, and increased susceptibility to diseases. Insufficient light during the seedling stage affects leaf and flower bud differentiation. Insufficient light during the fruit setting stage results in nutrient deficiency, small flowers, small ovaries, and easy flower and fruit drop. Insufficient light during the fruiting stage leads to a decrease in sugar content and poor fruit quality. Thick-skinned sweet melons have strict light requirements, while thin-skinned varieties have a wider range of light adaptability and are more tolerant to weak light conditions.

The duration of daylight significantly affects the growth and development of sweet melon. During the seedling stage, a daily light duration of 10 hours or more is beneficial for the accumulation of photosynthetic products and the differentiation of female flowers. This results in early flower bud differentiation, lower flowering positions, and a higher number of flowers. If the daily light duration is less than 8 hours, the plant's growth will be poor.

Different varieties of sweet melon have different requirements for the total hours of sunlight. Early-maturing varieties require 1 100–1 300 hours, mid-maturing varieties require 1 300–1 500 hours, and late-maturing varieties require over 1 500 hours.

004 What are the water requirements for sweet melon?

Sweet melon has a fast growth rate, abundant stems and leaves, and high transpiration, so it requires a large amount of water. It has been determined that a sweet melon seedling with three true leaves consumes 170 g of water per day, and a single sweet melon plant can consume up to 250 g of water in a day and night during the flowering and fruiting stage. The extensive transpiration of leaves helps regulate plant temperature, which is the basic function of sweet melon's heat tolerance and the main reason for high sugar accumulation. However, sweet melon has poor tolerance to waterlogging, and waterlogging can cause root damage due to oxygen deficiency, leading to plant death. Therefore, it is advisable to choose high and well-drained land for sweet melon cultivation and strengthen drainage and irrigation management.

Sweet melon has different water requirements during different growth stages. During the seed germination stage, sufficient water is needed, so the seedbed or substrate for sowing should be thoroughly watered. The seedling stage requires moderate water, but since the root system is shallow, it is necessary to keep the soil moist. The suitable soil moisture content is around 65% of the field's maximum water-holding capacity. During the vine elongation and flowering-fruiting stages, more water is needed, so the irrigation amount should be increased to ensure that the soil moisture content reaches around 70% of the field's maximum water-holding capacity. During the fruit development stage, the soil moisture should reach around 80%-85% of the field's maximum water-holding capacity, as insufficient water can affect fruit expansion. During the fruit ripening stage, the soil moisture should be lower, around 55%-60% of the field's maximum water-holding capacity. However, it should not be too low, as it can cause fruit cracking.

Sweet melon prefers dry air, and the suitable relative humidity is around 50%-60%. Therefore, sweet melons grown in dry air areas have higher sweetness and stronger aroma. High humidity weakens plant growth, affects fruit setting, reduces fruit flavor, and lowers overall fruit quality. It can also promote the occurrence of various diseases. Sweet melon adapts to higher air humidity before flowering and fruit setting, but its adaptability to high humidity decreases after fruit setting. For sweet melons grown in greenhouses, various measures should be taken to reduce the air humidity inside the greenhouse.

005 What are the soil requirements for sweet melon? How to select a suitable land for cultivation?

Sweet melon has a wide adaptability to soil conditions and can be grown in sandy soil, loamy soil, and clay soil. However, loose, deep, fertile, and well-drained loamy soil is considered the most suitable. Sandy loam soil has a fast temperature rise in early spring, which benefits seedling growth, early fruit ripening, and good fruit quality. Clayey soil generally has good fertility and strong water and nutrient retention capacity, making it stable for sweet melon growth in the later stages. When growing sweet melons in sandy soil, it is necessary to apply organic fertilizer and strengthen fertilizer and water management in the later stages to improve the soil's water and nutrient retention capacity.

The pH value of the soil for sweet melon cultivation should be between 6.0 and 6.8. It can tolerate mild salt and alkali conditions, and it can grow normally in soil with a pH of 8 to 9. A certain amount of salt content in the soil (total salt content below 0.615%) can promote plant growth, early maturity, and increase sugar content and other soluble solids, improving fruit quality.

Sweet melon is a warm-loving and light-demanding crop, so it is recommended to choose a sunny land with a higher elevation for cultivation. It should have convenient access to water sources and transportation. The previous crop is preferably millet, sorghum, corn, wheat, or other suitable crops. However, attention should be paid to the use of herbicides on the previous crop to avoid sulfonylurea herbicides, as they can have residual effects on the soil and affect sweet melon cultivation. It is best to prepare the land and apply fertilizers in autumn.

006 What are the nutrient requirements for sweet melon?

Sweet melon has a high nutrient demand, requiring approximately 2. 5 - 3. 5 kg of nitrogen(N), 1. 3-1. 7 kg of phosphorus(P_2O_5), and 4. 4-6. 8 kg of potassium(K_2O) per 1 000 kg of production. Sweet melon prefers nitrate nitrogen fertilizer, and excessive ammonium nitrogen content can affect photosynthetic efficiency and cause ammonia toxicity, resulting in decreased sugar content, green skin color in netted melons, and reduced marketability. Therefore, it is recommended to use nitrate nitrogen fertilizer whenever possible. The application of phosphorus fertilizer promotes root growth and flower bud differentiation, improving plant cold tolerance. Potassium fertilizer enhances plant disease resistance. In addition to nitrogen, phosphorus, and potassium, calcium and boron are also important for growth and development. Insufficient calcium not only affects sugar content but also damages fruit appearance, resulting in whitening of the fruit skin and coarse netting. Boron has a certain impact on sugar accumulation, and in boron-deficient areas, sweet melons may have reduced sugar accumulation and develop brown spots in the fruit flesh.

Sweet melon is sensitive to chloride ions and does not grow well in soils with high chloride ion content. It is not suitable to apply chloride-containing fertilizers such as ammonium chloride and potassium chloride, nor should chlorinated pesticides be used, to avoid unnecessary damage to the plants.

007 What are the growth and development stages of sweet melon? What are their characteristics?

Sweet melon is a plant with a wide variation in maturity among different varieties. The growth period of early-maturing varieties is around 65-70 days, while late-maturing varieties can take up to 150 days. Thin-skinned sweet melons have a shorter growth period compared to thick-skinned varieties. The time from seedling emergence to the first female flower opening is generally 48-55 days, and the variation in maturity mainly lies in the time from flowering and fruit setting to fruit maturity. The complete growth period of sweet melon can be divided into four stages: germination, seedling, vine elongation, and flowering-fruiting stages.

(1) Germination stage. This stage starts from seed imbibition, germination, emergence,

and the appearance of true leaves, which takes about 10 days. During this stage, growth mainly relies on the nutrients stored in the seed, and the absolute growth is small. The expansion of cotyledons, elongation of hypocotyls, and root growth are the main processes. The soaking time for seeds is usually 4-8 hours. Under optimal temperatures of 28-30 ℃, thin-skinned sweet melon seeds can germinate in about 12 hours, while thick-skinned sweet melon seeds germinate a bit slower, taking an additional 2 hours. The accumulated temperature required for seed germination is 170-180 ℃, with an effective temperature accumulation of 60-70 ℃. The growth center during the germination stage is the elongation of the hypocotyl and the establishment of underground organs.

(2) Seedling stage. This stage starts from the emergence of true leaves to the stage where the plant has four true leaves. It takes about 30 days. During this stage, the aboveground growth is slow, with the addition of one leaf every 5-7 days. The root system continues to expand, flower buds begin to differentiate, and the seedling structure forms. When the first true leaf appears, flower bud differentiation begins, and by the time the fifth true leaf appears, the main vine has differentiated more than 20 nodes, with over 130 young leaves, 27 lateral vine primordia, and over 100 flower primordia. The leaves, flowers, and vines related to cultivation have all been differentiated, and the seedling structure has taken shape. Under conditions of a daytime temperature of 25-30 ℃, a nighttime temperature of 17-20 ℃, and 12 hours of sunlight, flower bud differentiation is vigorous, and the quality of female flowers is high. The period with 2-4 true leaves is the peak period of flower bud differentiation. During this stage, the growth center gradually shifts from the root system to the shoot apex.

(3) Vine elongation stage. This stage starts from the appearance of the fifth true leaf until the first female flower opens, lasting for 20-25 days. The primary growth during this stage is the expansion of the nutrient organs. The root system rapidly expands horizontally and vertically, increasing nutrient uptake. Lateral vines continuously emerge and grow rapidly, with one leaf emerging every 2-3 days. Flower organs gradually develop and mature. The growth center during the vine elongation stage is at the shoot apex. The key to cultivation management is to ensure moderate nutrient growth without excessive elongation of the plant. Timely pruning is an important measure to adjust nutrient growth and reproductive growth.

(4) Flowering-fruiting stage. This stage starts from the opening of the first female flower until fruit maturity, lasting for 25-60 days. Thin-skinned sweet melons have a shorter fruiting period, while thick-skinned sweet melons have a longer fruiting period, resulting in significant differences in fruit maturity among different varieties. During this stage, the underground root system stops growing, the aboveground stems and leaves transition from vigorous growth to slower growth, and the fruits rapidly enter a period of vigorous growth followed by a gradual slowdown. The growth center shifts from the shoot apex to the fruits. This stage can be further divided into three periods.

Fruit setting period. This period lasts for 7–9 days from the opening of the female flowers to the stage where the fruit begins to lose its hairs. It is a transitional period when the plant shifts from nutrient growth dominance to reproductive growth dominance. The fruit starts to grow, but nutrient growth is still strong. When the fruit begins to lose its hairs, the fruit is about the size of an egg, and the hairs on the fruit surface start to disappear, indicating that the fruit has settled. At this stage, the seeds have formed rudiments. The main growth during this period is cell division.

Fruit expansion period. This period lasts for 10–25 days from the stage where the fruit begins to lose its hairs to the stage where the fruit reaches its final size. Early-maturing varieties take 13 – 16 days, mid-maturing varieties take 15 – 23 days, and late-maturing varieties take 19–26 days. During this period, fruit growth is the main focus, with daily increases in fruit diameter of 5–13 mm and weight increases of 50–150 g. At the "final size" stage, the fruit characteristics become more apparent, with the appearance of ridges, furrows, and netting, but the skin color has not changed, the flesh is still firm, the taste is not sweet, and there is no aroma. The main growth during this period is cell expansion.

Ripening period. This period lasts for 10–40 days from the "final size" stage to fruit maturity. Thin-skinned sweet melons take about 10 days, while thick-skinned varieties take 20–40 days. During this period, root, stem, and leaf growth tends to stop, and the fruit volume stops increasing, but the weight continues to increase. In addition to further nutrient accumulation, the main changes during this period are the conversion of internal stored substances in the fruit. At the same time, the fruit skin becomes more distinct, the flesh becomes sweeter and more aromatic, the texture becomes crisp or soft, and the seeds become fully developed.

For varieties that produce multiple fruits per plant, there is a continuous harvest period from the maturity of the first fruit until the end of harvest, which usually takes 10–25 days.

008 How many years can sweet melon be continuously grown in a plastic greenhouse under conventional production?

In Changtu County, Liaoning Province, sweet melon farmers usually cultivate sweet melon for only one year and then relocate and rebuild the greenhouse. This practice increases the cost of building greenhouses and digging wells, and it is only suitable for areas with abundant arable land.

Based on the author's production experience, it has been found that sweet melon can be continuously grown in a greenhouse for two years without experiencing disease outbreaks. However, in the third year of cultivation, the incidence of vine wilt disease reaches around 15%. If vine wilt disease is prevented in advance, it is possible to continue cultivation for three years.

The author's experience in continuous sweet melon cultivation is as follows: First, it is necessary to apply a large amount of organic fertilizer. For a 700 m^2 greenhouse, approximately 3 m^3 of chicken manure from two tractor loads was applied. This resulted in vigorous sweet melon growth and enhanced disease resistance. Secondly, it is important to have proper ventilation and humidity control. Maintaining low humidity conditions helps prevent leaf and fruit diseases.

009 Can sweet melon be continuously grown in a plastic greenhouse without grafting?

Sweet melon can be continuously grown in a plastic greenhouse without grafting. Mature techniques such as straw bioreactor technology, application of "Chongcha Bicke" (a soil conditioner), and the use of anti-continuous cropping microbial agents can be employed.

(1) Straw bioreactor technology straw bioreactor technology utilizes crop straw as raw material, mixed with specially made microbial strains, to rapidly decompose the straw and release large amounts of CO_2, heat, and disease-resistant microbial spores. This technology significantly increases the yield and improves the quality of crops, especially greenhouse fruits and vegetables, and significantly enhances economic benefits. Straw bio-reactors are divided into two types: internal and external. The internal type is further divided into in-row and inter-row internal types. For vegetable production, in-row internal bio-reactors are generally used.

This technique originated from Shandong and Hebei, and was introduced to Shenyang, Dalian, Benxi, and Chaoyang in Liaoning Province in 2006 for use in solar greenhouses and plastic sheds. By 2009, demonstrations were carried out in all 14 prefecture-level cities in Liaoning Province. The city of Benxi tested the technique on greenhouse melons in 2011. In 2014, Jin Shaohua, from Toudao Town, Changtu County, Liaoning Province, introduced the technology into greenhouse melon production, and achieved very good results.

Why does this technology enable continuous cropping of vegetable crops? Firstly, the microbial strains used in the bio-reactor include ones specially designed to control soil-borne diseases. The high-activity strains used in the reactor generate a large number of beneficial bacteria during fermentation, which inhibit and kill various pathogens. Secondly, the return of straw to the field significantly improves soil structure, greatly enhances soil permeability, greatly improves the soil's granular structure, increases the organic matter content in the soil, strengthens root growth, and enhances resistance.

The application of straw bio-reactor technology in greenhouse melon production not only increases the soil temperature by 2 – 8 ℃ at a depth of 10 cm and the greenhouse air temperature by 1 – 2 ℃, but also effectively mitigates the impact of early spring frost on greenhouse melon production, achieving disaster reduction and resilience. The application of this technology increases the yield of greenhouse melons by 30%, raises the sugar content by 0.6 Baume degrees, reduces pesticide use by 80%, and decreases fertilizer use by 30%.

(2) Application of rechabite in 2019, Liaoning Vocational College tested the use of Rechabite, produced by Chengde Panfeng Enzyme Bacteria Co., Ltd., in the production of watermelon and melon continuous cropping. Very good results were achieved, with no occurrence of plant death.

This product is propagated and produced using imported Japanese strains. It is a solidified active protein complex of beneficial microbial secretions. It contains rich antibiotics, bacteriostatic substances, growth factors, active enzymes, amino acids, and other active substances and trace elements. It has functions such as catalyzing the decomposition of organic matter, degrading residual pesticides and heavy metals in the soil, promoting crop root development, inhibiting soil-borne diseases, and reducing obstacles to continuous cropping. It also has the ability to resist continuous cropping, prevent lodging, promote root formation, inhibit the reproduction and growth of spore nematodes, and prevent root rot. It has significant effects on physiological states caused by deficiency and viruses, such as small leaves, yellow leaves, curled leaves, and dwarfism, and it can also quickly recover from crop pests, frost damage, and mechanical damage.

It can raise the soil temperature by 2-3 ℃, advance crop maturity by 5-7 days, and bring forward the market availability of vegetables and fruits by 7-10 days. It can increase crop yields by 10-20%, and even up to 30-40% for vegetables and fruits, and greatly improve the quality and taste of agricultural products.

Usage and dosage:

Soil mixing. When used as seed fertilizer or base fertilizer, Rechabite can be mixed with moist soil and applied around seeds or evenly spread in the field, the closer to the seed root system the better.

Seedling irrigation. Rechabite can be diluted with an appropriate amount of water and sprayed evenly with a spray bottle.

Root dipping. When transplanting crops, Rechabite can be diluted with an appropriate amount of water and used for root dipping before transplanting.

Fertilizer mixing. When used as seed fertilizer or base fertilizer, Rechabite can be mixed with chemical fertilizer or organic fertilizer and applied together, it should be used immediately after mixing.

The regular dosage is 1-2 bags/667 m^2, for severely recurring plots, 3-4 bags/667 m^2.

(3) Use of Anti-Continuous Cropping Micro-ecological Formulation.

The Anti-Continuous Cropping Micro-ecological Formulation (officially called Anti-Continuous Cropping Disease-resistant Microbial Agent) is a microbial formulation created by Zhongnong LvKang (Beijing) Biotechnology Co., Ltd., using the internationally advanced anti-continuous cropping and disease-resistant technology from China Agricultural University.

It is produced using efficient microbial strains such as Bacillus subtilis, Trichoderma, and Viscobroom, using modern microbial fermentation technology. The effective live bacteria count is ≥5.0 billion/g.

The function features of the vegetable-specific Anti-Continuous Cropping Micro-ecological Formulation include: regulating the microbial composition in the roots and body of the crops, effectively improving soil, restoring soil granular structure; increasing the number of beneficial bacteria, inhibiting pathogens, preventing the occurrence of soil-borne diseases, with significant anti-continuous cropping effect; it can produce a variety of physiologically active substances, enhance crop disease resistance and stress resistance; promote the growth of capillary absorption roots, improve vegetable quality, advance maturity, increase yield, and make the harvesting period relatively concentrated.

Specific usage methods include:

Base application, mix this product with fertilizer and evenly spread it into the ground for harrowing, or apply it in holes or trenches, using 2-4 kg per 667 m^2. It can also be used when transplanting, by dipping roots in 50 times diluted solution for more than 30 minutes, or in the early stage of the occurrence of continuous cropping disease, irrigating roots with 80-100 times diluted solution.

010 What are the harmful effects of soil acidification in greenhouse production? How can it be improved?

The normal pH value of neutral soil should be around 7.0, varying slightly depending on the region. If the same few types of vegetables are planted year after year in facility soil, and if more nitrogen fertilizer is applied and less manure is used, the soil will gradually acidify, the pH value will decrease, and green moss will appear on the soil surface. Soil acidification will exacerbate the occurrence of soil-borne diseases, with the incidence of disease increasing by approximately 14% - 18% for each unit decrease in soil pH value. Therefore, it is important to pay attention to the improvement of acidic soil.

If the soil pH is between 5.0-6.5, prevention techniques focused on reducing the use of chemical fertilizers and increasing the application of organic fertilizers should be adopted. Increasing the application of organic fertilizers is an effective means of soil improvement and preventing soil degradation.

If the soil pH is less than 5.0, treatment techniques centered on the use of acidic soil conditioners should be adopted. The Korean imported soil conditioner "Zetu" can be applied when plowing the soil.

011 How can supplemental lighting be provided in greenhouse production?

Supplemental lighting is necessary in greenhouse production, especially during the winter and early spring when daylight hours are short and the weather is cloudy. Insufficient light can significantly impact the photosynthesis and growth of light-demanding crops like sweet melon. To provide supplemental lighting, the following methods can be used.

Nighttime lighting. During the winter and early spring, supplemental lighting can be provided during the nighttime using artificial light sources. Lights can be turned on for 4-6 hours per night. In case of cloudy or rainy days, lights can be kept on throughout the day to increase the total light exposure.

Selection of plant growth lights. It is important to choose plant-specific growth lights rather than regular household lighting. Plant growth lights emit specific wavelengths of light that are beneficial for plant growth. Blue or pink light is commonly used for supplemental lighting. Lights with a power of 36 watts can be installed at intervals of 5 meters in the greenhouse.

012 What are the principles for choosing thin-skinned melon varieties for spring planting in plastic greenhouses?

The temperature and light conditions in spring greenhouses are suitable for the growth and development of thin-skinned melons. Therefore, the variety requirements for early-maturing cultivation of thin-skinned melons in greenhouses are not strict, and appropriate varieties can be selected according to local consumption habits and development trends.

(1) Since thin-skinned melons in Liaoning are mainly sold to Jilin and Heilongjiang, and a large part is also sold to cities in the province such as Shenyang and Dalian, the earlier they are on the market, the higher the price. The varieties planted have evolved from mid-maturing to early-maturing and ultra-early maturing. For example, they have developed from past varieties such as Golden Road, Fu'er series, and Red City series, to Princess series, and then to Wanjing series, etc.

(2) Due to the increase in consumption levels in large cities such as Shenyang and Dalian, high-end melons are becoming more and more popular in the market. Therefore, the selection of melon varieties has evolved from simply pursuing yield to select low-grade large melons, to pursuing high sugar, high-quality, high-grade, and small-medium sized melons. The market demands melons that are uniform in size, with a single melon weight of 0.3-0.5 kg. Farmers have come up with a good way to control the size of melons by increasing the planting density, which not only solves the problem of large melons being difficult to sell, but

also does not reduce the total yield. With the highest density of 6 000 plants per 667 m^2, both quality and yield are excellent.

(3) The market's demand for the sweetness of melons is getting higher and higher, so sweet-fleshed melons should be chosen.

(4) Choose varieties that are beautiful in appearance and are shiny, resistant to friction, and transportable. The skin color of the melon has evolved from green skin to yellow-green skin, then to yellow skin, and currently to yellow-white skin.

(5) The skin should be thin and the flesh thick, but not hard in texture. The melons should be sweet, crisp, contain appropriate water content, be resistant to storage, not easily collapse, and the surface should be smooth without grooves. The Fu'er series, Red City series, and Sweet Princess are all representatives in this regard.

013 What are some main varieties of thin-skinned sweet melon for spring cultivation in plastic greenhouses?

There are thin-skinned melon varieties suitable for production and popular with consumers in various places. Here are just a few varieties suitable for the Northeast region.

(1) Bixiang. This is a hybrid melon variety with disease-resistant factors newly bred by Daqing Ironman Agricultural Research Institute. It is suitable for cultivation in old melon areas and areas with high disease incidence. It takes about 21 days from setting fruit to maturity under high-temperature cultivation, and it has particularly strong disease resistance. It sets fruit easily, usually with 5–12 fruits per plant, grows rapidly, and comes to market in a concentrated manner. The fruit is oblong, with orange-yellow flesh. The mature melon is yellow-white and has faint green stripes. The single fruit weight is 400–500 g, and can reach up to 700 g, with very strong commercial characteristics. The sugar content is generally 16–18 degrees Brix. It is suitable for open field, greenhouse, and plastic film covered cultivation. The plant line spacing is 35 cm × 120 cm, sowed in double rows, topped at 4 leaves, leaving 3 lateral vines, and the tips of the lateral vines are pinched at 3 leaves. When the melon is mature, it must be covered with leaves to promote the color change of the melon surface. Apply more phosphorus, potassium fertilizer, and farmyard manure, but the use of pig manure is prohibited.

(2) Zuimei. Zuimei melon is a new generation of hybrid variety launched by the Daqing Ironman Agricultural Research Institute. It is currently the top variety of white-fleshed melon on the market. It matures 5–7 days earlier than similar white-fleshed melons, taking about 19 days from fruit setting to maturity. It mainly sets fruit on lateral vines, grows vigorously, has large leaves, and is high-yielding, high-sugar, beautiful, and attractive. It has strong disease resistance, quick growth, and a concentrated harvest period. The fruit is oblong, with shallow grooves, 5–7 fruits per plant, and can be planted in multiple crops. The single fruit

weighs 400–500 g, is uniform, has excellent commercial characteristics, and is resistant to storage and transport, providing a very considerable economic benefit.

It is suitable for cultivation under greenhouse and plastic film cover. The plant line spacing is 35 cm × 120 cm, sowed in double rows, topped at 4 leaves, leaving 3 lateral vines, and the tips of the lateral vines are pinched at 3 leaves. When the melon is mature, it must be covered with leaves to promote the color change of the melon surface. Apply more phosphorus, potassium fertilizer, and farmyard manure, but the use of pig manure is prohibited.

(3) Wanjin Zaotemei. This new early maturing yellow-white round melon variety was bred by Jilin Kefeng Seed Industry Co., Ltd. It matures about 21 days after fruit setting. It has a high early yield, grows rapidly, changes color quickly, is resistant to low temperatures and weak light, and easily sets fruit on both primary and secondary vines. Each plant produces 8–10 fruits, with each fruit weighing around 500 g. The melon has a bright and delicate color, is shiny and attractive, and has good marketability. The melon flesh is sweet and crispy, and deliciously sweet. It has a strong outer aroma, strong comprehensive resistance, is resistant to transportation, has stable performance, and is a disease-resistant variety for melon and vegetable bases.

(4) Wanjin Zaocuitian. Bred by Jilin Kefeng Seed Industry Co., Ltd, this is an extremely early maturing variety, maturing about 5 days earlier than similar varieties. It has a high early yield, grows rapidly, changes color quickly, is resistant to low temperatures and weak light, and easily sets fruit on both primary and secondary vines. Each plant produces 8–10 fruits, with each fruit weighing around 450 g. The melon has a bright and delicate color, is shiny and attractive, and the melon flesh is sweet, crispy and not mushy, very delicious. It has a sugar content 19% higher than the Fei series, a strong outer aroma, strong comprehensive resistance, is resistant to transportation, highly resistant to wilt and downy mildew, has stable performance, and is a new disease-resistant variety for melon and vegetable bases that can withstand multiple cropping(Figure 1).

(5) Cuibao. The plant has a robust growth, resistance to wilt, vine wilt, white powder, downy mildew, leaf spot and other diseases. It has good disease resistance and anti-adversity, strong resistance to low temperature and weak light, does not age early, and does not die. Mainly sets fruit on secondary vines, each plant can yield 6–9 fruits, and has a strong continuous fruiting ability, maturing in about 26 days from fruit setting. The weight of a single fruit is 400–600 g, the flesh is 2.8 cm thick, sugar content is 16–18 baume degrees, the fruit is nearly round like an apple, deep gray-green, shiny exterior, green flesh, elegant appearance, uniform, resistant to storage and transportation, excellent quality, exceptionally crispy and sweet in taste.

(6) Zhengguan No. 2. Produced by Shenyang Jinyi Xuelong Seed Industry Co., Ltd. The plant grows vigorously, is resistant to low temperature and weak light, has good maturity,

Figure 1 Wanjin Zaocuitian

and the fruit is light green, high pineapple shape, green flesh, sweet taste, young melon swells quickly, an average of 4-5 melons per plant, each melon weighs 300-400 g, sugar content is 18-20 baume degrees, is resistant to transportation. With proper management, the yield can reach about 4 000 kg per 667 m^2, suitable for winter greenhouse and large shed hanging vine cultivation. After trial cultivation, it is superior to Cuibao type melon in all aspects and is a standout in the new generation of green skin and green flesh melons. It can be planted in all regions of the country with thin-skinned melons.

(7) Ganlu 19. Bred by the Vegetable Institute of Liaoning Academy of Agricultural Sciences. This variety has strong fruit setting ability, the fruit has a good inhibitory effect on the occurrence of side branches, does not need frequent pruning, and the fruit can still swell, eliminating the need for support in later stages. The weight of a single melon is 500-600 g, and the yield per plant can reach 4-5 kg. The fruit has a high content of soluble solids, which is particularly suitable for simplified cultivation.

(8) Sweet Princess. An early-maturing thin-skinned melon hybrid, the plant grows vigorously and is easy to set fruit. Both the secondary and tertiary vines can set fruit, with a fruit development period of about 28 days. The fruit is pear-shaped, the mature fruit skin is yellowish-white with a light green halo, the aroma is strong, the sugar content reaches 18 baume degrees, it is disease-resistant, high-yielding, and has strong adaptability.

014 What are the main varieties of thick-skinned melons for spring planting in greenhouses?

There is a wide variety of thick-skinned melon varieties, too numerous to mention.

(1) Elizabeth. Imported hybrid from Japan, with strong disease resistance, especially strong tolerance to low light, and good quality. The fruit is spherical, with a single fruit weight of 600−800 g, the skin is thick yellow, glossy, and without netting, the flesh is thick and white. The sugar content is 13−15 baume degrees, and it can reach up to 17 baume degrees. It has a fragrant taste and excellent flavor. The whole growth period is about 90 days, about 40 days after flowering, the fruit skin gradually turns from deep green to deep yellow, and it is not easy to fall off after maturing.

(2) Beauty. A hybrid produced by Shanghai Shimanfeng Seed Industry Co., Ltd., it is a medium-early maturing white-skinned, green-fleshed, thick-skinned melon. It has good fruit setting ability under low temperature and high fruit setting rate. The fruit is high-round and the surface is smooth. The weight of a single fruit is about 2.0 kg. The sugar content is usually 15−18 baume degrees. The flesh is jade green, tender and juicy. The fruit matures about 40 days after flowering, does not fall off the stem, and is resistant to storage and transportation. During the high-temperature season, attention should be paid to cooling, moisturizing, and timely prevention and control of diseases (Figure 2).

Figure 2 Beauty

(3) Ruby No. 6. A large-fruited high-grade netted melon variety bred by Beijing Jiaoxue Seedling Technology Development Co., Ltd. in 2012. It has been trial planted in many places in China such as Hebei, Shandong, Hainan, Zhejiang, Xinjiang, Liaoning and Southeast Asian countries. It is suitable for protected cultivation, and can also be cultivated in the open field in the northwest. The plant grows vigorously, with large leaves and thick vines, strong disease resistance, high resistance to powdery mildew, etc., resistant to low

temperature and weak light, and likes high temperature, strong light, and high temperature difference. It is easy to set fruit, late-maturing, generally matures 50 – 55 days after flowering, high-round shape, generally a single fruit weighs 2 500–3 000 g, up to 5 000 g, grey-green skin, tough, with fine, protruding, stable, uniform, exquisite netting. The flesh is orange-red, bright, crisp and sweet, about 5 cm thick, generally has a sugar content of 16–19 baume degrees, up to 20 baume degrees or more, with a high commercial rate.

(4) Honey King 998. Produced by Weifang Luyu Seed Industry Co., Ltd., it is a large-fruited Western melon variety with thicker leaves and significantly improved resistance. The fruit development period is around 50 days, the fruit is oval, the skin base color is blackish green, the netting is beautiful, the flesh is orange-red, about 4.5 cm thick, the flesh is fine and crisp, the central sugar content is about 17 baume degrees, the taste is refreshing and sweet, and the quality is excellent. The weight of a single fruit is over 3.5 kg, with a yield of about 5 000 kg per 667 m^2, and it is resistant to storage and transportation.

(5) Yaolong No. 25. Produced by Hainan Fuyou Seedling Co., Ltd. It grows vigorously, is resistant to wilt disease, the melon is oval, the single melon weight is 1.5–2.5 kg, the melon skin is gray-green, with fine netting covering the whole melon, the flesh is red, hard and crisp, without acid taste, the soluble solids content is 14%, the growing period is 85–95 days, from fruit setting to harvest is 40–50 days.

2. Greenhouse Spring Melon Seedling Cultivation Technology

015 How to prepare the traditional heating nursery for sweet melon seedlings in early spring?

To ensure safe production and prevent soil-borne diseases that can cause sweet melon seedlings to die, the selection of the nursery site is crucial. Generally, greenhouses are used for seedling cultivation, so the greenhouse film should be put on before the freeze in the previous year, and enough nutrient soil for seedling cultivation should be prepared. Farmers who cannot put on the film must do so before February 15 of the following year.

For a greenhouse that is 80 meters long, three stoves should be installed for heating (Figure 3). The greenhouse film is covered with a grass curtain.

Figure 3 Installing a Heating Stove in the Greenhouse

Some farmers use solar greenhouses for seedling cultivation. In fact, cultivating seedlings in a solar greenhouse is not as good as in a traditional greenhouse, because the environmental conditions of a solar greenhouse differ greatly from those of a traditional greenhouse. When the seedlings are transplanted to the greenhouse, they need to go through an adaptation process due to the change in environmental conditions. However, seedlings grown in a traditional greenhouse are robust due to the large temperature difference between day and night. Towards the end of the seedling cultivation period, heating is gradually reduced until it is stopped altogether, making the environmental conditions of the seedlings identical to those of the greenhouse where they will be transplanted. This results in the seedlings adapting quickly after transplantation, speeding up their recovery.

016 Can sweet melon seedlings be cultivated in a plastic greenhouse in early spring without heating?

Jin Shaohua from Toutao Town, Changtu County, Liaoning Province, has innovated a method of cultivating sweet melon seedlings in a plastic greenhouse in early spring without heating, namely by using multi-layer covering for insulation and straw bio-reactor bed for increasing temperature. The method involves setting up a central arched shed in the greenhouse, and a small arched shed inside the central arched shed, which is covered with insulation felt(Figure 4). The seedling bed uses a warm bed set up based on the principle of straw bio-reactor. This kind of warm bed is covered with 10 cm thick rice straw, sprinkled with the strain used for straw bio-reactor, then sprayed with water, covered with plastic film, and the seedling tray is placed on top.

Picture 4 Multi-layer Covered Warm Bed

This method of seedling cultivation can save the trouble and investment of firing stoves and the investment in electric heating beds, and it is energy-saving and environmentally friendly. According to measurements, the minimum soil temperature of the hole tray on the warm bed is 7 ℃ higher than that of the soil without the warm bed.

017 How to prepare seedling soil for traditional sweet melon seedling cultivation?

The nutrient soil should be fertile, loose, moisture and nutrient retaining, and free of pests and weed seeds. It can be made from field soil, paddy soil, soil from onion and garlic gardens, river bay soil, furnace ash, and well-rotted livestock and poultry manure. The best choice for field soil is from soybean stubble fields, and it should be free of pesticide residues. The ratio of field soil to manure is 7: 3. When preparing, crush and sieve the soil. For each cubic meter of nutrient soil, add 150 g of fungus, insect, and poison triple clear", mix well, pile up for 5 days, then add 1 kg of phosphorus and potassium bacterial agent, 800 g of dipotassium phosphate, 0. 5 kg of seedling mother agent, 1 kg of biological potassium fertilizer, and 120 g of active root-promoting seedling agent, mix well and prepare for potting.

018 What are the wrong practices farmers use when preparing seedling soil?

In visiting farmers, I found some improper practices when preparing seedling soil.

First, using compound microbial fertilizer and Fungus, Insect, Poison Triple Clear together. The compound microbial fertilizer contains beneficial biological bacteria, while Fungus, Insect, Poison Triple Clear contains fungicides that can kill or harm beneficial biological bacteria, making the microbial fertilizer ineffective. Therefore, these two cannot be used together.

Second, using fungicides to disinfect soil from corn fields. Fields that have been growing corn for many years are basically not infected with vegetable pathogens, so there is no need to use fungicides for disinfection. Using fungicides is wasteful and increases chemical pollution. Therefore, when taking soil from fields and mixing it with fully decomposed organic fertilizer to prepare seedling soil, there is no need for disinfection.

You can mix compound microbial fertilizer into the prepared agricultural fertilizer one week in advance to promote the fermentation of the agricultural fertilizer, and turn it over once after one week. The fermented agricultural fertilizer should be mixed into the field soil one or two days before potting. The field soil should be disinfected with a fungicide at least 3 days in advance. In this way, the fungicide will not harm the bacteria in the microbial fertilizer when preparing the seedling soil.

019 How to disinfect sweet melon seeds?

Seeds may carry viruses and bacteria, so it is essential to disinfect them. The basic disinfection method is hot water soaking. Before soaking, the seeds must be selected and then sun-dried for one day for disinfection. When soaking, first soak the seeds in normal temperature water for 0.5 hours, then dry them and soak them in hot water.

The key to hot water soaking treatment is to grasp three points: First, wet the seeds with normal temperature water to change the dormant bacteria on the seed surface to an active state. Only bacteria in an active state are afraid of scalding. Second, the water temperature should be between 55 ℃ and 60 ℃. 55 ℃ is the lethal temperature for most bacteria. The starting water temperature is sufficient in production, but the temperature is lower in winter and spring, and the water temperature drops quickly. Some people do not add hot water in time, or do not add hot water at all, which causes the water temperature to drop quickly, lower than the lethal temperature, or does not reach the lethal time. Third, the water temperature above 55 ℃ needs to be maintained for 10-15 minutes. 10 minutes is the lethal time at the lethal temperature. Therefore, to achieve good results in hot water soaking, especially the last two points, you cannot guarantee the effect if one of them is not achieved. The inactivation temperature of most plant viruses is between 55 ℃ and 70 ℃. Therefore, hot water soaking has an inhibitory effect on most plant viruses.

To ensure the water temperature, first, do not process too many seeds at once, and second, use two containers, one large and one small, with the small one for hot water soaking. Mix 2 parts boiling water with 1 part normal temperature water, and the temperature should be about 55-60 ℃. If the temperature is at the lower limit, add some boiling water to make the water temperature reach or approach 60 ℃. Pour the prepared hot water into the large container, then place the small container with the seeds inside, and then pour hot water into the small container and stir while observing the change in water temperature. Place the container for hot water soaking in 60 ℃ hot water, and the water temperature will drop slowly.

Another method is to treat the seeds with a fungicide. You can soak the seeds in a 500 to 700 times dilution of 70% methylthiophanate-methyl wettable powder for 1 hour, then rinse them and soak them in clean water for germination. This can effectively kill or inhibit various pathogens carried by the seeds, such as Fusarium wilt.

Now, some commercial seeds have been coated before leaving the factory, and there is no need to disinfect such seeds.

020 How to germinate sweet melon seeds?

Place the disinfected seeds in water at a temperature of 25-30 ℃, and add 5ml of Aiduosho and 5ml of Antikuning per 1 kg of water. Soak for 6-8 hours, stirring once every hour to ensure that the seeds are evenly medicated. Finally, rinse with clean water, drain, and put the seeds in a gauze bag for germination.

It is best to use a constant temperature germination box or germination room for germination.

Farmers or cooperatives generally do not have constant temperature conditions for germination. For safe germination and mass seed germination, it is advocated to use a hot water bag for germination. The hot water bag is a plastic bag filled with hot water at 36 to 37 ℃, which is then placed in a large basin. The basin is placed on a heated traditional Chinese bed (Figure 5). The gauze bag containing seeds is placed on the hot water bag, and a wet towel and quilt are covered on top. During the germination period, it is necessary to frequently check the temperature and moisture. If the temperature is low, add hot water, turn the seeds approximately every 3–4 hours, and wash the seeds every 8 hours to change the air and ensure the seeds are evenly heated. With careful management as described, the seeds should sprout in about 12–15 hours.

Figure 5 Hot Water Bag Germination Method

After germination, in order for the seedlings to adapt to the temperature inside the shed and the ground temperature, it is advocated to use low-temperature sprout training. When the seedlings grow to 2 – 3 mm, the temperature should be reduced to around 25 ℃. Before sowing, the temperature should be gradually reduced to around 15 ℃ to enhance the sprout's resistance to adversity. This prevents the seedlings from growing too tall in high temperatures, thereby achieving the goal of uniform and robust seedlings.

021 What preparations need to be made before sowing sweet melon seeds?

Before sowing, preparations for the seedbed, nutrient pots, and seedling trays should be made.

(1) Seedbed preparation. For every 667 m^2 of cultivation area, a seedbed area of 3–4 m^2 is needed.

(2) Nutrient pot preparation. A week before sowing, set up the nutrient pots. It's best to use new nutrient pots. The size of the pots should be 10 cm × 9 cm or 8 cm × 9 cm, filled to 9/10 full. Place them on the seedbed. To increase the ground temperature, you can lay weeds or grass curtains underneath the seedbed. After arranging the nutrient pots, water them thoroughly, cover with a ground film to increase the ground temperature, and then wait for sowing.

(3) Seedling tray preparation. If you're using seedling trays, you can choose a 50-hole seedling tray. New seedling trays do not need to be disinfected, but old trays must be disinfected before reuse to avoid infection from residual bacteria from previous crops. The best disinfection method is to soak the seedling tray in a 100-fold solution of 40% formalin for 15 to 20 minutes, then seal it with a film for 7 days. After uncovering, rinse with clean water and it's ready for use.

022 How to sow sweet melon seeds?

The sowing time in the central and northern parts of Liaoning is from February 23 to March 5. For seedlings grown in plastic greenhouses, a stove should be lit to increase the temperature the day before sowing. On the day of sowing, the nutrient pots should be sprayed with a small amount of water again, with a fungicide in the water. Use 0. 25 kg of 50% carbendazim with 100 kg of water. After the water seeps down, use a thin stick to make a shallow hole in the center of the pot. Each pot should be sown with one seed, the seed should be laid flat with the sprout facing down, and then covered with 1 to 1. 5 cm thick nutrient soil. When covering the soil, a pesticide should be added, the formula is 60 kg of prepared nutrient soil and 20 g of seedling fungicide. After covering the soil, cover it with a layer of clean fine river sand about the thickness of a piece of paper, which can effectively prevent seedling diseases such as damping-off. After covering with river sand, cover with a ground film to prevent water evaporation, and place a small arch shed on the bed surface. This method not only increases the ground temperature but also increases the air temperature.

For substrate sowing, first spray water to moisten the substrate, then fill it into the seedling tray, scrape it flat with a wooden board, then spray water again, press a shallow pit with another seedling tray, and put one seed into the shallow pit(Figure 6), then cover it with substrate and place it on the seedbed.

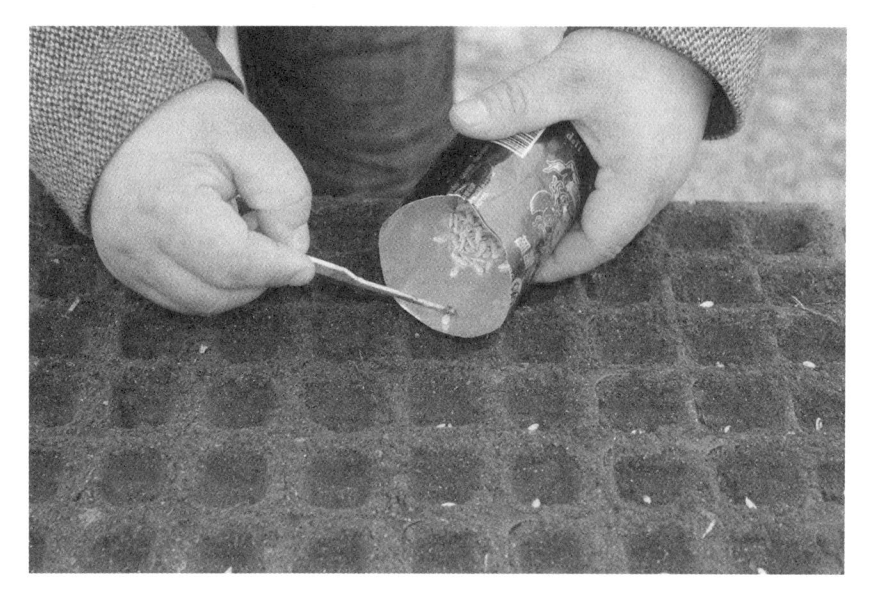

Figure 6 Manual Seed Sowing in Seedling Tray

023 How to manage sweet melons after sowing?

(1) Post-sowing pre-seedling management. Management should pay attention to three points. Pay attention to the phenomenon of seedlings being baked by high temperatures produced by strong sunlight at noon under the film. Therefore, before the seedlings emerge, the temperature of the nutrient pots under the film should be checked frequently. It can reach 30-34 ℃ at noon, but it cannot reach 40 ℃. High temperatures that could bake the seedlings must be strictly prevented. Pay attention to the phenomenon of dry buds caused by insufficient water in the nutrient pots after sowing, which leads to missing seedlings. Pay attention to the need to heat the stove when the night temperature is below 13 ℃.

After sowing, the daytime temperature of the pre-seedlings is 28-35 ℃ and the nighttime temperature is maintained at 18-20 ℃. After three days and nights, most of the seedlings can mound the soil.

(2) Post-emergence management. After most of the seedlings have emerged from the soil, the film should be removed in time, and the temperature should be lowered. The daytime air temperature is 20-22 ℃, and the nighttime temperature is 16-18 ℃ to prevent the emergence of tall seedlings. About 7-10 days after sowing, the seedlings are uniform and the cotyledons are flat. At this time, the seedbed should be dry rather than wet. To prevent seedling diseases, spray Green Heng No. 2 once after the seedlings are uniform, and spray it

again one day before planting. When the melon seedlings grow true leaves, the temperature in the shed should be appropriately increased, which is conducive to rapid growth during the seedling stage. The daytime temperature is generally controlled at 28 - 30 ℃, and the nighttime temperature is controlled at 20-22 ℃, otherwise, it will affect the growth of the seedlings. For seedbeds covered with small arch sheds, when the seedbed temperature reaches 35 ℃, the arch shed film should be uncovered to allow the melon seedlings to receive more sunlight, ventilate the humid air in the seedbed, and prevent the occurrence of diseases.

In terms of water management, do not water if not dry. If there is indeed a lack of water, choose a sunny morning to lightly water with warm water around 20 ℃.

Stop spraying water 5 to 7 days before planting, gradually increase the ventilation, start hardening off the seedlings, and gradually make the melon seedlings adapt to the cold shed environment to improve the survival rate after planting.

Management should achieve three preventions and two attentions.

Strictly prevent bad weather attacks, prevent the grass curtain from being lifted by strong winds at night, and have countermeasures for long-term cloudy and sudden sunny weather. Strictly prevent the serious harm of sulfur dioxide and carbon monoxide produced by heating stoves to humans and seedlings. Strictly prevent the destruction of seeds by rodents after sowing.

Pay attention to the fire hazard caused by strong wind weather. Pay attention to the cold damage that may be caused by short-term hardening off before planting.

Through meticulous scientific management, strong seedlings with 3 leaves and 1 heart can be cultivated in 25 days, ready for planting.

3. Field Management of Melons

024 How to manage temperature and fertilizer water from planting to flowering in greenhouse melons?

After planting, close the greenhouse to raise the temperature and promote seedling recovery. The temperature during the day is between 25-35 ℃, and it should not be lower than 15 ℃ at night. Straw mats are placed around the greenhouse at night. After seedling recovery, the temperature during the day is 25-35 ℃, at night it is 15-18 ℃, and the soil temperature should be 20-25 ℃. For greenhouses with small arch sheds, uncover during the day and cover at night. As the outside temperature rises, the small arch shed can be removed, and ventilation can be appropriately increased to reduce the temperature.

Water the seedlings once about 7 days after planting. If the base fertilizer is sufficient and the soil moisture is appropriate, there is no need to top dress or water until before fruit setting. Appropriate seedling squatting promotes root growth.

The period of cell division in the melon embryo tissue after the pollination of the melon female flowers is the sensitive period of the melon to water. If there is sufficient water, the cells divide normally and the melon can set; if there is not enough water, it will cause the cells to stop dividing, and the melon embryo will not develop or enlarge, resulting in dead melons. Dead melons are dull (Figure 7), while normally developing melon embryos are glossy. Therefore, the seedling recovery water must ensure the needs of the melon embryo tissue cell division, otherwise, near flowering, water must be irrigated again to prevent the

Figure 7 Dead Melons in Sweet Melon

occurrence of dead melons, affecting fruit setting and yield.

025 What are the common pruning methods for greenhouse thin-skinned melons?

The goal of spring greenhouse thin-skinned melons is early maturity, so the fruit is set on lateral vines, and generally 3 fruits are kept per plant, each fruit should have 6–8 true leaves as functional leaves. There are two commonly used pruning methods.

(1) "4321" pruning method. The main vine is topped after leaving 4 true leaves, retaining the 3 lateral vines that grow out of the axils of the upper 3 true leaves, and each lateral vine is topped after leaving 3 leaves. Each lateral vine produces 3 secondary vines, and each secondary vine is topped after leaving 2 leaves. Each secondary vine produces 2 tertiary vines, and each tertiary vine is topped after leaving 1 leaf. The vines that grow out of the cotyledons are also all removed.

(2) Three lateral vine pruning method. The main vine is topped after leaving 4 true leaves, retaining the 3 lateral vines that grow out of the axils of the upper 3 true leaves, and each lateral vine is topped after leaving 7 leaves. All the secondary vines on the lateral vines are removed. The vines that grow out of the cotyledons are also all removed.

026 How to specifically prune the greenhouse spring thin-skinned melons using the "4321" pruning method?

(1) Topping the main vine. Topping the main vine after leaving four true leaves. Generally, the tip can be pinched as soon as the 4th true leaf is exposed, and the branch that grows at the cotyledon is also pinched off. The branches at the cotyledon are very small when the main vine is pinched, but they must be pinched off early, otherwise they will grow bigger later.

(2) Keeping lateral vines. The main vine grows 4 lateral vines, and 3 are kept. Generally speaking, the lateral vine that grows from the first true leaf often does not grow well, so it should be removed, leaving the upper 3 lateral vines.

(3) Topping the lateral vine. The lateral vine is topped after leaving three leaves. However, when topping, it cannot be done mechanically, we need to check if there are female flowers at the axil of the true leaves on the lateral vine. If there are no female flowers at the axil of the three leaves on the lateral vine (this situation is not uncommon), then we need to change the fruit setting to the secondary vine. The way to do this is to leave the first leaf at the base of this lateral vine, cut off the lateral vine from its front, and let the secondary vine at the first leaf expand, replacing this lateral vine.

A common mistake in production when topping the lateral vine is to mistakenly consider the leaves on the main vine as the first leaf of the lateral vine. In this case, when topping, only two leaves are actually left on the lateral vine, which should be noted.

(4) Topping the secondary vines. Topping the secondary vines after leaving two leaves. When topping the secondary vines, you need to see if the melon on the lateral vine is set. If it is set, you can top it normally. If it is not set, you need to choose the secondary vine that can open the female flower first to keep, and top this secondary vine after leaving three leaves, and remove the other two secondary vines on this lateral vine. This is to let this secondary vine replace the lateral vine that could not set the melon.

(5) Topping the tertiary vines. Topping the tertiary vines after leaving one leaf.

027 How to retain melons in a greenhouse in spring? How to determine if the young melon is set?

Retaining melons is one of the key techniques in melon cultivation. The ideal number of melons to keep is one per lateral vine, that is, three per plant. If a lateral vine has set two melons at the same time and they are of similar size, then four melons can be kept per plant.

How to determine whether the young melon is developing normally and can be set? It depends on whether the melon fetus has a gloss. If it is glossy, it indicates that the melon is growing normally and can be set. If it is dull and no longer develops, it is a dead melon.

028 What should be noticed in the three lateral vine pruning method for thin-skinned sweet melons?

The three lateral vine pruning method is relatively simple to operate. When topping the secondary vine, it is necessary to check whether there are female flowers on the lateral vine. If you cannot see the female flowers on the lateral vine when topping the first secondary vine, but there are female flowers on the secondary vine, then the part of the lateral vine ahead of the secondary vine joint is removed, and this secondary vine is allowed to grow, replacing the lateral vine to set the melon(Figure 8).

Figure 8 The Secondary Vine Replacing the Lateral Vine to Set the Melon.

029 How to prune and retain melons in suspended vine cultivation?

For sweet melon cultivation using suspended vines, generally single vine pruning is used, and one melon is retained per plant. That is, the main vine is not topped, all lateral vines below the 10th leaf are removed, and lateral vines are retained from the 10th leaf onwards to see if they have a melon. Generally, there will be a melon, and when the female flowers bloom, treatment with a fruit setting agent or pollination with bees is used. Choose one with a good melon shape, leave two as backups, and top the vine leaving 2 leaves before the melon. If the melon on the lateral vine of the 10th leaf is set and has a good shape, then the lateral vines on the 11th and 12th leaves are all removed. If the melon on the lateral vine of the 10th leaf is not set or the melon shape is not good, then the best melon from the lateral vines of the 11th and 12th leaves is retained, and all other lateral vines are removed. If the greenhouse temperature and moisture management are good, you can see at the time that the melon is set. You can directly keep one melon, and the melon will grow rapidly without a lack of nutrients. When the main vine grows to 25-30 leaves, it is topped, generally based on the horizontal wire used to tie the suspension rope. When the growing point reaches the iron wire, it is topped.

If it is a small melon variety, more melons are retained. That is, starting from the 9th leaf, first retain 3 melons, and alternately retain 3 lateral vines, with 1 melon retained on each lateral vine, and all other lateral vines between the 3 lateral vines retaining the melons are removed. You can keep two or three more melons going up, but this depends on whether the growth of the plant can meet the growth needs of the extra retained melons. If it can't meet the needs, don't keep more. This is the pruning method for the small-sized melon series of Green Treasure. There is also a method for pruning small melons, which is to continuously keep four or five lateral vines from the 9th leaf, with one melon retained on each lateral vine. This method has a bit smaller melon count because the melon setting is concentrated, and the management level needs to be high, the nutrients need to be supplied, the water needs to keep up, and the temperature needs to be appropriate.

In production, if there are not enough seedlings, double vine pruning can also be used, with the main vine topped at the 4th leaf, and two lateral vines are retained to grow. It also retains melons from the 10th leaf upwards, and the vine is topped leaving 2 leaves before the melon.

3-5 days after the melon is set, a sterile bag specifically for sweet melons is put on, and it is removed 7 days before maturation. The bagged melon has a more delicate, smooth, and clean skin, and the commercial price is higher. If a fruit enlargement agent is sprayed, it should be 13-15 days after the melon is set. If it is too early, the melon may be deformed, and if it is too late, there may be cracks on the melon.

030 What commercial fruit-setting agent should be used for sweet melons?

Artificial pollination of sweet melons is time-consuming and laborious, so producers generally use a fruit-setting agent. Production practice has shown that Sweet Melon Meiling (produced by Liaoning Anshan Huaxin Agricultural Technology Co., Ltd.) is a fruit-setting agent preferred by producers in Liaoning Province. It is a water-soluble fertilizer containing trace elements and is a substitute for chloropyrimidine and fruit-setting agents, without chloropyrimidine. The melons grown with it can be harvested earlier, have a high fruit-setting rate, expand quickly, do not make the melon bitter when the concentration is high, and have a good taste when the melon matures early.

Sweet Melon Meiling must be diluted with water immediately before use, and spray the petals and melon embryos of the female flowers that will open on the day and the next day. When the temperature is 14-25 ℃, dilute with 500-750 g of water per bottle; when the temperature is 26-31 ℃, dilute with 750-1 000 g of water per bottle; do not use when the temperature is above 31 ℃, cool down before use. For those who plant varieties like Golden Consort, Jing Sweet, and other medium and large melons with a single fruit weight of more than 400 g, or when the melon seedlings are vigorous, or during the flowering period when there is low temperature and heavy rain, it is necessary to spray the petals and melon embryos on the day of flowering, and the amount of water for dilution must be ensured between 500-750 g, otherwise, it is easy to cause low fruit-setting rate and problems with dead and rotten melons.

031 What are the advantages of bee pollination for sweet melons in a controlled environment?

Bee pollination can significantly increase the vitamin C content of sweet melon fruits, increasing it by 73.93% compared to chloropyrimidine treatment, and 6.12% compared to manual pollination. The quality of single melons pollinated by bees is not significantly different from those pollinated by chloropyrimidine, but it is 6.90% higher than those pollinated manually (Wu Qianxing, Liu Yong, et al., 2013). The maturity period of sweet melons pollinated by bees is shorter than that of manual pollination (Yan Debin, Niu Qingsheng, etc., 2012), indicating that bee pollination can mature earlier. The fruits of sweet melons pollinated by bees contain seeds. The seeds produce endogenous hormones, which make the sweet melons uniform in size, neat in shape, and free of malformed melons. Bee pollination is environmentally friendly, reducing chemical pollution in the melon, after all, fruit-setting agents contain chemical growth regulators. Bee pollination also reduces human labor input.

The experience of bee pollination of sweet melons in Hebei Province believes that bumblebee pollination is suitable for sweet melons in a controlled environment, and bee pollination is suitable for watermelons in a controlled environment. Each box of bees can be reused for 2–3 pollination cycles. The fruit-setting rate can reach more than 95%, the rate of malformed fruits can be reduced by 5%, and the sugar content can be increased by 0.5%–1.5%.

032 How to manage bee pollination for sweet melons in a controlled environment?

If insect nets are used for ventilation, only one box of bees should be placed in each greenhouse or large shed, and the bee box should be placed in the middle of the shed, usually 20–40 cm above the ground. If the large shed does not use insect nets, because the large shed has ventilation on both sides, it is convenient for bees to enter and exit, and one box of bees can be placed in 2–3 large sheds, and the bee box is 40 cm above the ground. The adaptation period for bees is 2–3 days, and they should be put in the shed 2–3 days before the peak flowering period. They start activity outside the box when the temperature is above 12 ℃. The adaptation temperature of bumblebees is 2 ℃ lower than that of bees. The suitable temperature for bees to pollinate in the greenhouse is 18–30 ℃, and the temperature for bee colony reproduction is 34–35 ℃, so during the pollination period, the daytime temperature in the shed should be maintained at 18–30 ℃. This range of temperature is suitable for the growth of sweet melons, the reproduction of bee colonies, and the pollination activities of bees.

The pollinating bee colony should be checked every 3–5 days, timely feeding syrup and pollen to ensure sufficient food in the hive. Regularly reward feeding with syrup soaked in sweet melon flowers in the evening to improve the pollination efficiency of bees visiting sweet melon flowers. During the pollination period, do not spray pesticides that are toxic to bees.

033 How to manage sweet melons during the fruit expansion period in a greenhouse?

(1) Fertilizer and water management. When the sweet melon embryo grows to the size of an egg yolk and enters the expansion period, it is the second sensitive period to water. Adequate fertilizer and water should be provided to ensure the rapid expansion of the melon. If the fertilizer and water are insufficient, the melon will not grow large and will not reach the size of this variety's commercial melon, affecting the yield. You can apply a balanced

nitrogen-phosphorus-potassium fertilizer+chitosan once at the beginning of melon expansion, and then apply a high-potassium fertilizer+chitosan twice with water.

(2) Temperature management. The fruit expansion period of the spring crop of sweet melons in the greenhouse coincides in May, so the vents do not need to be closed at night, and ventilation is achieved day and night. But a daytime temperature of 28 - 32 ℃ is preferred. Higher daytime temperatures are conducive to photosynthesis and also help to increase the day-night temperature difference, which is beneficial for the development of fruits and the accumulation of sugar.

(3) Vine control management. Vine control management is very important during the melon expansion period. If the vine and leaves grow too vigorously, that is, the nutritional growth is too strong, it will affect the melon expansion. The tip of the heavy grandson vine left by the "4321" pruning method is pinched after one leaf, and the vines that grow out later should be pinched in time. Regardless of which pruning method is adopted, the vines should still be pinched in time after the melon expands to control the vigorous growth. If the growth of the vines is left unchecked, the vines will grow rapidly, not only consuming nutrients, but also causing large humidity inside the melon due to the dense branches and leaves, causing the melon to breed diseases.

034 How to increase the sweetness of sweet melons during the maturation period in a greenhouse?

(1) Leaf protection management. The first fruit of the greenhouse sweet melon enters the maturity period after setting, and leaf protection management is very important. Good leaves lead to good photosynthesis, making the melon sweet.

Spring melon leaves in a plastic greenhouse are not prone to diseases, but aphids are prone to occur in the later stage of fruit setting(late May and thereafter). If prevention and control are not timely, aphids will spread rapidly, causing the leaf function to decline severely, making the melon not easy to mature and decreasing the sweetness. Therefore, it is necessary to first pay attention to the occurrence of aphids and prevent and control them in time. Secondly, foliar fertilization can be carried out, spraying 0. 3% potassium dihydrogen phosphate, etc., which can prevent premature aging, promote photosynthesis and carbohydrate transportation, and help to increase sweetness.

(2) Large temperature difference management. After the first fruit of the sweet melon is set, the ventilation should be reduced during the day to keep the temperature as close to 28-35 ℃ as possible, and then the vents should be opened to the maximum at night to increase the day-night temperature difference and promote sugar transformation and accumulation.

(3) Water control management. On the premise of the normal growth of sweet melons, the soil moisture should be appropriately controlled when the fruit enters the maturity period,

and watering should be stopped 10 days before harvesting, otherwise the melon will not be sweet.

035 What causes melon cracking? How to prevent it?

Melon cracking often occurs during the young melon expansion period(melon girl period) and fruit maturity period(six or seven mature), and there are mainly three reasons:

(1) Relative drought during the fruit growth period. That is, during the young melon expansion period or fruit maturity period, the air humidity is small, the fruit epidermis ages or basically forms, and then suddenly watering or raining, the fruit cells suddenly enlarge, and the epidermis growth rate cannot keep up with the melon flesh growth speed, causing melon cracking.

(2) Improper use of plant growth regulators. When using fruit setting agents, if the concentration is too high or the amount is too much, young melons will soon crack(Figure 9). Overuse of gibberellin in the later stage of fruit also causes a large amount of melon cracking.

Figure 9 Young Melon Cracking

(3) Lack of calcium and boron during the fruit expansion period. Over-application of nitrogen, phosphorus, and potassium fertilizers in the fertilization process inhibits the absorption of micronutrients such as boron and calcium, leading to thinning of the fruit skin cell wall, which can easily cause melon cracking.

Once melon cracking occurs, there is no cure, and the marketability declines severely. So prevention is key, and the prevention methods are:

Use fruit setting agents as required, and it is best to use bee pollination. Watering should be done according to the principle of seeing dry, seeing wet. Ensure sufficient moisture before setting fruit, timely watering during expansion to promote fruit expansion, and avoid watering heavily after expansion. Water and fertilize 7 - 8 days after setting fruit, preferably using biological organic fertilizers, and apply 1 kg of boron fertilizer per 667 m^2 with water. During the rapid growth period of young fruit and the middle and later stages of fruit growth, avoid drastic environmental changes and excessive drought. If it is dry, water it lightly first, and then gradually increase the amount of water when the plant adapts. Try not to water or only water a small amount 7-10 days before picking, and avoid heavy flooding. Protect the leaves near the melon to prevent the melon from being directly exposed to sunlight, and it is better to use a paper bag for bagging. Spray a 500-fold amino acid calcium solution, or high-efficiency meilin calcium solution, or 0. 3% calcium chloride+0. 2% boric acid solution 5-7 days before the fruit matures after setting. It is not recommended to use an expansion agent. If you use an expansion agent, be cautious and stop using it in the later stage of fruit growth.

036 How to determine the maturity of melon fruit?

The quality of muskmelon is directly related to the maturity of the fruit. Only timely harvested melons can guarantee the unique sweet quality of the variety. At present, fruit maturity is mainly determined by experience, but some physiological indicators can also be used.

(1) According to the number of days of fruit development. Generally, the development days of thin-skinned melons are 25-35 days, early-maturing thick-skinned melons are 35-40 days, mid-maturing varieties are 45-50 days, and late-maturing varieties are as long as 65-90 days. According to the number of days of fruit development of the planted variety, a marking method is used to mark the date on the day of flowering. When the maturity period of the same batch of fruit arrives, pick 1-2 melons for verification first. Once mature, you can harvest in order. However, this method is only applicable to thick-skinned melons with a single melon left per plant, and thin-skinned melons are generally not used.

(2) Observe the appearance characteristics of the fruit. The fruit skin is bright and fully expresses the inherent color characteristics of this variety, indicating maturity. The yellow-skinned variety only changes color completely when it is mature, and the netted variety is marked as mature when the net on the fruit surface hardens.

(3) Hand press method. Press the melon surface near the fruit umbilicus lightly with your finger. If it starts to soften, it is a ripe melon. This method is quite accurate when applied to soft flesh varieties.

(4) Smell the melon navel. Those who emit the unique strong fragrance of this variety are marked as mature, and the immature raw melons do not emit fragrance.

(5) Check if the melon stem falls off. Some varieties, such as honeydew melon and

yellow egg, fall off the melon stem when the fruit is ripe and are easy to pick. However, not all varieties fall off the melon stem when the fruit is ripe, especially thick-skinned melons, which often specifically choose varieties that do not fall off the melon stem when ripe, so the falling off of the melon stem cannot be used as a judgment.

037 How to harvest muskmelons?

Timely harvesting of muskmelons is crucial, and it is the final measure to ensure the sweetness of muskmelons. Generally, muskmelons are fully developed and mature, with high sugar content and good taste. The total sugar content of muskmelon fruit continues to increase during the maturation process, especially the sucrose content increases sharply a few days before full maturity. Therefore, fully mature muskmelons are the sweetest.

For local sales, it is best to harvest when it is ninety percent ripe, while for sales to other places, harvest when it is eighty percent ripe. It is advisable to use scissors to harvest. Thin-skinned muskmelons should leave a short fruit stem, and thick-skinned muskmelons should cut the fruit stem into a T-shape with scissors to protect the water in the melon from loss through the wound and extend the storage period.

4. Prevention and Treatment of Diseases and Pests in Muskmelons

038 What are the common symptoms of cold damage in muskmelons? How to prevent cold damage?

Muskmelons are thermophilic crops and are extremely intolerant to cold. When the temperature is below the low temperature limit of the plant tissue, some damage symptoms will appear. Cold damage often occurs at the front bottom of the greenhouse and on both sides of the shed. Acute cold damage has the following common symptoms:

(1) Patches. When the plant is below 5 ℃ for more than 24 hours, the cell tissue will suffer acute damage, manifested as water-stained patches on the leaves and stems, mainly on the leaf edges. If the damage time is longer, the damaged part will show completely necrotic dry patches, and severe ones will wilt and die.

(2) Degreening. When the temperature is below 7.5 ℃ for more than 48 hours, there are no water-stained patches on the leaf surface, but a few days later the leaf tissue degreens, and irregular light green patches or dry patches appear on the leaf surface, still mainly on the leaf edges.

(3) When the root temperature is below 13 ℃, no more root hairs will occur, the absorption function will decrease, the nutrition will be poor, and the growth of the plant will be inhibited. If the ground temperature remains below 13 ℃ for a long time, the root color will turn brown, and severe cases will cause root rot or root rot caused by fusarium infection. Even if the temperature rises afterwards, the root system will not be able to produce new roots. If it is cloudy for a long time and then sunny, the plants will wilt in large areas, and all will wither and die a few days later.

Prevention measures:

(1) Do not plant too early. Planting too early can easily encounter late cold weather. The minimum temperature in the greenhouse should reach the lower limit of the temperature suitable for the melons at the time of planting, and should not be too low.

(2) Set a foot apron. The front foot and both sides of the greenhouse are lined with old thin films on the inside of the greenhouse to form a double-layer film, which has the effect of buffering cold air.

(3) Moderate. Moderate low-temperature management should be carried out the day

before cooling, lowering the highest and lowest temperatures by 1-2 ℃ under the premise of ensuring the basic physiological needs of melons, which can help melons adapt to the low temperature environment in advance and reduce physiological obstacles caused by sudden temperature drops.

(4) Covering with straw. When a cold wave hits, wrapping the greenhouse with straw at night can effectively prevent seedlings at the edge of the shed from being damaged.

(5) Temporary heating. For those who do not use straw bio-reactor technology, if the lowest temperature in the greenhouse can drop to 5 ℃ in the early morning, it is best to use temporary heating to maintain the temperature above 5 ℃, but pay attention not to overheat. It is recommended to use a greenhouse heating block, using 3-5 blocks per 667 m^2 area according to the cooling situation, and ignite when the greenhouse temperature is 7-8 ℃ in the first half of the night, which can not only increase the temperature by 4-6 ℃, but also increase the application of carbon dioxide gas fertilizer. You can also use coal, fuel or electric heating as a heat source, but you should pay attention to fire prevention.

(6) Spray nutrient cold resistance. In winter or late spring cold weather with low temperatures, the root absorption ability is weak, and spraying photosynthetic micro-fertilizer on the leaves can supplement the deficiency symptoms caused by the root's inadequate absorption of nutrients. Spraying rice vinegar on the leaves can inhibit bacteria and drive away insects. When mixed with sugar and superphosphate, it can increase the sugar content and hardness of the leaves and improve cold resistance.

(7) Spray Bi Hu cold resistance. Bi Hu is a pure natural high-tech biological product developed by German scientists based on the wonderful phenomenon of plant chemosensation (also known as allopathic life) in nature and the principles of ecological biochemistry over 30 years. It contains more than 30 kinds of plant active substances such as natural plant endogenous hormones, flavonoids and amino acids, forming a unique plant growth complex balance regulation system, which can activate multiple activities of plants through system induction, make plants flourish, root system developed, significantly improve plant stress resistance(frost, drought, waterlogging, soil compaction, salinity, etc.) and pest and disease resistance, increase yield and improve quality. It is a new type of compound balanced plant growth regulator. It can be sprayed 1-2 days before the cold wave hits.

(8) Use Ice Guard antifreeze. Ice Guard" antifreeze is known as plant warm underwear. Ice Guard is a high-end antifreeze brand under Shanghai Hanhe, specializing in providing efficient and safe frost protection for crops, reducing damage to crops caused by extremely cold weather. Ice Guard antifreeze comes from Norway. Its core ingredient comes from natural cold-resistant brown algae in the cold sea near the Arctic Ocean, which contains the anti-freezing immune factor DCT, which can quickly activate the plant's anti-freezing immune function and has a unique anti-freezing cumulative immune characteristic.

Even if DCT is continuously applied to crops during the non-frost period, it can enable crops to gain an accumulated resistance to frost, improve crop resistance, achieve proactive defense effects, greatly enhance the adaptability of crops to sudden frost disasters, and prevent frost damage. Ice Guard antifreeze has good preventive and reparative effects. One week before the arrival of snowy weather, root irrigation and foliar spraying can be carried out on crops to achieve good frost resistance.

039 What diseases are common in the seedling stage of melons?

The main diseases of melon seedlings include damping-off, wilt, and anthracnose.

(1) Damping-off. Young seedlings are affected, with water-stain-like lesions appearing at the base of the lower hypocotyl near the soil surface. The affected area then turns yellow-brown, dries up and shrinks into a line, and the seedling collapses before the cotyledons wilt. Sometimes diseased seedlings appear the same as healthy ones, but they are prostrate on the soil surface and cannot stand upright. Upon closer examination, the base of the stem has already shrunk into a line. The disease spreads quickly in the seedbed. It initially affects only a few seedlings, but within a few days, it causes a large number of seedlings to collapse. This disease often occurs in the early stages of seedling growth when the soil temperature in the seedbed is low (10 - 15 ℃), the soil moisture is high, or during cold spells, low temperatures, and continuous rainy weather, and insufficient light, which can easily cause damping-off.

(2) Wilt. Young seedlings develop elliptical dark brown lesions on the stem. In the early stages, diseased seedlings wilt at noon and recover in the morning and evening. The diseased part gradually sinks, expands, and wraps around the stem, dries up, and the plant dies but remains upright. Seedlings with mild disease only form brown lesions at the base of the stem, grow poorly, but do not die. Wilt often occurs in seedbeds with higher temperatures or in the middle and later stages of seedling growth. In rainy weather, the humidity of the seedbed is high, which can cause seedlings to grow excessively, exacerbating the spread of wilt.

(3) Anthracnose. Anthracnose can occur throughout the entire growth period of melons. When young seedlings are affected, the disease mainly affects the cotyledons, with round or semi-circular lesions appearing on the leaf edges, which are brown and have black spots and pale red sticky substance on them. The pathogen relies on rainwater or irrigation water to spread, so the leaves near the soil surface are the first to be affected. High humidity is the main cause of disease, which can occur at temperatures between 10–30 ℃, but it is usually severe at 95% high humidity and a temperature of 24 ℃.

(4) Bacterial leaf blight. It is common in grafted seedlings with pumpkin rootstocks and mainly affects the leaves. Initially, water-soaked spots appear near the stomata on the leaf

edges, which later expand into light brown irregular lesions with a halo around them; severe cases form a V-shaped large brown spot. When there are many lesions, the entire leaf dies. The pathogen is a bacterium that likes a low-temperature and high-humidity environment; the optimal disease occurrence environment is at a temperature of 8–25 ℃ and a relative humidity of more than 95%.

040 How to comprehensively prevent and control diseases in the seedling stage of melons?

(1) Seedbed soil disinfection. Both wilt and damping-off in the seedling stage of melons are soil-borne diseases. Therefore, the nutrient soil should be disinfected 15–30 days before seedling cultivation.

(2) Cultivate robust seedlings. Insulation work should be done on the seedbed at night to prevent cold wind and low temperature attacks and avoid seedlings from freezing. During the day, ventilation should be strengthened to reduce the temperature and humidity of the seedbed, make the seedlings grow stronger, prevent excessive growth, and enhance disease resistance. Melon seedlings do not like moisture as much as cucumber seedlings, so pay attention to the frequency of watering. As long as the air humidity is not high, leaf diseases such as anthracnose will not generally occur. When the seedlings have about 3 true leaves, foliar fertilization can be carried out to enhance the disease resistance of the melon seedlings, and substances such as fulvic acid can be selected.

(3) Disease prevention during grafting seedlings healing period. It is important to prevent bacterial leaf blight during the grafting seedlings healing period. First, the air humidity should not exceed 95%, and second, the entire seedbed should be sprayed with a 600 times solution of 50% methamidophos wettable powder after grafting.

(4) Drug prevention and control. When the seedbed begins to sporadically appear wilt and damping-off seedlings, the diseased plants must be immediately pulled out along with the substrate or soil clumps. At the same time, select 70% iprodione 3 000 times+Bi Hu, 30% ether fungicide suspension 2 000 times+Bi Hu, 70% methyl tobutin wettable powder 800 times+Bi Hu for spraying, and pay attention to the rotation of the three drugs.

041 Can melons contract root-knot nematode disease? How to prevent and control it?

Root-knot nematode disease is a common disease in cucurbits, solanaceous fruits, and legume vegetables. The affected plants produce bead-like root knots on the root hairs and lateral roots or form swollen deformities of varying sizes, which are initially milky white and later form a light brown color. Crop rotation cultivation of melons will not cause root-knot nematode disease. Generally, it does not occur in the second year of heavy cropping, but it may occur in the third year, so prevention must be taken into consideration. The most effective prevention method is high-temperature fumigation during the summer fallow period.

(1) High-temperature fumigation prevention and control method. The suitable temperature range for the onset of root-knot nematodes is 15 – 35 ℃, and they grow and develop best at 25–30 ℃, which is most conducive to infection. They are rarely active below 10 ℃ or above 40 ℃. They die within 4 hours at 42 ℃ and within 10 minutes at 55 ℃. Therefore, fumigation can begin in early July.

The suitable soil relative humidity for root-knot nematodes is 40% – 70%. In dry or excessively wet soil, their activity is inhibited. Therefore, a combination of dry and wet fumigation methods is used. Firstly, base fertilizer should be applied to prepare the cultivation ridges, and the ground film should be covered. Then, the vents should be closed to seal the greenhouse, and it should be dry fumigated for 15 days. After the dry fumigation is finished, a large amount of water should be poured to ensure thorough watering, and then the greenhouse should be fumigated again for 15 days. If it is rainy and cloudy, the fumigation time should be appropriately extended to ensure effectiveness.

Please note that if you plan to apply biofertilizer for the next crop, you should apply the fertilizer and prepare the ridges after fumigation, otherwise, the high-temperature fumigation can harm the bio-organisms.

(2) Biopesticide prevention and control method. Avermectin, produced by the fermentation of Streptomyces avermitilis, is the main biological product for the prevention and control of root-knot nematodes. Avermectin can interfere with the neurological activities of nematodes, causing them to become paralyzed and die. Because it is easily degraded and does not accumulate in the environment, many commercial formulations with avermectin as the active ingredient have been developed and widely used in the prevention and control of root-knot nematodes.

The Vegetable Practice Base of Liaoning Vocational College applied avermectin combined with chitosan to control root-knot nematodes, achieving very good results. After chitosan enters the soil, it significantly promotes the proliferation of beneficial bacteria such as nitrogen-fixing bacteria, cellulose-decomposing bacteria, lactic acid bacteria, and actinomycetes, and inhibits harmful bacteria, thus enhancing resistance. Pure chitosan should be used. Liaoning Vocational College used Chitosan Feng produced by Dalian Dongyan Technology Development Co., Ltd. The specific method is: 1 000 times dilution of 1.8% avermectin and 500 times dilution of chitosan feng. At the time of planting, apply 500–1 000 g of the medicine solution to the root of each plant (depending on the size of the plant), and spray the leaves after planting. During the growth period, soil irrigation with the same concentration can be done once more.

According to research by the Institute of Plant Protection of the Chinese Academy of

Agricultural Sciences, the use of abamectin in combination with the Sr18 biocontrol agent can achieve a control effect of over 80%. This can reduce the amount of abamectin used, with 1 333.4 mL of abamectin per 667 square meters in combination with 6 666 mL of Sr18 per 667 square meters, the control effect is equivalent to using 2 000 mL of abamectin alone per 667 square meters.

(3) The Nematode Control Microbial Agent is a biological agricultural product made by Lvkang (Beijing) Biotechnology Co., Ltd. using the internationally leading nematode control technology results from the China Agricultural University. It uses efficient bacterial strains such as Bacillus and Actinomycetes and modern microbial fermentation technology. Its functional features include preventing and controlling nematode diseases, effectively improving soil, increasing soil organic matter content, relieving soil compaction, and increasing yield and efficiency.

After being applied to cucumbers in Saarqin Town, Donghe District, Baotou City, Inner Mongolia in 2017, the root knots of cucumber roots were significantly reduced and the yield increased by nearly 50%, with the seedling pulling period delayed by one month.

(4) Chemical drug prevention and control methods, such as using soil treatment agents like Mirelon, Weibai Mu, and cyanamide (drug fertilizer) for soil fumigation treatment, are also one of the effective methods to prevent root-knot nematodes. During the growth period, Bayer Lu Fuda can be used for flushing.

In addition, it is necessary to adopt a reasonable cultivation system. The rotation of melons and non-cucurbit vegetables, and the interplanting of amaranth and crown daisy during the gap between the two main crops in summer, can help reduce the incidence of root-knot nematodes.

042 How to prevent and control melon powdery mildew?

Melon powdery mildew is caused by the infection of the fungus Podosphaera xanthii in the subphylum Ascomycota. At the early stage of powdery mildew, white powdery spots appear on the leaf surface, which gradually enlarge and coalesce, finally covering the entire leaf, leading to leaf death and loss of photosynthetic function. The fungus prefers warm and humid environments, with suitable temperatures for disease occurrence between 10 and 35 ℃, and relative air humidity between 45% and 95%. The fungus has a very wide range of humidity adaptation, and can germinate with relative humidity above 25%. Therefore, as long as the fungus is present, powdery mildew can easily occur. The disease is most susceptible during the mature plant stage.

(1) Green prevention and control methods for powdery mildew.

Apply more farm fertilizers. Based on many years of production practice observations, melons are less likely to develop powdery mildew when they receive more farm fertilizers and the plants grow robustly.

Facility disinfection. Before greenhouse cultivation, 250 g of sulfur powder and 500 g of sawdust, or 250 g of 45% bactericidal fumigant, are used for each 100 m^3 space, lit at various points, and fumigated overnight with the space sealed.

High-temperature spraying with baking soda. Powdery mildew fungi are afraid of high temperatures and alkalis, so you can spray baking soda at high noon. From 11: 30 to 12: 00, spray a 500-fold solution of baking soda, with a large amount of water, so that the leaves are soaked in the liquid. After spraying, close the vents to raise the temperature to 35-40 ℃ for 45 minutes, then gradually open the vents to lower it to 30 ℃.

(2) Chemical prevention and control methods for powdery mildew.

Start spraying medication at the early stage of the disease, using a 40% difenoconazole emulsion at 6 000-8 000 times dilution, spray about once every 10 days, and spray 2-4 times consecutively, the comprehensive prevention effect can reach over 90%. Note that if the concentration of difenoconazole is too high, it will have a significant inhibitory effect on crop growth. It should be used strictly according to requirements, and the number of uses should not exceed 4 times in the same growing season to avoid the development of drug resistance and a decrease in drug efficacy.

In addition, imidazoles such as prochloraz, morpholines such as tridemorph, and methoxyl acrylic esters such as kresoxim-methyl and fenoxanil can be alternated with difenoconazole.

After melons are settled, it is best not to use the spraying method to prevent black spots from appearing on the melons, affecting their commercial quality. A bactericidal fumigant can be used to fumigate the shed. When the humidity is high, a dusting machine can be used to spray powder, using Bacillus subtilis as the agent, with a dosage of 100 g/667 m^2.

043 How to prevent and control bacterial angular leaf spot in melons?

Melons can be infected throughout the whole growth period, mainly harming the leaves, but also harming stems and fruits. The infection starts from the lower leaves and gradually develops upward. In the early stage of the disease, small water-soaked spots appear on the leaves, gradually enlarging and restricted by the leaf veins to form angular or irregular shapes. Sometimes yellow-white fluid oozes out of the back of the affected leaf, and the diseased leaves turn yellow-brown and dry in the later stage. The disease spots become brittle and easily crack and fall off. The disease spots on stems and fruits initially appear water-soaked and concave, and carry a large amount of milky white mucus. The disease spots on the fruit surface are prone to rot, the cracks extend inward to the seeds, causing the seeds to carry the bacteria.

The bacteria overwinter in diseased seeds or in diseased residual tissues and become the initial infection source for the second year. The bacteria prefer warm and humid conditions, with a disease occurrence temperature range of 4–40 ℃, and the most suitable temperature is 20–34 ℃. The relative humidity is over 95%, and it is prone to occur from the flowering and fruiting stage to the middle and late harvest stage. The size of the disease spots is related to the air humidity. If the saturation humidity at night is more than 6 hours, large disease spots will be produced. If the humidity at night is less than 85%, or the saturation humidity time is less than 3 hours, small disease spots will be produced.

The main prevention and control methods are:

(1) Seed disinfection treatment. Soak the seeds in 55–60 ℃ warm water for 15 minutes.

(2) Control humidity. Ventilate and exchange air in time, water should be lightly and frequently, and carried out in the sunny morning, and ventilate and dehumidify in time. On rainy days, as long as the temperature is not lower than 12 ℃, moderate ventilation and air exchange should be carried out.

(3) Clean the garden. Timely remove old diseased leaves, remove diseased residues after harvest, and bury or burn them deeply.

(4) Chemical prevention and control. If it spreads quickly, use imidazole quinoline copper+chlorbromuron+fluorescent pseudomonas to quickly control it. Another recommended drug is 20% thiophanate-zinc suspension, which is highly efficient and low toxic, with characteristics such as high activity, broad-spectrum bactericide, advanced formulation, safety for crops, and combined protection and systemic bactericidal treatment.

044 How to prevent and control bacterial fruit spot in melons?

Bacterial fruit spot in melons is a typical seed-transmitted bacterial disease, caused by Acidovorax citrulli, with characteristics of rapid onset, fast transmission speed, and strong outbreak. It has become one of the main diseases affecting melon production.

Seedlings can be infected, forming water-soaked spots on cotyledons, then extending to the base of the cotyledons, presenting strip-like or irregular dark green spots. In severe cases, they will develop into black-brown necrotic spots along the main vein. In mature plants, the spots on the leaves are round to polygonal, initially showing a V-shaped water-soaked edge, and then thinning in the middle, the spots dry up. The back of the spots overflows with white bacterial pus, after drying, it forms a thin layer and shines. In severe cases, multiple spots merge into large spots, the color deepens, mostly turning brown to black-brown. The stems and vines are damaged, often forming depressed spots and producing bacterial pus, leading to vine rot. When the air is dry, a white powdery substance forms on the affected parts of the stem and vine, which the farmers call water rot vine. In severe cases, the leaves wither, and the whole field is affected, appearing as if it is burned. The fruit is infected, often infecting young fruits less than 3 weeks old. Small water-soaked spots first appear on the upward-facing skin of the fruit, gradually turning brown and slightly sunken, and later often cracking and turning brown.

At the early stage of the disease, it is only limited to the fruit skin. After entering the mid-stage of the disease, the bacteria can extend to the fruit pulp alone or together with saprophytic bacteria, causing the fruit pulp to become water-soaked and rotten.

The pathogenic bacteria mainly overwinter on the surface of seeds and diseased residues in the field, becoming the initial infection source for the following year. The bacteria mainly invade through wounds and stomata. The disease is transmitted over long distances by the transportation of infected seeds, and both the seed surface and embryo can carry bacteria. The bacterial pus overflowing from the spots is spread and spread by wind and rain, insects, and farming operations, causing multiple re-infections.

The main prevention and control methods are:

(1) Seed disinfection. Use Tsunami100: water at a 1: 80 ratio to disinfect the seeds. The ratio of disinfectant to wet seeds is 2: 1. Pour the seeds into the container with the disinfectant and soak for 15 min. During soaking, constantly stir the seeds to ensure thorough disinfection. After the time is up, remove the seeds and rinse them twice with water and then spread them on a plastic mesh for drying, and turn them over manually from time to time.

Or, the seeds are sterilized with 70 ℃ constant temperature dry heat for 72 hours, or soaked in 55 ℃ warm water for 25 minutes, rinsed with clean water and dried, and germinated for sowing within 2 days; or soaked in 40% formalin 150 times for 1. 5 hours, cleaned and rinsed before germination and sowing; or treated with 200 times the dose of Bactericide No. 1 (patented by the Institute of Plant Protection, Chinese Academy of Agricultural Sciences) for 1 hour, thoroughly washed with water and then germinated for sowing. This has a good preventive effect on the fruit spot disease in the seedling stage.

(2) Agricultural prevention. Cultivate seedlings in disease-free soil to ensure that the seedlings are disease-free. Reasonable irrigation, preferably using drip irrigation or reducing water pressure to prevent flooding, pay attention to ventilation and humidity reduction, shorten the dew time on the plant surface, carry out farming operations after the dew is dry, and timely prevent and control pests.

If there are many diseases on the cotyledons, remove them in time. When the melon has two leaves and one heart, you can choose a sunny day to use disinfected scissors to cut off the cotyledons and take the cotyledons out of the shed for treatment. After harvesting, remove the diseased plants in time and take them out of the shed for treatment.

(3) Chemical prevention. Prevent and control with agents such as Galenon, Kasugamycin 3 000, Tianjin, Thiomycin, Tetramycin, and Allicin before or at the early stage of the disease. It should be noted that young melons are relatively sensitive to copper preparations, so be sure to reduce the amount of copper preparations to prevent phytotoxicity.

045 How to prevent and control vine wilt disease in melons?

In recent years, vine wilt disease has become a major disease of melons. Vine wilt disease can occur during the entire growth period of melons, but it usually happens after planting in production. Its level of harm is much higher than that of melon wilt disease and blight.

Vine wilt disease can harm leaves, stems, vines, fruit, etc. When leaves are infected, it often starts from the leaf edge, forming V-shaped black-brown necrotic spots. As the disease progresses, black spots grow on the spots, and the spots are prone to crack when the air is dry, sometimes ring patterns appear on the spots. Most stem and vine infections form water-soaked spots near the stem and vine nodes, which gradually turn gray-white and even light reddish-brown necrosis and crack with the progress of the disease, then produce amber gummy substance(cracking and gummy substance are typical symptoms of vine wilt disease), in the later stage, the stem and vine rot until the seedlings die. When the fruit is infected, water-soaked small spots appear on the surface, and then expand into round, dark brown sunken spots. On some varieties of fruit, the surface of the spots often cracks in a star shape, and the inside is cork-like dry rot.

The pathogen of melon vine wilt disease, the asexual type is the Fusarium of melon, and the sexual type belongs to the genus Ascochyta. The pathogen overwinters and survives the summer with mycelium, conidiophores, and ascomata in plant disease residues on the ground, in the soil and in unripe manure, or attached to the surface of the shed, and can also be transmitted through seeds. Under suitable conditions in the following year, the pathogen is spread by wind and rain, and irrigation water, and after the spores germinate, they invade from the host's young roots, stomata, hydathodes, and wounds, forming the initial infection source in the field.

The vine wilt pathogen likes warm and humid environments, and the development temperature range of the pathogen is 5-35 ℃, and the most suitable disease temperature is 20-30 ℃, and relative humidity above 85%. Humidity has a greater impact on the occurrence and prevalence of the disease than temperature. Within the suitable temperature range for disease occurrence, if there are many rainy days, large amounts of irrigation, and high humidity in the field, it is easy to cause an epidemic of the disease.

Prevention and control methods:

(1) Soil disinfection. Using the summer leisure period for high temperature steaming treatment in the greenhouse can effectively eliminate the pathogens in the greenhouse and reduce the number of pathogens.

(2) Seed disinfection. Producers cannot ensure that seeds are not carrying the pathogen, so seed disinfection before sowing is an essential step. You can soak the seeds in warm water,

or soak the seeds in 500–700 times liquid of 70% methyl thiophanate wettable powder, wash and then sow, which can effectively kill or inhibit various pathogens carried by the seeds.

(3) Strengthen field management. Apply sufficient mature organic fertilizer, increase the application of phosphorus and potassium fertilizer, cultivate robust plants, enhance the plant's disease resistance; the greenhouse should be ventilated and dehumidified in time, use under-membrane drip irrigation or dark irrigation, reduce the chance of membrane condensation, cultivate on high ridges, reasonably dense planting, increase plant permeability, close the ventilation port in time when it rains; timely remove diseased plants from the greenhouse, reduce the number of pathogens; choose sunny mornings for branch pruning.

(4) Rational rotation. In places with conditions, rotation with gramineous plants can effectively reduce the incidence of vine wilt disease. For example, in Changtu County, Liaoning Province, a melon-producing area, because there is much arable land, farmers build new greenhouses on the open field to grow melons every year.

(5) Chemical control. The vine wilt pathogen generally invades through wounds and cracks left after pruning, so it is necessary to spray medicine on the wounds and vines in time. At the early stage of the disease, choose 10% prochloraz water-dispersible granules 1 200 times liquid, or 25% azoxystrobin suspension 1 500 times liquid, or 22. 5% difenoconazole suspension 1 500 times liquid for spraying, spray once every 5 – 7 days, and continuous spraying for 3–4 times. For severely diseased stems and vines, use a brush dipped in 10% prochloraz water-dispersible granules 300 times liquid to apply on the diseased spots, especially promoting wound healing at the gumming area. Pyraclostrobin and methyl thiophanate have developed drug resistance and should not be used as the first choice of drugs.

046 How to prevent and control melon wilt disease?

The pathogen of melon wilt disease is Fusarium oxysporum, also known as black rot, which is one of the major diseases of melons and occurs throughout the country, especially in greenhouses where crops are continuously grown.

The wilt disease mainly affects the roots and stems. In the early stages of the disease, water-soaked dark green sunken spots appear, and then the base of the stem softens and shrinks, the diseased part is rough and longitudinally cracked, and amber gum-like substance often overflows from the surface. When it is humid, a pink mold layer often grows on the diseased part. When the diseased stem is longitudinally cut, the vascular bundles are discolored, impeding the normal water supply of the plant and causing wilting.

The pathogen can overwinter in the field and in uncomposted organic manure, and it can also overwinter on seeds and greenhouse frames. The pathogen is mainly spread by rainwater, irrigation water, and insects. When the environmental conditions are suitable, the pathogen invades from wounds on the roots, natural cracks, or root crowns, and can also enter from cracks at the base of the stem. After the germination of infected seeds, they can invade the seedlings. Greenhouse frames, farm tools, and underground pests can also spread the disease.

The pathogen prefers warm and humid environments. The disease occurs at temperatures ranging from 4-34 ℃, with the optimal disease occurrence temperature being 24-28 ℃. Temperatures above 35 ℃ can inhibit the occurrence of the disease.

Prevention and control methods:

Rotate with non-cucurbit vegetables to avoid planting cucurbit vegetables in successive crops. In greenhouses where wilt disease is severe, water and drought rotation can be implemented.

Apply fully composted organic manure to avoid fertilizer carrying the pathogen. Control the amount of nitrogen fertilizer and increase the application of phosphorus, potassium, and trace element fertilizers.

Dehumidify and increase the temperature to create unfavorable conditions for the pathogen. The temperature in the greenhouse can be increased to 35-38 ℃ at noon.

During transplanting and the early stage of disease, irrigate the roots 2-3 times with a 500 times dilution of Bacillus subtilis SN16-1.

047 How to prevent and control melon downy mildew?

The pathogen of melon downy mildew is Pseudoperonospora cubensis. The disease occurs throughout the country, with a more severe occurrence in summer and autumn open-field cultivation. It is also prone to occur in greenhouse cultivation with high air humidity.

The disease mainly affects leaves and can occur from seedling stage to mature plant stage. When leaves are infected, they first produce water-soaked spots on the leaves, which turn pale brown as the spots expand. In humid conditions, a gray-purple mold layer is produced on the back of the leaf spots. The disease spreads from the lower leaves to the upper ones, initially producing water-soaked spots, which then become irregular pale yellow spots restricted by the leaf veins. Several spots can merge into small patches. In severe cases, the spots join together, and the entire leaf turns yellowish-brown and withers, curling up.

The pathogen prefers warm and high humidity conditions. The suitable temperature range for disease occurrence is 10-30 ℃, with an optimal daily average temperature of 15-22 ℃, and a relative humidity of 90%-100%.

If there are cucumbers near the melon field, the downy mildew on the cucumbers can infect the melons. Therefore, in terms of spatial layout, greenhouse melons should be kept away from greenhouse cucumbers, and open-field melons should be kept away from greenhouse and field-grown cucumbers and melons.

On the field level, various measures should be considered to reduce field humidity and create ecological conditions unfavorable for pathogen infection, such as trellis cultivation. Avoid excessive fertilization to prevent overgrowth and overcrowding of melon seedlings. Water moderately and increase ventilation. If the film of the facility is damaged, it should be repaired in time to prevent rain leakage.

When the central diseased plants are found, they should be treated promptly. Options include using a 72% mancozeb, a 69% acylamine manganese zinc, or a 60% fluorine manganese zinc wettable powder at 600 times dilution, or a 10% win green 2 000 times liquid, evenly spraying 60 kg of solution per 667 m^2, spraying once every 5-7 days, and treating 2-6 times depending on the disease situation.

When pulling out melon seedlings, burn or deeply bury the diseased residues to reduce the source of reinfection.

048 How to prevent and control melon root rot disease?

Root rot is caused by the infection of the fungus Fusarium solani cucurbitae, which is a common disease in melons. It is prone to occur in heavy clay soil, during the seedling stage, and before the seedlings are established after transplanting. The disease can also occur if uncomposted manure is used, or for other reasons that cause root damage, and the pathogen invades through the wounds. High temperature and high humidity are favorable for the disease occurrence.

The disease mainly infects the roots and root neck. Initially, the diseased part appears as water-soaked spots, which later turn yellowish-brown and rot. When the humidity is high, a small amount of white mold is produced on the surface of the diseased root neck, the vascular bundles turn brown, but they don't develop upward, which distinguishes it from wilt disease. As the disease progresses, the diseased part slightly shrinks and thins, and the leaves of the plant or seedling gradually turn yellow from the bottom up, wilt, and eventually die. Finally, the diseased part completely rots, leaving only filamentous vascular bundle tissues.

Prevention and control methods:

Rotate with non-cucurbit vegetables for more than 3 years.

The manure and base fertilizer used for seedling cultivation should be fully composted. Use high ridge cultivation to prevent flooding.

For chemical prevention and control, spray 50% carbendazim 500 times solution, or 45% propineb suspension 1 000 times solution, or 25% difenoconazole emulsifiable concentrate 1 500 times solution at the early stage of the disease.

049 How to prevent and control melon anthracnose?

Anthracnose, caused by the fungus Colletotrichum lagenarium, can affect the leaves, vines and melons, and can cause widespread death in severe cases. It is one of the main diseases causing melon rot.

When a leaf is infected, it initially develops small water-soaked spots. These spots enlarge to form round or irregular light brown lesions with reddish-brown edges. When wet, a pink, sticky substance is produced on the leaf surface.

When the vine and petiole are infected, they first develop yellow-brown, slightly sunken streaks. As the disease progresses, the lesions surround the vine and petiole, causing everything above the infected area to die.

When melons are affected, they initially appear water-soaked. This later develops into a yellow-brown, slightly sunken oval lesion. When wet, a pink, sticky substance is produced on the surface, and the infected area eventually cracks.

Anthracnose develops most rapidly at temperatures of 22-27 ℃ and a relative humidity of around 90%. The fungus is spread by rainwater and flowing water, and can also be carried by seeds.

Prevention and control methods include the following:

Crop rotation. In heavily infected areas, rotate with non-melon crops for more than 3 years.

Seed treatment. Treat seeds with hot water soaking method.

Strengthen management and adopt a formula fertilization method. Apply fully decomposed organic fertilizer, level the land to prevent water accumulation, drain water promptly after rain, plant reasonably, and remove weeds in a timely manner. During the disease period, remove diseased melons, avoid water accumulation in the field, increase ventilation in the shelter, reduce air humidity as much as possible, and control the disease.

Chemical control. Fungicides such as carbendazim, polyoxin, metyltobuzin, mancozeb, and anthrax magnesium can be used. Spray once every 7-10 days, and can be sprayed 2-3 times consecutively. The best effect is achieved by alternating the above drugs.

050 How to prevent and control aphids?

The aphids that occur on melons are mainly peach aphids and melon aphids. In the season when the greenhouse temperature is high, aphids can reproduce a generation every 4 to 5 days, so it is advisable to catch them early in prevention and control, and eliminate them when they first appear in spots or patches. At this time, chemical pesticides are not needed for prevention and control. If the treatment is late and the greenhouse is full of aphids, it is not easy to treat them with chemical pesticides. Therefore, field management should pay attention to melon leaves. If they curl downward or find ants crawling, they should check for aphids. If found early, the aphids haven't spread yet, and can be controlled by manual crushing or washing urine mixture method.

(1) Plant lure aphid method. In places where melons cannot be planted on the edge of the plastic greenhouse, some rapeseed and pakchoi are planted to attract aphids, then the aphid-infected vegetables are pulled out and processed, which can avoid or alleviate the damage of melons by aphids.

(2) Manual crushing method. At the initial stage, there are few plants with aphids, which can be crushed by hand. The central plant must be searched leaf by leaf, and a sign must be made for the damaged plants, and crushed continuously for 3 days, because the aphids are small and there will be omissions, at least for 2 consecutive days.

(3) Wash urine mixture prevention and control method. The ratio of laundry detergent, urea, and water is generally controlled at 1: 4: 100. It must be ensured that the laundry detergent is dissolved during preparation. Aphids are generally concentrated on the back of the leaves and the tips. They must be sprayed evenly and thoroughly. The central plants at the initial stage are best sprayed with a handheld atomizer leaf by leaf, so that the insect body can be sprayed to form a liquid film, in order to achieve the prevention and control effect, and spray again every 2 days to kill the aphids that were not sprayed last time.

051 How to identify and control thrips in cucurbits?

Thrips in cucurbits belong to the Thripidae family of Terebrantia order, also known as palm thrips, brown-yellow thrips, and melon thrips. It harms a variety of vegetables such as cucurbits, solanaceae, and legumes, and is a common thrips pest on greenhouse vegetables.

Adult thrips are about 1 mm long and yellow. The eggs are elongated oval, 0. 2 mm long, and pale yellow. There are four instars of nymphs, which are white or pale yellow.

In the north, adult thrips overwinter in greenhouses. Adults are attracted to the yellow and tender green parts of plants, they move quickly, are good at jumping, and are afraid of strong light.

Adults and nymphs rasp and suck the sap of the heart leaves, buds, and young cucumbers, causing the heart leaves to not open normally, the growth point to shrink and turn black, and cluster growth to occur. The leaves show gray-white elongated spots. The fuzz on the young cucumbers damaged turns black and becomes deformed, and the mature cucumbers have rough skin, yellow-brown spots, or are covered with rust.

Prevention and control methods:

Agricultural prevention and control. Timely removal of weeds and strengthened management to promote vigorous plant growth can reduce damage.

Color board trapping. Set up yellow boards containing plant-derived attractants in the field, one per 20–30 square meters, to lure and kill adult thrips.

Chemical prevention and control. Regular inspections are necessary, and when 2–3 thrips are found on the heart leaves, use medicine for prevention and control, spray once every 7–15 days. The first choice is ethyl thiram suspension (Ailvshi) with high biological activity, wider insecticidal spectrum, better quick-acting, longer-lasting effect, and no phytotoxicity, then alternate with spinetoram suspension (Muwangte), cypermethrin suspension (Caixi), and dinotefuran water-dispersible granules (Gengmeng) to prevent drug resistance. The timing of medication is also important, and if the timing is not right, the effect is not good. Since thrips are not active in strong light, it is advisable to apply medicine before 9 am and after 4 pm. Thrips are sensitive to organosilicon, so organosilicon can be added when preparing the medicine, focusing on spraying the back of the leaves.

052 How to control greenhouse whitefly?

The greenhouse whitefly is a species in the Aleyrodidae family, commonly known as the little white moth. It can be found across the country, especially in the protected areas of the northern region. Its main hosts are various vegetables such as cucumbers, solanaceous fruit, beans, and also flowers, with over 900 types of hosts in total.

The adult whitefly is 1.0–1.5 mm in length, pale yellow, with white wax powder covering the wings. The eggs are oval and 0.22–0.26 mm long. They are light green when first laid and turn black before hatching. The nymphs go through 4 instars, are long and oval, and the mature nymphs are about 0.5 mm long, pale yellow or yellow-green, and semi-transparent.

Whiteflies cannot overwinter in the open fields of the north, but survive the winter in various stages in greenhouses or continue to cause damage. Under the conditions of northern facility production, they occur in more than 10 generations a year, with overlapping generations.

Adults have a strong attraction to the color yellow and prefer tender leaves for egg-laying.

With piercing and sucking mouthparts, both adults and nymphs cluster on the back of leaves to suck sap, causing the leaves to fade, turn yellow, wilt, and even causing the entire plant to die. They secrete honeydew, which seriously contaminates the leaves and fruit, often leading to the occurrence of sooty mold disease, making sweet melons lose their commercial value. In addition, they also transmit viral diseases.

Control methods:

Agricultural control. Before using the nursery and production greenhouses, clean up the weeds and remains of previous crops and thoroughly fumigate. Pruning and removing the leaves and vines can prevent and control the pest.

Crop rotation. For seriously infested facilities, before planting greenhouse melon in spring, avoid planting solanaceous fruit and bean vegetables in the previous crop, choose crops that are not suitable for the occurrence and damage of whiteflies, such as celery that is resistant to low temperature, to cut off the food chain supply of host crops and reduce the base number of overwintering pests.

Yellow board to lure adults. In the early stages of infestation, hang yellow boards containing plant-derived attractants for whiteflies in the field, 1 board per 20 - 30 m^2, to attract and kill adults.

Biological control. In the early stages of whitefly infestation in the facility, release Aphelinus abdominalis at a ratio of 1: (2 - 4) to whiteflies, once every 7 - 10 days, continuously for 3 times, can effectively control the early damage of whiteflies.

Chemical control. Use pesticides in a timely manner during the early stage of spotting" infestation. Spray the pesticide when the morning dewf is not dry, which makes it easier to spray onto the whiteflies. Pesticides that can be selected include imidacloprid emulsion, fenpropathrin emulsion, bifenthrin emulsion, and imidacloprid wettable powder.

Chapter 5 Q&A on Practical Techniques for Vegetable Disease and Pest Control

Manual of Horticultural Techniques

园艺技术手册

001 What is the difference between cucumber downy mildew and cucumber gray mold?

Cucumber downy mildew occurs from seedling to mature plant stage, with local infection mainly damaging the leaves. During the seedling stage, the front of the cotyledons produce light yellow or yellow-brown spots, which produce gray-brown mold when damp and then dry out. During the mature plant stage, the disease usually starts from the southern end of the greenhouse, and the plant develops upwards from the middle. In the early stage of the disease, small spots the size of needle tips appear on the back of the leaves, which are more visible in the morning, and the front of the leaves is not obvious. Later, the spots gradually enlarge, the front begins to show symptoms, the spots turn yellow-brown, and the range is limited by the veins to form polygons. Under humid conditions, a purple-black or gray-brown sparse mold layer appears on the back of the spots. When severe, the diseased leaves curl upwards, the spots merge into one, and the leaves dry out. Under high temperature and dry conditions, the spots stop developing and dry out, and no mold layer is produced on the back of the leaves(Figure 1).

Figure 1 Cucumber Downy Mildew

Cucumber gray mold mainly harms petals, young cucumbers, leaves and stems. The pathogen often invades from decayed female flowers, causing the petals to turn brown and wilt, and in severe cases rot, producing a gray-brown mold layer. It extends to the young cucumber, forming a watery spot at the navel, and the head or entire cucumber rots. Rotten cucumbers, rotten flowers and leaves and stems come into contact, causing the leaves and stems to get sick. Generally, leaf spots start from the leaf tips, initially becoming water-soaked, and then light gray-brown. Large, obvious ring spots often form on the leaves, with yellow-brown edges. When the disease is severe, the lower part of the stem becomes water-soaked and softens, then turns yellow-brown and constricts. Under high humidity, a thick gray mold layer grows densely, and when the spots wrap around the stem, the entire plant dies(Figure 2).

Figure 2 Cucumber Gray Mold

002 How to prevent and control cucumber downy mildew?

(1) Choose disease-resistant varieties.

(2) Adopt high mound film mulching and drip irrigation cultivation to increase soil temperature and reduce humidity.

(3) Regulate the temperature and humidity in the greenhouse. During the fruiting period, you can use the method of ventilating in the early morning to reduce humidity, closing the greenhouse at sunrise to increase temperature, and the temperature can rise by 6-7 ℃ per hour, and the relative humidity drops by about 20%. When the temperature reaches 28 ℃, ventilate slightly, and when it reaches 32 ℃, ventilate greatly, so that the highest temperature does not exceed 35 ℃. Ventilate greatly in the afternoon, control the temperature at 20-25 ℃, relative humidity 60%-70%, and control the temperature at 13 ℃ at midnight. If the minimum temperature at night is above 12 ℃, you can ventilate all night. You also need to ventilate on rainy days, and be careful to prevent rain splashing. Watering should be done on sunny mornings. After watering, close the greenhouse to increase the temperature to 38-40 ℃, and ventilate after 1 hour.

(4) Timely remove diseased leaves, control watering, and ensure that the condensation time in the greenhouse does not exceed 2 hours. Release carbon dioxide in the morning, close the stomata of the cucumber, and prevent the pathogen from invading.

(5) Drug prevention and control.

Spray in the afternoon on sunny days. You can choose 52. 5% inhibitory net water dispersion granules 1 500 times liquid, or 72% clove wettable powder 600-800 times liquid+100 grams of sugar+50 grams of urea, spray once every 7 days, continuous prevention and control 3–4 times.

On rainy days in the greenhouse, use the fumigation method, and close the greenhouse in the evening. Use 200–250 grams of 45% hundred fungicides per mu, evenly distributed in 4–5 places under the back wall of the greenhouse, and fumigated overnight after being ignited by a dark fire, ventilated and ventilated the next morning, and fumigated once every 7 days. Pay attention to the placement position should not be under the seedlings.

In the early stage of the disease, you can also use the dusting method. In the evening on rainy days, use a duster to spray 5% hundred fungicides or 7% anti-mold spirit powder, use 1 kg per mu each time, and apply it once every 9–10 days.

(6) Rotate medication. Pay attention to rotating drugs to prevent the pathogen from developing drug resistance.

003 How to prevent and control cucumber gray mold?

(1) Strengthen cultivation management. Adopt measures such as high mound film mulching, drip irrigation, and increasing ventilation to increase greenhouse temperature and reduce humidity.

(2) Seed treatment. Soak seeds in 55 ℃ constant temperature water for 20–30 min, then germinate and sow.

(3) Handle disease residues. In the early stage of the disease, promptly remove diseased flowers, diseased fruits, diseased leaves or pull out diseased plants, bring them out of the greenhouse in plastic bags, bury them deeply or burn them to reduce the spread of the pathogen.

(4) Regulate the temperature and humidity in the greenhouse. The indoor temperature is raised to 31 – 33 ℃, and ventilation starts when it exceeds 33 ℃. The temperature is maintained at 20–25 ℃ in the afternoon, and the vents are closed when it drops to 20 ℃. The night temperature is kept at 15–17 ℃. If the greenhouse temperature exceeds 30 ℃ for 2–3 hours a day, it can effectively inhibit the growth and spread of the pathogen.

(5) Set up a carbon dioxide generator. Release carbon dioxide at regular intervals in the morning to supplement the lack of carbon dioxide in the greenhouse. Create a high-temperature and relatively low-humidity ecological environment to inhibit the growth and spread of the pathogen.

(6) Rotation. Rotate for 2 – 3 years with non-cucumber vegetables (such as leafy vegetables) on plots with heavy disease.

(7) Drug prevention and control.

Spray before planting with 50% speed spirit wettable powder 1 000–1 500 times liquid or 50% push sea wettable powder 1 000 times liquid. In the early stage of the disease, use 70% hundred fungicide wettable powder 600 times liquid, 40% methomyl 600–800 times liquid,

50% methyl tobutin 500 times liquid to spray on sunny mornings. Ventilate in time after spraying to lower humidity.

Use 10% speed spirit smoke, 45% hundred fungicide smoke, 15% gram spirit smoke, 200-250 grams of medicine per mu for fumigation.

004 What are the symptoms of cucumber black star disease?

This disease can infect plants from the seedling stage to the late growth stage, with young tissues being more susceptible. After the cotyledons of seedlings are infected, nearly circular yellow-white spots appear, the whole leaf wilts, and the plant stops growing. Half-expanded new leaves produce nearly circular pale yellow or dark green spots, which later develop star-shaped perforations. The perforations have a yellow halo around the edge, which can distinguish them from the perforations of angular spot disease and anthracnose. The vein tissues necrotize, causing the leaf tissues around the disease to wrinkle. After the stem is damaged, green water-stain-like stripes along the stem axis and depressed cracks with gum exudation appear. In severe cases, the diseased area turns brown and rots. The growth point is infected for 2-3 days, the top rots and stops growing, forming a bald seedling, and lateral buds sprout below. When the fruit is infected, brown scab-like depressions appear on the disease area, semi-transparent gum-like substance oozes out, turns amber, the tissue stops growing, causing cucumber deformity. When the humidity is high, a gray mold layer grows on the surface of the disease area (Figure 3).

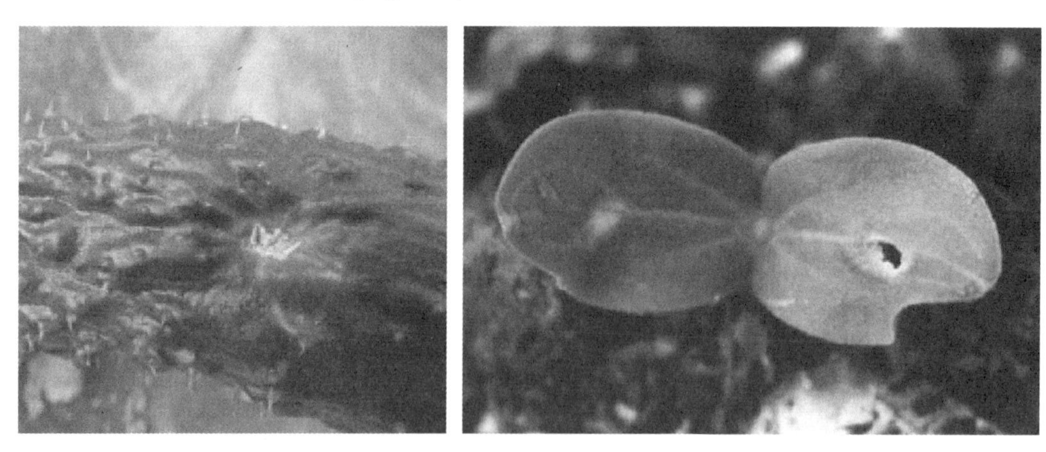

Figure 3 Cucumber Black Star Disease

005 How to prevent and control cucumber black star disease?

(1) Strengthen plant quarantine. Strictly implement origin quarantine and transportation

quarantine when transporting seeds to avoid the introduction of diseased seeds.

(2) Select disease resistant varieties. Such as Jinchun No. 1, Jinyou No. 31, Jinyou No. 32, Zhongnong No. 7, Zhongnong No. 13, Nongda No. 14, Changchun Dense Thorn, etc.

(3) Seed treatment. Soak the seeds in constant temperature water at 55 ℃ for 15 minutes, then germinate and sow. Mix the seeds with 0.3%–0.4% of 50% carbendazim, the medicinal powder evenly adheres to the seed surface, sow the seeds on the same day or the next day after mixing.

(4) Fumigate with sulfur powder. Use high-raised bed mulching cultivation, after plowing before greenhouse planting or seedling raising, mix 2.5 kg of sulfur powder with an equal amount of dry sawdust per mu, ignite for fumigation for 24 hours, and sow or plant the seedlings 7–10 days later.

(5) Regulate the temperature and humidity in the greenhouse. Do not flood during the production period. Close all vents every morning, let the temperature rise to 33–35 ℃ before suddenly opening the top vents for ventilation and dehumidification. When the temperature drops to 30 ℃, close the vents again, ventilate again when it exceeds 33 ℃, and perform high-temperature ventilation 2–3 times every morning. In the afternoon, ventilate to lower the temperature to 25–28 ℃, which can greatly reduce the humidity, especially reduce the night humidity, which is conducive to high yield, good control of humidity, and reduce the night dew time.

(6) Clean up the fields. After harvesting, clean up the frame materials and disease tendrils, destroy the diseased plants or bury them deeply.

(7) Chemical prevention and control. In the early stage of the disease, spray with 50% carbendazim wettable powder 600 times liquid for prevention and control, or use fumigation No. 3 for fumigation at night once; in the middle and late stages of the disease, you can choose 40% Xinxing wettable powder 8 000 times liquid, 40% Xinxing 10 000 times liquid, 50% Xingguang 600 times liquid, and 50% Puhaiyin wettable powder 1 000 times liquid for foliar spray, use medicine once every 5–7 days, and continue spraying 2–3 times depending on the disease. Rotate the medicines, strictly control the concentration of the medicine solution to prevent drug damage.

006 What are the symptoms of cucumber bacterial angular leaf spot disease?

Cucumber bacterial angular leaf spot disease mainly harms the leaves, leaf stalks, tendrils, and fruits. It can harm both seedlings and mature plants, with the greatest damage to the leaves of mature plants. In the seedling stage, it appears on the cotyledons producing circular, water-soaked, slightly sunken spots, which then enlarge and turn yellow-brown. In the mature plant stage, there are fresh green water-soaked small spots on the leaves, which then spread around, forming multi-angular gray-brown or yellow-brown spots due to the limitation of leaf veins. The back of the leaf spot has an oily appearance, and later in dry conditions, the center of the spot gradually dries up and finally falls off forming a hole. When the air is humid, the spot produces bead-like white bacterial pus, which dries up into a white film. The spots on the stem and leaf stalk are also water-soaked, almost circular, then extend longitudinally along the stem groove into a short stripe, with white bacterial pus on the surface, leaving white traces after drying. The fruit produces oil-soaked dark-colored sunken spots, which extend outward in irregular or contiguous forms, rot after exuding white bacterial pus, and emit a foul smell(Figure 4).

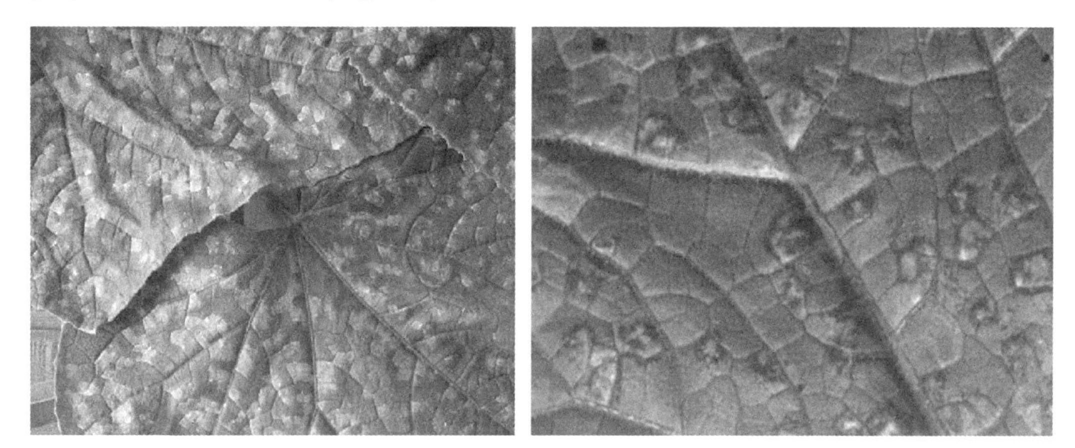

Figure 4 Cucumber Bacterial Angular Leaf Spot Disease

007 How to prevent and control cucumber bacterial angular leaf spot disease?

(1) Strengthen quarantine. Control the transportation of seeds carrying bacteria and carry out seed quarantine.

(2) Seed treatment. Soak the seeds in 55 ℃ water for 20 minutes, or soak them in new plant mold for 1 hour, then soak in clear water for 3 hours and germinate and sow.

(3) Seedling soil treatment. Mix the soil used for planting onions, garlic or leeks and fully decomposed organic manure at a ratio of 6∶4, sieve it, add 2 kg/m^3 of diammonium phosphate, 5 kg/m^3 of wood ash, 100 g/m^3 of 50% carbendazim, 100 mL/m^3 of 40% phoxim, and mix evenly.

(4) Regulate the temperature and humidity in the greenhouse(the same method as the prevention and control of cucumber downy mildew). The greenhouse should use a non-drip film shed to reduce the dew time on the cucumber leaves as much as possible. Watering should use the under-film dark watering technology, water less to reduce humidity, reduce the

dew time in the shed, and control the disease.

(5) Chemical prevention and control. In the early stage of the disease, spray with 1 000 times liquid of agricultural streptomycin sulfate wettable powder, or 1 000 grams/mu of 50% bactericidal clear powder, which has a high prevention effect. If it occurs simultaneously with downy mildew, use 250 times liquid of 70% metalaxyl copper or 600 times liquid of 50% Ridomil copper wettable powder for prevention and control, and continue for 2-3 times.

008 What is the difference in symptoms between cucumber wilt disease and cucumber vine wilt disease?

Cucumber wilt disease can occur at any stage of cucumber growth, and the typical symptom is wilt of the plant. When young seedlings are infected, the cotyledons wilt or the stem base becomes brown and constricted, causing the whole plant to wilt and collapse. When the disease is more serious after flowering, the lower leaves turn green in the early stage, with a network of bright yellow stripes along the veins, wilting at noon and recovering in the morning and evening, gradually developing to the upper leaves, until the whole leaf turns yellow, wilts, and the whole plant dies. The roots are brown and decayed, the base of the stem is constricted, sometimes longitudinally split, often with gelatinous exudations, pink moldy objects appear in damp conditions, and brown vascular bundles can be seen when the lower stem is cut open, which can be distinguished from other wilt diseases such as blight, vine wilt, and sclerotinia(Figure 5).

Figure 5 Cucumber Wilt Disease

Cucumber vine wilt disease mainly damages the leaves and vines of cucumbers. The disease spots on the vine often occur at the base of the stem. In the early stage of the disease, oil spot-like elliptical small spots are formed. Later, the disease spots expand to a grayish-white color, sometimes exuding amber-colored resinous gelatinous material, and when severe, the diseased parts rot and the stems and leaves above the diseased parts wilt. In the later stage, the diseased stems shrink, the epidermis longitudinally splits like a mess, exposing the vascular bundles. The leaves often start from the leaf edge and form a yellow-brown "V" or semi-circular slightly ringed disease spot inward, and small black spots are scattered in the later stage, which are the conidial bodies of the disease. Sometimes, the disease can also invade from the petals and stigma, causing fruit yellowing and shrinkage (Figure 6). The difference between this disease and wilt disease is that the vascular bundles of the diseased stems do not change color, and the difference between anthracnose and blight is that small black spots can be produced on the disease spots.

Figure 6　Cucumber Vine Wilt Disease

009　How to prevent and control cucumber wilt disease?

(1) Select disease resistant varieties. Cucumber wilt disease is a common soil-borne disease, so when planting, you must choose varieties with strong disease resistance, which can greatly reduce the probability of disease.

(2) Seed treatment.

Soak the seeds in 55 ℃ warm water for 15min, and continue stirring, then germinate and sow.

Soak the seeds in 500 times dilution of 50% carbendazim wettable powder for 1 hour, rinse with clean water, then germinate and sow.

(3) Grafting with black seed pumpkin as rootstock. When the cucumber seedlings are 7–10 cm high and the first true leaf is the size of a coin, and the pumpkin seedlings have unfolded their cotyledons and the true leaves are about 5 mm long, it is the suitable period for grafting. Currently, side grafting method is mostly used. During the operation, avoid opening the shed film at will to prevent the humidity inside the shed from decreasing. Water-filled basins should be set up in the shed or trenches should be opened outside the seedbed to water, in order to keep the humidity in the shed above 95%. The temperature in the shed should be maintained at 25 – 28 ℃. If the temperature in the shed is too high, the temperature can be lowered by shading with a shading net. During grafting, prepare tools and operation tables. Specifically, first use a specially made bamboo stick knife to pick off the growth point of the pumpkin seedlings, then use a blade to cut diagonally from 0.5 cm below the cotyledon node, with a cut length of 0.6 cm, and the cut depth reaches 1/2 of the root neck. The cucumber seedlings are cut diagonally from 2 cm below the cotyledon node, with a cut length of 0.6 cm, and the cut depth reaches 3/5–2/3 of the root neck (note that the cut depth of the cucumber seedlings must exceed 1/2 of the root diameter). After the two cucumber seedlings are cut, hang the cut of the cucumber seedlings on the cut of the pumpkin seedlings, align them, press the true leaves of the cucumber on top of the two cotyledons of the pumpkin, and clamp them firmly with a grafting clip. When clamping, make sure that the cucumber seedlings are on the inside and the pumpkin seedlings are on the outside, so that the roots can be cut off. During the grafting operation, pay attention that the blade should be clean and not contaminated with mud and water to avoid bacterial infection; the operation should be careful, and the hand strength should be light, and the cucumber seedlings should not be damaged, causing tissue necrosis of the cucumber seedlings.

(4) Chemical control.

Soil disinfection. For continuous cucumbers with slight disease, before planting, use methamidophos or carbendazim and other drugs to prepare a 1: 100 ratio of medicated soil, with a dosage of 1–1.5 kg per mu, spread in the planting ditch, or use 50% carbendazim wettable powder 4–5 kg per mu when planting, mixed into the planting hole.

After the disease occurs, use methamidophos, carbendazim, apron and other pesticides 400 times liquid to irrigate the roots, once every 8–10 days, which can control the spread. In the early stage of wilt disease, use 500 times liquid of 50% carbendazim wettable powder, 800 times liquid of 70% methamidophos wettable powder to irrigate the roots, once every 7–10 days, continuously irrigate 2–3 times.

010 How to prevent and control cucumber vine wilt disease?

(1) Seed treatment.

Soak the seeds in 55 ℃ warm water for 15 min, and continue stirring, then germinate and sow. Save seeds from disease-free plants.

(2) Strengthen cultivation management. Adopt high ridge mulching drip irrigation cultivation to reduce humidity and control relative humidity below 80%. Also, use temperature and humidity control technology in combination with the prevention and control of downy mildew.

(3) Remove disease residues. In the early stage of the disease, thoroughly remove diseased leaves and vines and bury them deep outside the shed.

(4) Chemical control.

Spray. In the early stage of the disease, use 600 times liquid of 75% chlorothalonil wettable powder, 500 times liquid of 65% mancozeb wettable powder, 500 times liquid of 50% methamidophos wettable powder, 500 times liquid of 50% carbendazim wettable powder to spray. Spray once every 6–7 days, and spray 2–3 times in a row.

Fumigation. Use 45% chlorothalonil smoke agent, 250–300 grams per mu, placed in 5–6 places, lit from the inside to the outside, seal the shed overnight. Fumigate once every 6 days or so, and fumigate 2–3 times in a row. It is best to alternate fumigation and spraying.

Disease prevention. After cucumber planting and seedling recovery, spray the ground around the plant with 600 times liquid of 30% metalaxyl or 800 times liquid of 38% metalaxyl copper bacteria ester to effectively prevent the occurrence of vine wilt disease.

011 What causes the new leaves of cucumber to turn yellow, and in severe cases, the upper leaves turn yellow-white? How to prevent and treat it?

The new leaves of cucumber turn yellow, and in severe cases, the upper leaves turn yellow-white(Figure 7). This is a symptom of iron deficiency in cucumbers. The soil pH for cuc umber cultivation should be between 6 and 6. 5 to prevent the soil from becoming alkaline. Water management should be monitored to prevent excessive dryness or wetness. An emergency measure is to spray the leaves with a 0. 1% to 0. 5% solution of ferrous sulfate or citric acid iron in 100 mL/L of water. Alternatively, apply a 50 mL/L solution of chelated iron to the soil at 100mL per plant.

Figure 7 Cucumber Iron Deficiency

012 What causes cucumbers to grow into "big head" cucumbers? How to prevent and treat it?

The end of the fruit, close to the part where the flower falls off, swells, while the middle is particularly thin, mostly turning into curved fruit. This is caused by the following three reasons:

(1) The female flower of cucumber is not fully fertilized, and seeds are only formed at the tip of the cucumber, resulting in the accumulation of nutrients at the tip. This leads to an especially large fruit flesh tissue at the tip, forming a "big belly" cucumber head.

(2) The plant lacks potassium, and nitrogen fertilizer is supplied excessively.

(3) Cucumber lacks water during the early growth period, which slows down cell growth. However, a large amount of water supply in the later period leads to rapid cell development and can result in a "big belly" cucumber(Figure 8).

Figure 8 Cucumber "Big Head"

How to prevent and control the "big head" disease?

(1) Prevention is the primary goal. During the growth process of the plant, enhance the photosynthetic assimilation function of the leaves, do not forcefully remove leaves, pay attention to the light conditions of the lower leaves, and maintain the growth vigor of the plant.

(2) Balanced water supply. Do not lack water in the early stage due to growth control, and do not supply too much water in the later stage due to rapid growth.

(3) Reasonable temperature control. The temperature should not be too high, with 25–28 ℃ during the day and 13–15 ℃ at night.

(4) Ensure nutrient supply.

Spray an appropriate amount of trace fertilizer in time. Lack of boron can easily cause poor development of young cucumbers, "waist" cucumbers, and "big head" cucumbers. An appropriate amount of boron fertilizer should be supplemented during the management process. Boron can be applied using a 800-fold dilution of Boron, sprayed once every 7–10 days, for 2–3 consecutive times, which can effectively alleviate the problem.

The deformed cucumbers that have already appeared should be picked early, so that nutrients are concentrated on the growth and development of normal young cucumbers.

(5) Prevent insects such as bees from flying in.

013 What causes the leaves of cucumber to turn dark green in the early stages of growth, and then develop brown spots in the later stages?

During the seedling stage, the leaves are dark green, hardened, dwarfed, and small. The leaves slightly stick up, and after planting in the open field, growth stops and fruit ripening is delayed. Because of low temperatures, even if there is sufficient phosphorus in the soil, it is difficult to absorb phosphorus, and phosphorus deficiency can easily occur. In the early stages of growth, the leaf color is dark green, and brown spots appear in the later stages. The normal phosphorus content in leaves is between 0.2% and 0.4%, and if it is below 0.2%, it indicates phosphorus deficiency(Figure 9).

Figure 9 Cucumber Phosphorus Deficiency

When the total phosphorus content in the soil is below 30mg/100 g soil, in addition to applying phosphorus fertilizer, the soil should be improved in advance and fused phosphorus fertilizer should be applied. When the total phosphorus content in the soil is below 150 mg/100 g soil, the effect of applying phosphorus fertilizer is good, and sufficient phosphorus fertilizer and organic fertilizer should be applied during the seedling stage. Cucumber seedlings especially need phosphorus, and for seedling nutrient soil, an average of 1 000–1 500 mg of pentoxide phosphorus can be applied per liter.

014 What causes the edges and veins of cucumber leaves to turn yellow and the cucumber to not grow large? How to solve it?

In the early stages of cucumber growth, the leaf edges show slight yellowing, first at the leaf edges, then between the veins. In the middle and later stages of growth, the middle leaves show the same symptoms. As the leaves continue to grow and curl outward, the leaves slightly harden, the cucumber vines are a little short, the enlargement is poor, and potassium deficiency symptoms appear(Figure 10).

Figure 10 Cucumber Potassium Deficiency

Ensure sufficient potassium fertilizer supply, especially during the middle and later stages of growth, potassium fertilizer should not be lacking. Use organic fertilizer more often.

015 What are the symptoms of melon root-knot nematode disease?

The melon root-knot nematode mainly harms the roots, with no obvious symptoms above ground at the initial stage. In severe cases, the plants are dwarfed and growth stops. The leaves lose their gloss, and wilt during high noon temperatures. They can recover to normal when watered sufficiently in the morning or evening. The leaves turn yellow from the bottom upwards. As the disease progresses, wilting cannot be recovered until the plant dies. The lateral roots and fibrous roots are covered with white tumors of different sizes, called root-knots. They turn brown to dark brown in the later stages of growth, and sometimes the surface cracks. Cutting open the root-knot reveals milky white nematodes(Figure 11).

Figure 11 Cucumber Root-Knot Nematode

016 How does melon root-knot nematode disease occur? How to prevent and control it?

The melon root-knot nematode belongs to the Meloidogyne genus of the Heteroderidae family. It overwinters in the soil as eggs or second-stage juveniles attached to diseased residues. The second-stage juveniles can move in the soil for active transmission and enter from the root crown. They can also be passively transmitted by diseased soil, seedlings, and irrigation water. They are mostly distributed in the soil layer of 5–30 cm. They are almost inactive above 40 ℃ or below 10 ℃, suitable for soil with pH4–8. In greenhouses, about 10 generations can occur in a year. Generally, high and dry terrain, loose soil, and low salt content are suitable for nematode activity, and the disease is severe in continuous cropping fields.

(1) Soil treatment. Replace the nursery soil with field soil. Steam disinfection. In modern large greenhouses, steam pipes are laid, water is poured on the tilled land, and then covered with plastic film. High-pressure steam is introduced, and the temperature of the 20 cm soil layer reaches 60 ℃, which can be achieved after 30 minutes.

(2) Strengthen cultivation management. Use disease-free soil for seedling cultivation and cultivate healthy seedlings. If diseased plants are found during transplanting, they should be removed in time. Reasonable fertilization and watering can reduce the occurrence of diseases. After harvest, the soil should be deeply plowed and turned. Deep plowing 50 cm, raising high ridges 30 cm, irrigating water in the ditch, covering the soil with film, and closing the greenhouse or large shed for 15–20 days, the prevention effect is more than 90% after high temperature or flooding in summer.

(3) Remove diseased residues. After cucumber harvest, remove diseased residues promptly, especially the residual roots, which should be incinerated.

(4) Crop rotation. Rotate with nematode-resistant vegetables such as leeks, onions, and garlic.

(5) Chemical control. You can choose 10% Kxylphos, 3% Mirel, 5% Yishubao and other granules, 3–5 kg per mu, evenly spread and then turned into the soil, or use the above drugs to open trenches on both sides of the planting row; or with the planting hole, use 1–2 kg of medicine per mu, mix the soil after applying the medicine to prevent the roots from coming into direct contact with the medicine.

017 What are the symptoms of cucumber sclerotinia disease?

Cucumber sclerotinia disease can occur from the seedling stage to the mature plant stage. Most of the diseases occur 5–30 cm above the ground. The fruit is mainly infected by the fungus invading the residual flowers, forming water-soaked spots on the navel. Soft rot, producing white cottony mycelium, sometimes produces a gel-like substance, and later forms black sclerotia. After the stem is infected, it shows soft rot, densely covered with white mycelium, sometimes accompanied by gum exudation. In the early stage of infection, the leaves above the infected part show wilting around noon, and recover in the morning and evening. In the later stage of infection, it does not recover, and finally the whole plant dies. Leaf infection is due to the contact of the infected part with the diseased remains of the fruit, forming irregular yellow-brown spots on the leaves, which later expand to form holes. The front is yellow-brown, the edges are dark brown, and the back of the leaf spot is light gray; in the later stage of infection, the whole leaf is like being scalded by water, and when the humidity is high, dense white mycelium is produced. The edge of the leaf can form a "V" shaped spot(Figure 12).

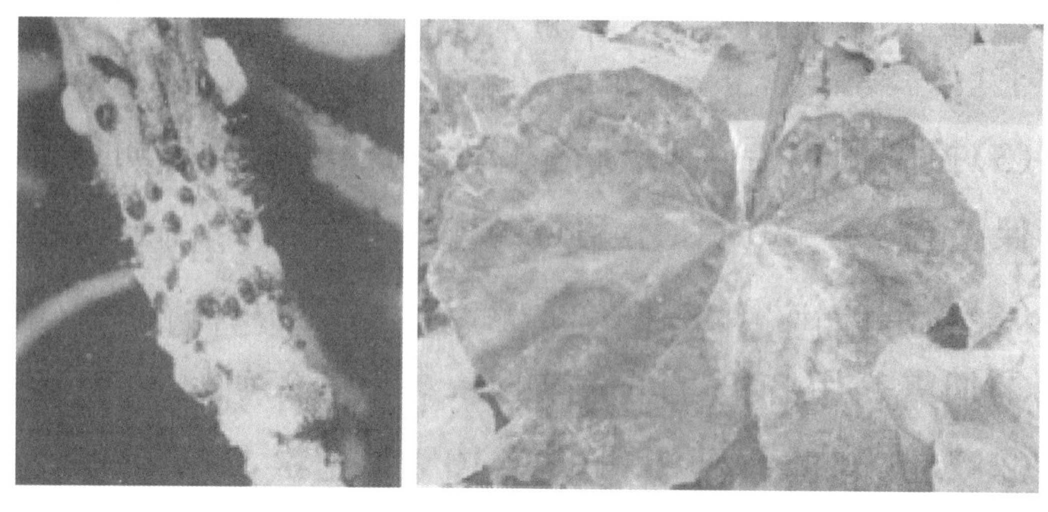

Figure 12 Cucumber Sclerotinia Disease

018 What are the occurrence pattern of cucumber sclerotinia disease? How to prevent and control?

The pathogen of cucumber sclerotinia disease is a fungus of the genus Sclerotinia in the sub-phylum Ascomycota, which is highly saprophytic. The pathogen overwinters in the soil or mixed with seeds in the form of sclerotia. Sclerotia can absorb water and germinate, grow out of the ascocarp, and after maturation, eject ascospores to infect and cause disease, or can be directly infected by mycelium. Mycelium can grow at 0–30 ℃, and the optimal temperature is about 15 ℃. Low temperature and high humidity are conducive to the occurrence of the disease, and the relative humidity below 85% is not conducive to the growth of mycelium, and the disease is light.

(1) Clean up the fields. Remove the diseased remains, deeply turn the soil, apply 50 kg of quicklime per mu, bury the sclerotia below 30 cm, irrigate and soak the soil, and seal the shed film tightly around the shed, and press it tightly to raise the soil temperature in the shed to above 60 ℃, and fumigate the shed for 10 days.

(2) Seed and soil disinfection. Soak seeds in 50 ℃ water for 10 minutes, or use 40% pentachloronitrobenzene to make medicinal soil before sowing. Use 1 kg of medicine per mu, add 15–20 kg of fine soil, and sow after applying medicinal soil.

(3) Control the temperature in the shed. In the early spring and autumn and winter seasons, the shed should be closed in the morning to raise the temperature, and the temperature should be kept at 28–30 ℃, not exceeding 32 ℃.

(4) Pesticide prevention and control.

Fumigation. Use 30% bactericidal smoke agent, 250 g per mu.

Dusting. Use 5% bactericidal powder or 10% mek powder, 1 kg per mu, use medicine once every 7–10 days, apply 3–4 times continuously depending on the disease.

Spray. At the beginning of the disease, use 50% speed kill wettable powder 1 500 times solution, 50% agricultural ling water dispersible granule 1 000 times solution, 40% sclerotinia net wettable powder 800 times solution spray, spray once every 7 – 10 days, continuous spray 2 – 3 times, you can also use 50% speed kill wettable powder 50 times solution to apply to the infected part.

019 What are the symptoms of cucumber bacterial edge blight? How can it be prevented and treated?

All parts of the cucumber plant can be affected by bacterial wilt. Small water-soaked spots form near the leaf edge stomata, which later expand into irregular pale brown spots surrounded by halos. In severe cases, large water-soaked wedge-shaped spots form, expanding from the leaf edge to the middle of the leaf. The lesions on the petioles, stems, and tendrils are brown and water-soaked. The fruit is often infected from the pedicel, forming lesions, yellowing, wilting, and becoming rigid after dehydration. In high humidity, bacterial pus oozes from the diseased parts(Figure 13).

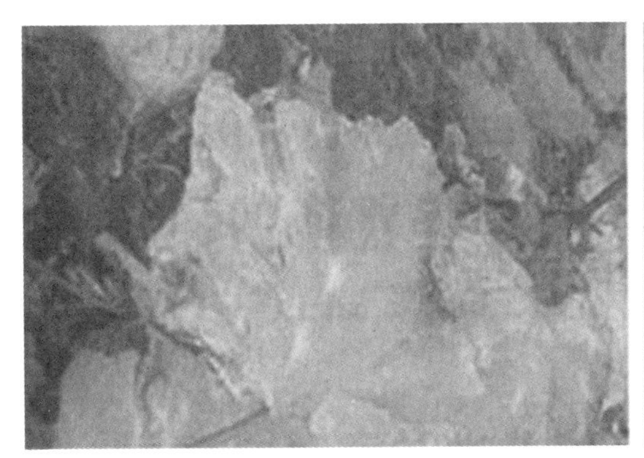
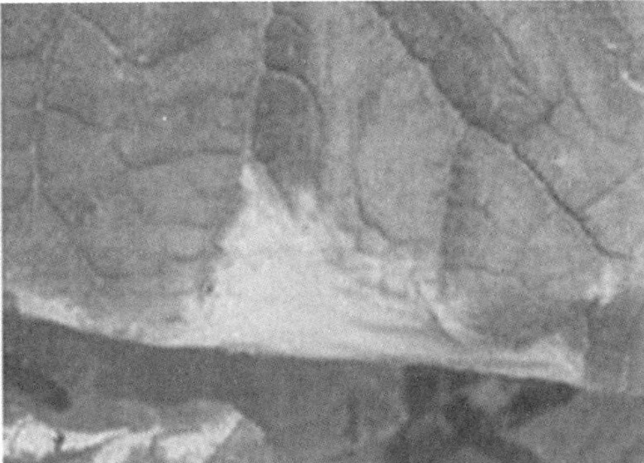

Figure 13 Cucumber Bacterial Edge Blight

The pathogen of cucumber bacterial edge blight is a bacterium from the genus Pseudomonas. The pathogen overwinters in the soil or adheres to seeds with the diseased remains. It spreads and spreads by air flow, running water, and field operations. The bacteria invade from natural openings such as leaf edge stomata. It is mainly affected by changes in humidity and dew on the leaf surface. When the humidity in the shed is high, especially when the night temperature drops, the relative humidity is above 70% for 7 – 8 hours, the water vapor in the shed will condense on the cucumber leaves or stems to form dew on the leaf surface. Long dew time and easy exudation of stomata at the leaf edge can easily cause the disease to spread.

(1) Seed treatment. Soak seeds in 55 ℃ warm water for 15 minutes, or in 100 times acetic acid solution for 30 minutes, or in 40% formalin 150 times solution for 1.5 hours, or in 500 times solution of 1 million U agricultural streptomycin for 2 hours, then rinse and germinate for sowing.

(2) Rotation. Implement a rotation with non-cucurbit crops for more than 2 years.

(3) Reduce the source of bacteria. Clean the soil and use sterile soil for seedlings.

(4) Pesticide prevention and control. Use 30% copper oxychloride wettable powder 500 times solution, 60% copper ethophosphorus aluminum wettable powder 500 times solution, 14% copper ammonia complex water solution 300 times solution, 50% methoxy copper wettable powder 600 times solution, 2% chunlei mycin water solution 400 - 750 times solution, 77% can kill wettable microgranule 400 times solution, 70% bacteria 500 - 600 times solution spray.

020 What are the main symptoms of tomato wilt disease? How can it be prevented and treated?

The disease often occurs in the field during the flowering and fruiting period of tomatoes. In the early stage of the disease, the leaves of the middle and lower part of the plant wilt at noon and return to normal in the morning and evening. After several days of repetition, the wilting gradually worsens and the leaves no longer recover. The diseased plant leaves wilt, turn yellow, and finally dry up. When the diseased stem is cross-sectioned, the vascular bundle of the diseased part is brown. In high humidity, there is often a pink mold layer at the base of the dead stem (Figure 14).

Figure 14 Tomato Wilt Disease

(1) Soil treatment. Use new seedbeds for seedlings, and disinfect the soil when using old beds. For soil disinfection, use 8-10 g of 50% mancozeb per square meter, mixed with 10-15 kg of dry fine soil to make medicinal soil, and use medicinal soil to underlay and cover when sowing.

(2) Seed treatment. Soak seeds in 52 ℃ warm water for 30 minutes. Or soak seeds in 0.1% copper sulfate 1 000 times solution for 5 minutes, or mix seeds with 50% thiram at 0.3%-0.5% of the seed weight.

(3) Pesticide prevention and control. When sporadic diseased plants are found, they should be removed in time, the planting hole should be filled with quicklime and covered with soil and treaded, and fungicides should be sprayed to the root area and surrounding soil, such as 50% mancozeb wettable powder 500 times solution, 50% methyl tobujin wettable powder 500 times solution, 10% double effect spirit water solution 200 times solution, etc. Or use 70% dichlorvos wettable powder 500 times solution to irrigate the roots, each plant is irrigated with 0.3-0.5 kg of medicinal solution, every 7-10 days, continuously irrigated 2-3 times.

021 How to prevent and control tomato blossom end rot?

(1) Select disease resistant varieties. In areas where the disease often occurs, varieties with round fruit shape, smooth and thicker skin can be selected.

(2) Reasonable fertilization. Increase the application of organic fertilizer, avoid partial application of nitrogen fertilizer. Top dressing mainly uses nitrogen and potassium fertilizers, and pay attention to the combination of fertilizer and water. Top dressing times: fruiting fertilizer during the first fruit expansion period, and fruiting fertilizer during the second and third fruit expansion periods. Apply 3-4 times.

(3) Properly supplement calcium elements. After entering the fruiting period, the method of spraying fertilizer outside the root can be adopted. One month after the tomato fruiting, it is the key period for calcium absorption. Can spray 1% calcium metaphosphate, 0.5% calcium chloride plus 5mg/kg naphthylacetic acid, 0.1% calcium nitrate and Aiduo Collect 6 000 times liquid, or Green Fenwei No. 3 1 000-1 500 times liquid. Starting from the initial flowering period, spray once every 10-15 days, continuously spray 2-3 times, can reduce the occurrence of blossom end rot.

022 What is tomato leaf mold? What is the prevention and control method?

Tomato leaf mold mainly damages leaves, but also stems and fruits. Initially, round or nearly round pale yellow spots are formed on the back of the leaves, and the front is discolored. The affected area on the back of the leaf has a light white and then turns to a brownish-brown mold layer. When there are many disease spots, they merge into a piece, the leaves turn yellow and curl, dry up, the plants prematurely pull seedlings, and in severe cases, the front of the leaves will also appear brown. When the fruit is infected, black hard concave disease spots are often formed around the fruit pedicle(Figure 15).

Figure 15 Tomato Leaf Mold

The pathogen of tomato leaf mold is a fungus of the genus Cladosporium in the sub-phylum Ascomycota. The pathogen overwinters on the surface of the soil in the form of mycelium in the diseased remains, or spores attached to the surface of the seeds and mycelium lurking in the seeds. The most suitable temperature for the disease is 21-25 ℃, and the suitable relative humidity is above 95%. The humidity is too high in the protected area, ventilation is poor, light is insufficient, the plant growth is weak, and the disease is heavy. Especially on rainy days, it is prone to disease.

(1) Seed treatment. Soak the seeds in 52 ℃ water for 30 minutes.

(2) Crop rotation. A 3-year rotation should be implemented in the heavy disease greenhouse.

(3) Strengthen cultivation management. The focus is on controlling temperature and humidity and strengthening water management. The tomatoes in the greenhouse and the big shed should control watering, and ventilate and dehumidify in time. Reasonably plant densely, increase organic fertilizer, and enhance the disease resistance of the plant. Keep warm in the early stage, strengthen ventilation in the later stage, reduce humidity, and timely remove the old leaves at the bottom of the plant.

(4) Pesticide prevention and control. In the early stage of the disease, medication should be used in time. Can choose to use 65% mancozeb wettable powder 600 times liquid, 50% metalaxyl wettable powder 700 times liquid, 70% methotrexate wettable powder 800-1 000 times liquid, 75% chlorothalonil wettable powder 500 times liquid, 50% sulfur suspension 700-800 times liquid spray.

023 How does tomato early blight occur? What are the disease symptoms?

Tomato early blight can harm all parts of the tomato. The leaf spots are nearly round, gray-brown, with concentric rings, surrounded by yellow halos, and when severe, the spots coalesce and the leaves dry up. The stem disease often produces round or oval spots at the branching point, which are gray-brown with concentric rings. When the spots surround the stem, it can cause the plant to fall over, and the damage to seedlings is particularly severe. The fruit spots are dark brown, slightly sunken, with rings. When it is warm and humid, a black fluff-like mold layer is produced on the spots(Figure 16).

The pathogen of tomato early blight is a fungus of the genus Alternaria in the subphylum Ascomycota. The pathogen mainly overwinters on seeds and diseased residues, becoming the initial source of infection in the field in the following year. It is spread by rainwater, airflow, and irrigation water, and invades directly from the leaf stomata, wounds or host epidermis. The disease often occurs in hot and humid weather. Continuous cropping, insufficient base fertilizer, excessive density, and weak plant growth plots are prone to severe disease.

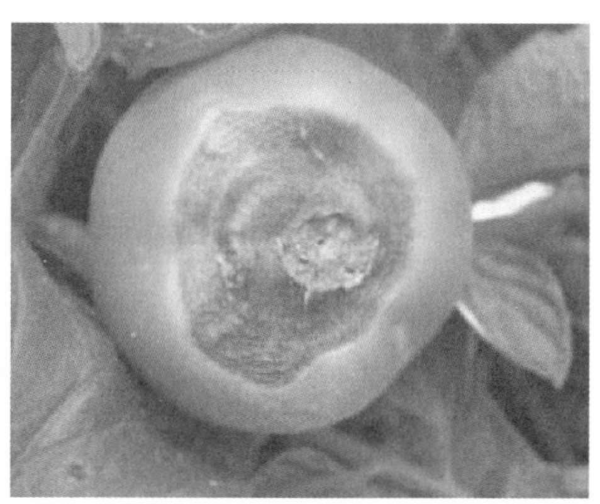

Figure 16 Tomato Early Blight

024 How to prevent and control tomato early blight?

(1) Select disease resistant varieties. Varieties such as Delia, Otley, and Triumph 158 are more resistant.

(2) Crop rotation. Rotate with non-solanaceous vegetables for more than two years.

(3) Strengthen field management. Reasonably dense planting, sufficient base fertilizer, and increased application of phosphorus and potassium fertilizers. Control the temperature and humidity in the shed, ventilate in time to prevent excessive humidity and high temperature.

(4) Seed disinfection. Soak the seeds in 50 ℃ water for 20 minutes, then soak them in cold water for 3–4 hours. Or soak the seeds in cold water for 4 hours, then soak them in 1% copper sulfate solution for 10 minutes, then soak in 1% soap water for 5 minutes, then wash, germinate and sow.

(5) Pesticide prevention and control. Start using medicine during the seedling stage, after transplanting, before the disease occurs, you can choose 50% mancozeb wettable powder 500 times solution, 65% zinc dimethyldithiocarbamate wettable powder 500 times solution, 75% chlorothalonil wettable powder 500 times solution, 50% mancozeb wettable powder 500 times solution, 70% manganese zinc dimethyldithiocarbamate wettable powder 500 times solution, 40% metalaxyl 400 times solution, 77% kasugamycin 600 times solution, spray once every 5–7 days, continuous spray 2–3 times. The stem spot can be scraped off first, and then applied with a 10–fold diluted 2% agricultural resistance 120 solution.

025 What are the conditions for the occurrence of tomato late blight? What are the disease symptoms?

Tomato late blight mainly harms leaves, fruits, and stems. When seedlings are infected, the leaves show dark green water-soaked spots, which expand into large spots with indistinct edges, and spread to the leaf veins and stem, causing the stem to thin and turn black-brown, eventually causing the plant to wilt or fall over. Under high humidity conditions, the diseased part produces a white mold layer. Leaf infection often starts from the leaf tip and edge, initially as dark green irregular water-soaked spots, then turns brown, and when humid, white mold grows at the junction of the diseased and healthy parts on the back of the leaf, and the entire leaf rots and can spread to the petiole and main stem. The spots on the stem are dark brown, slightly sunken, with unclear edges. The main disease of the fruit occurs on the green fruit, the spots are initially oil-soaked dark green, then turn dark brown to brown, the spots are irregular cloud-like, slightly sunken, with obvious edges. The diseased fruit is harder at the beginning and does not rot. When the humidity is high, a small amount of white mold grows on it, and the diseased fruit rots quickly(Figure 17).

The pathogen of tomato late blight is a fungus of the genus Phytophthora in the subphylum Oomycota. The pathogen mainly overwinters in the soil with diseased residues in the form of mycelium, and can also harm and overwinter on tomatoes cultivated in greenhouses in winter. The sporangia produced on the central diseased plant are spread by air flow and rainwater. The pathogen likes low temperature and high humidity conditions, the temperature is 20–25 ℃, and the relative humidity above 85% is conducive to the occurrence of the disease. The disease is more likely to occur when the plant grows vigorously, the terrain is low-lying, the drainage is poor, and the field humidity is too high.

 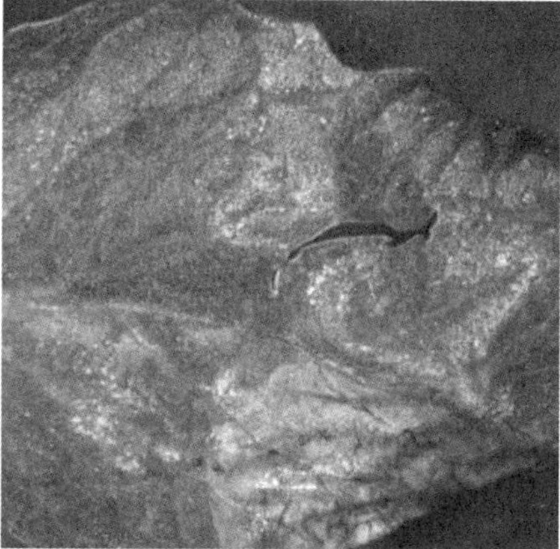

Figure 17 Tomato Late Blight

026 What are the prevention and control measures for tomato late blight?

(1) Select disease-resistant varieties.

(2) Crop rotation. Practice rotation with cruciferous vegetables for more than 3 years, and avoid planting adjacent to potatoes.

(3) Cultivate disease-free strong seedlings. The soil for seedlings must be strictly selected from soil that has not been planted with solanaceous crops, and it is advocated to cultivate disease-free strong seedlings with nutrition bowls, nutrition bags, and hole trays.

(4) Strengthen field management. Apply sufficient base fertilizers, implement formula fertilization, avoid partial application of nitrogen fertilizers, and increase phosphorus and potassium fertilizers. Ventilate in time to reduce humidity. Reasonable pruning, topping, and branching reduce nutrient consumption and promote the growth of the main stem.

(5) Pesticide prevention and control. It can be prevented and controlled by fumigation in the evening. Each acre is treated with 45% bactericidal smoke agent 200–250 grams, or 5% bactericidal dust agent, 1 kg per acre, one time every 9 days; at the beginning of the disease, timely remove diseased leaves and fruits and spray in time. It can be sprayed with 72.2% prolic water 600–700 times liquid, 64% alum fungicide wettable powder 500 times liquid, 70% ethyl phosphorus aluminum manganese zinc wettable powder 500 times liquid, 72% clotrimazole wettable powder 600–800 times liquid, 75% bactericidal clear wettable powder 500 times liquid.

027 What is tomato bacterial wilt? How to prevent and control?

Tomato bacterial wilt, also known as bacterial wilt, is a bacterial disease. The disease symptoms mainly appear on the leaves and stems of mature plants. The top, bottom, and middle leaves of the diseased plant wilt successively, the symptoms are obvious at noon, and they recover after evening. A few days later, it quickly spread to the whole plant, wilted and died without recovery. Because the plant still remains green after death, only the leaf color becomes lighter, so it is called wilt. The stem of the disease shows brown patches, the skin of the middle and lower part of the stem is rough, and often produces adventitious roots. When the diseased stem is cut open, the vascular bundle turns brown. After cross-cutting, white bacterial pus is squeezed out by hand, which can be distinguished from fungal wilt (Figure 18).

Figure 18 Tomato Bacterial Wilt

The pathogen of tomato bacterial wilt is a bacterium in the genus Pseudomonas. The pathogen mainly overwinters in the field with diseased residues, and is spread by irrigation water and agricultural operations, and invades from the wounds of the roots or stems of the host. The soil temperature is high, the humidity is high, and the air temperature rises rapidly, which is most likely to induce wilt. In addition, tomato continuous cropping, lack of potassium fertilizer, and root damage are all conducive to the occurrence of the disease.

Prevention and control techniques should be adopted, focusing on strengthening cultivation management, selecting disease-resistant varieties, and supplemented by pesticide prevention and control.

(1) Seed treatment. Select disease-resistant varieties and disinfect seeds.

(2) Adjust soil acidity. The wilt pathogen is suitable for growth in slightly acidic soil, and a suitable amount of lime can be spread in combination with soil preparation to make the

soil slightly alkaline.

(3) Reasonable fertilization. The use of fully matured organic fertilizer, wood ash or bio-bacterial fertilizer can alleviate the disease. And pay attention to the reasonable combination of nitrogen, phosphorus and potassium fertilizers, and appropriately increase nitrogen and potassium fertilizers.

(4) Pesticide prevention and control. At the early stage of the disease, use 72% agricultural streptomycin sulfate soluble powder 3 000–4 000 times liquid or new plant mycin 4 000 times liquid, 50% bacterial wilt wettable powder 500 times liquid, 77% can kill wettable powder 600 – 800 times liquid irrigation. Each plant irrigates about 0. 25 kg of medicated water, irrigates once every 7 days, and irrigates 2–3 times in a row. At the same time, it can be sprayed with 200 mg/L of streptomycin sulfate.

028 What are the symptoms of tomato virus disease?

The symptoms of tomato virus disease(Figure 19) mainly include the following types.

(1) Floral leaf type. It mainly occurs in the upper leaves of the plant, characterized by uneven leaf color, alternating yellow and green, leaf color fading, leaf surface wrinkling, plant dwarfing, and fewer and smaller flowers.

(2) Fern leaf type. It mainly occurs in the middle and lower leaves of the plant. The plant is dwarfed, the young leaves are slender and narrow, the upper leaves become linear, the middle and lower leaves roll up, and the fruits are fewer and smaller.

(3) Striped spot type. Stems, leaves, and fruits can all be affected. When the leaves are affected, they show mottled or tea-brown spots, and the veins undergo necrosis. The stem initially has dark green sunken stripes, which later develop into dark brown. The fruit surface of the diseased fruit shows dark brown slightly sunken stripes or irregular spots, deformed, and the diseased fruit is easy to fall off.

(4) Curly leaf type. The leaf surface turns yellow and the edges of the leaves curl up.

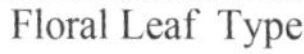
Floral Leaf Type

Curly Leaf Type

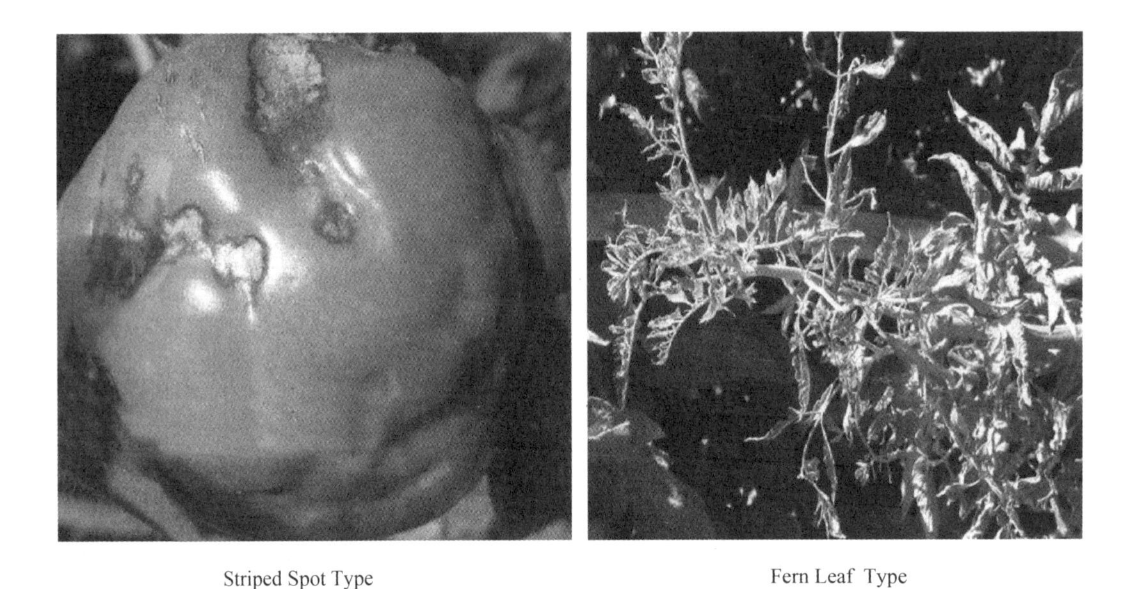

Striped Spot Type　　Fern Leaf Type

Figure 19　Tomato Virus Disease

029　What is the occurrence pattern of tomato virus disease? How to prevent and control it?

The pathogens of tomato virus disease are Tobacco Mosaic Virus and Cucumber Mosaic Virus. They are obligate parasites that survive in living plants. The Tobacco Mosaic Virus can overwinter on various plants, and seeds can also carry the virus, becoming the initial source of infection. The virus is transmitted through sap contact, entering from the micro-wounds of the host. The fruit debris attached to the tomato seeds can also carry the virus. Disease residues in the soil, overwintering host residues in the field, and weeds in the field can all become the initial sources of infection for this disease. The Cucumber Mosaic Virus is mainly transmitted by aphids, and sap can also transmit it. Various agricultural operations in the field can become transmission routes for the virus. When there are many weeds around the greenhouse and heavy aphids in the greenhouse, the virus disease can also worsen. High temperature and dry environment conditions are conducive to the occurrence and spread of the disease, and small day-night temperature differences and poor ventilation in protected areas are also conducive to the occurrence of the disease.

(1) Use disease-free seeds. Soak seeds in 60 ℃ water for 15 minutes before sowing, then soak them in 35 ℃ water for 4-6 hours. Or soak seeds in clean water for 3-4 hours, then soak them in a 10% trisodium phosphate solution for 30 minutes, then rinse with clean water, dry and germinate for sowing.

(2) Cultivation measures. Select disease resistant varieties. Rotate with non-solanaceous vegetables.

(3) Prevent and control virus-transmitting insects. Timely prevention and control of aphids, using the method of silver film to drive away aphids to reduce the aphids from migrating into the greenhouse, preventing them from spreading the virus further.

(4) Chemical control. Before the disease occurs or at the beginning of the disease, a 20% virus A wettable powder 700–1 000 times liquid or a bactericidal star 3 000 times liquid can be used for spraying prevention, spraying once every 7 – 10 days, and continuously spraying 2–3 times.

030 What are the symptoms of pepper infection with viral disease?

The main manifestations of pepper virus disease(Figure 20) are as follows.

Striped Type　　Yellowing Type

Flower leaf Type　　Deformed Type

Figure 20 Pepper Virus Disease

(1) Flower leaf type. The diseased leaves show mottled spots of varying shades of green, the leaf surface is wrinkled and deformed, the plant growth is slow, and the fruit becomes smaller.

(2) Yellowing type. The leaves turn yellow and flowers and fruits are easily shed.

(3) Striped type. The leaves show brown or dark brown necrotic spots, which gradually extend to the side branches and stems along the veins, causing them to turn brown, necrotic, and appear stripe-like spots, necrotic spots, or tip dieback.

(4) Deformed type. The leaves become thin and fern-like, the plant is dwarfed, and branching increases.

031 How does chili virus disease occur? How to prevent and control it?

There are more than 10 pathogens of chili virus disease, 7 of which have been found in China, including cucumber mosaic virus, chili mottle virus, tobacco mosaic virus, potato Y virus, potato X virus, etc. The virus overwinters in cultivated vegetables, vegetable plants and weeds, and is transmitted and infected by insects, leaf friction, pruning branches and other agricultural activities. Aphids feeding is the main route of disease transmission. High temperatures, drought, poor plant growth, continuous cropping, and untimely control of aphids are all conducive to the occurrence of the disease.

(1) Breeding(use) of resistant varieties. Breeding(use) of resistant varieties is the most effective method to control chili virus disease. There are some sweet(spicy) pepper varieties that are relatively resistant and tolerant to chili virus disease, such as Medium Pepper 220, Medium Pepper Y371, Single-Born Chaotian Ying Shan Hong Super Long Line Pepper Zhuo Yue, etc. Different regions should choose resistant varieties according to local climatic conditions and planting methods.

(2) Clean the garden. Thoroughly eradicate weeds and overwintering vegetable roots in the field.

(3) Crop rotation. Rotate with non-solanaceous vegetables for more than 2 years.

(4) Control virus-transmitting insects. Timely control of aphids.

(5) Seed treatment. Treat dry seeds at a constant temperature of 70 ℃ for 72 hours to inactivate the virus, or soak them in a 10% trisodium phosphate solution for 40–50 minutes, wash them, germinate them and then sow them, or soak them in a 0.1% potassium permanganate solution for 30 minutes, then wash them and germinate them.

(6) Chemical control. At the beginning of the disease, spray with a 500 times solution of 20% virus A wettable powder or a 1 000 times solution of 1.5% plant disease spirit emulsion, or a 3 000 times solution of fungicidal star to prevent the disease. Spray once every 7–10 days, and continue for 2–3 times.

032 What are the symptoms of chili blight?

Chili blight is a devastating disease in chili production. After the plants are infected, they wilt rapidly and can cause widespread death(Figure 21).

Figure 21 Chili Blight

Chili seedlings and mature plants can both be affected by chili blight. When seedlings are infected, water-soaked dark green soft rot appears at the stem base, and the seedlings fall over. The leaf lesions are dark green, with a clear boundary between healthy and diseased tissues, and they expand rapidly. When the humidity is high, white mold can be seen on the diseased parts. The stem is often infected at the base, initially showing dark green water-soaked spots, which gradually turn into brown streaks. The branches and leaves above the diseased part gradually wilt, and the diseased stem is easily broken. The fruit lesions are dark green and water-soaked, soft rot, and brown. When the humidity is high, the surface of the diseased fruit is densely covered with white mold. The diseased fruit can fall off or dry out and hang on the branches as dark green stiff fruit.

033 What is the occurrence pattern of chili blight? What are the prevention and control methods?

The pathogen of chili blight is a fungus from the genus Phytophthora in the phylum Oomycota. The pathogen overwinters in plant residues, soil or seeds as oospores and thick-walled spores. Oospores can survive in the soil for 2-3 years. Oospores can germinate and directly infect the stem base. Germinating infected seeds can directly infect seedlings. After the disease occurs, the sporangia produced on the diseased part can be spread by wind, rainwater, agricultural tools, etc. Wounds on the plant are conducive to the invasion of the blight pathogen. The disease is prone to occur under conditions of continuous cropping, excessive nitrogen fertilizer, insufficient phosphorus and potassium fertilizer and trace elements, the use of unripe organic fertilizer, high temperature of 30 ℃ and high humidity.

(1) Select good varieties resistant to disease.

(2) Crop rotation. Implement a 3-year rotation with beans, melons or onions and garlic.

(3) Strengthen cultivation management. Apply fertilizer reasonably, and apply nitrogen, phosphorus, and potassium fertilizers together for topdressing. Diseased plants should be promptly uprooted and taken out of the field for centralized burning or deep burial. Lime powder should be sprinkled on the disease pit for disinfection. Residues should be cleared in time after harvest.

(4) Seed disinfection. Pre-soaked seeds are treated in 55 ℃ warm water for 15 minutes, or soaked in a 1% copper sulfate solution for 5 minutes, or soaked in a 600 times solution of 72.2% Pulk water for 15 minutes, washed and seeded or germinated.

(5) Chemical control. When the first diseased plants are found, they should be promptly uprooted and sprayed with pesticides for control. Options include a 600 times solution of 75% Ridomil WP, an 800 times solution of 25% Metalaxyl WP, a 500 times solution of 64% Kasugamycin WP, a 600 times solution of 75% Carbendazim WP, a 0.3% copper sulfate solution, an 800 times solution of 77% Kasugamycin WP, a 600-800 times solution of 72.2% Pulk water, a 600 times solution of 70% Difenoconazole WP, or a 1:1:200 Bordeaux mixture. These should be used alternately, once every 5-7 days, for 2-3 consecutive times.

034 What is chili leaf blight? How to prevent and control it?

Chili leaf blight mainly harms the leaves. In the early stage of leaf disease, scattered brown spots appear, quickly expanding into circular or irregular disease spots, the center is gray-white, the edge is dark brown, the diameter varies from 2 to 10 mm, the center of the disease spot necrotizes and is easy to perforate, and the diseased leaves are easy to fall off (Figure 22).

The pathogen of chili leaf blight is a fungus of the genus Monilinia in the subphylum Ascomycota. It overwinters with mycelium or conidia on disease residues or attached to seeds, and is spread by air flow. The application of unripe farmyard manure or old seedbed for seedling cultivation, the seedbed cannot be ventilated in time after the temperature rises, and the temperature and humidity are too high, which is conducive to disease occurrence. Improper field management, excessive application of nitrogen fertilizer, vigorous plant growth in the early stage, or standing water in the field are all conducive to disease occurrence.

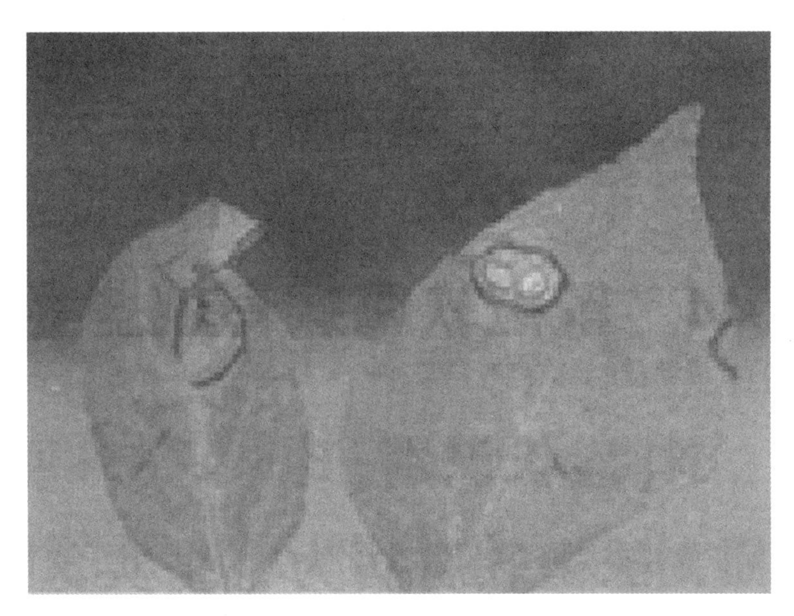

Figure 22 Chili Leaf Blight

(1) Cultivate strong seedlings. Use mature organic fertilizer to prepare nutrient soil, pay attention to ventilation during the seedling cultivation process, and strictly control the temperature and humidity of the seedbed. Strengthen management, apply nitrogen fertilizer reasonably, increase the application of phosphorus and potassium fertilizer, spray foliar fertilizer, and pay attention to tillage and soil loosening after planting.

(2) Rotation. Implement rotation and timely remove disease residues.

(3) Chemical prevention and control. Spray 64% bluestone wettable powder 500 times liquid, 50% carbendazim wettable powder 600 times liquid, 50% polyoxin wettable powder 500 times liquid, 50% benomyl wettable powder 1 000 times liquid, 75% methyl tobujin wettable powder 600 times liquid, 50% pyrimethanil wettable powder 1 500-2 000 times liquid, 70% chlorothalonil wettable powder 800-1 200 times liquid, 40% DuPont FuXing emulsifiable oil 8 000-10 000 times liquid, 50% thiophanate-methyl wettable powder 500-1 000 times liquid at the beginning of the disease, spray once every 7 days, and spray 2-3 times in a row.

035 What are the symptoms of chili soft rot?

Chili soft rot harms fruit. It starts from the wound, initially showing water-soaked dark green spots, which later turn dark brown. The keratin of the fruit skin is not infected by bacteria, it remains intact and unbroken, the interior of the fruit is completely rotten, and emits a foul smell. The diseased fruit falls off or stays on the branch, the remaining fruit gradually dehydrates and turns gray-white or white, leaving only the fruit skin, hollow, and looks like a lantern(Figure 23).

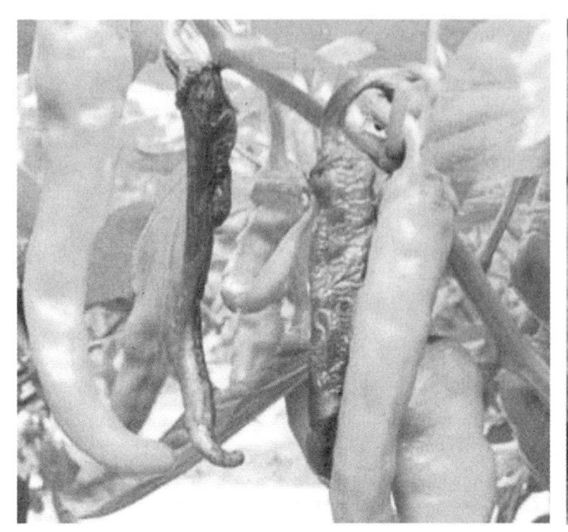

Figure 23 Chili Soft Rot

The pathogen of chili soft rot is a bacterium of the genus Erwinia carotovora. The pathogen mainly overwinters in the soil, inside vector insects, and on various host plants through disease residues, enters through wounds by splashing irrigation water or shed film drips, and can also be spread by vector insects and air flow. The optimal growth temperature is 25-30 ℃, the highest is 40 ℃, the lowest is 2 ℃, the lethal temperature is 50 ℃ for 10 minutes, the suitable pH is 5. 3-9. 3, and the optimal pH is 7. 3. It is easy to spread in greenhouses with many boring pests, continuous rainy weather, and high humidity.

036 What is the prevention and control method for chili soft rot?

(1) Rotation. Rotate with non-Solanaceae and Brassicaceae vegetables for more than 2 years.

(2) Strengthen cultivation management. Use high ridge cultivation and timely clean up diseased plants in the field.

(3) Prevent mechanical damage. Avoid mechanical damage as much as possible during transportation and storage, and pick out damaged and wounded fruits in time.

(4) Chemical prevention and control. Spray 72% agricultural-grade streptomycin sulfate soluble powder 4 000 times liquid, Xinzhimeisu 4 000 times liquid, 50% copper soap fatty acid wettable powder 500 times liquid, 77% Kasugamycin 101 wettable micro-granule powder 500 times liquid, and 14% copper amino acid water agent 300 times liquid at the beginning of the disease.

037 What are the symptoms of powdery mildew in peppers? What are the preventive measures?

Powdery mildew mainly affects the leaves of peppers. Both old and young leaves can be infected. At the initial stage of the disease, the front of the leaf shows fading green spots which gradually expand into unclear fading green-yellow patches. The back of the diseased area produces a powdery white substance. When severe, the disease spots are densely covered, the white powder increases rapidly, the entire leaf turns yellow and falls off easily. During the epidemic of the disease, all leaves of the plant fall off, leaving only the tender leaves at the top. The fruits are small, severely affecting yield and quality(Figure 24).

Figure 24 Powdery Mildew in Peppers

The pathogen of powdery mildew in peppers is a fungus of the Ascomycota subphylum Erysiphales. It overwinters in the soil with the leaf litter and is mainly spread by air flow in the field. The disease is prone to occur and spread under conditions of about 25 ℃, 50%-80% relative humidity and dim light. Plants that are overgrown or weak are most susceptible to the disease.

(1) Clean the garden. Remove diseased leaves as soon as they are discovered, bury or burn them in a concentrated manner. Remove plant residues after harvest to reduce the source of the fungus.

(2) Strengthen temperature and humidity control. Pay attention to controlling fertilizers and water, promote strong plant growth, and increase disease resistance. Control the temperature to prevent the greenhouse humidity from being too low and the air from being too dry.

(3) Seed disinfection. Soak the seeds in 55 ℃ warm water for 15 minutes, or soak the seeds in 0.1% potassium permanganate solution for 15-20 minutes.

(4) Fumigation disinfection of the greenhouse. For the seedling greenhouse and planting greenhouse, 7 days in advance, use 0.25 kg of sulfur powder and 0.5 kg of sawdust per 100 square meters, ignite in several places, fumigate and close for 1 day and night.

(5) Chemical prevention and control. At the initial stage of the disease, spray the medicine. You can use 75% chlorothalonil wettable powder 500 times liquid, 20% tricyclazole emulsion oil 2 000 times liquid, 20% rust ning wettable powder 1 500 times liquid, 70% methyl tobujin wettable powder 800 times liquid, 40% sulfur suspension 500 times liquid. 45% chlorothalonil fumigant can also be used for fumigation, fumigate once every 7 to 10 days, continuously 2 to 3 times.

038 What are the symptoms of Chili scab disease?

Chili scab disease can occur from seedling to mature plant stage. When seedlings are infected, the cotyledons produce water-stained white spots, which later turn into dark depressed lesions. Infected seedlings lose all their leaves and die. During the mature plant stage, the disease mainly affects leaves, stems, and fruits. After the leaves are infected, numerous circular or irregular black-green to yellow-brown spots appear in the early stage, sometimes with rings. The back of the leaf is slightly raised, blister-like, and the front is slightly depressed. When the stem is affected, the lesions are irregular streaks or patches, which later cork and bulge. When the fruit is affected, it shows round or oblong dark green lesions, about 0.5 cm in diameter, with slightly raised edges and a rough surface, causing fruit rot. When the humidity is high, bacterial fluid overflows in the middle of the canker (Figure 25).

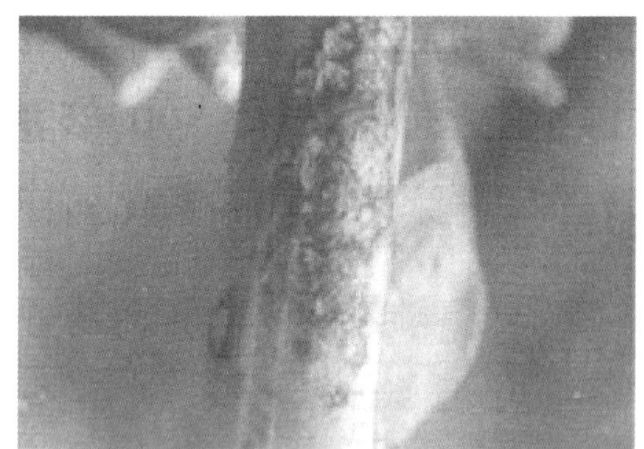
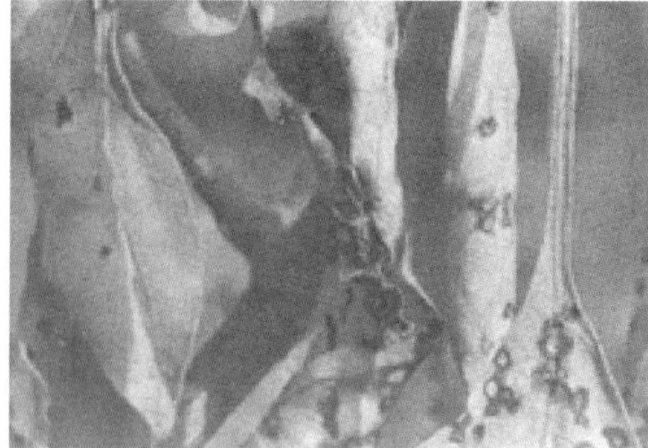

Figure 25 ChiliScab Disease

039 How to prevent and control Chili scab disease?

The pathogen of Chili scab disease is a bacterium of the genus Pseudomonas. The bacterium overwinters in the soil with the diseased residues or adheres to the seeds. It is spread by irrigation water, air flow, insects, and farming operations, and invades through stomata and wounds. The suitable temperature for the development of the bacterium is 27–30 ℃, and the bacterium can be killed at 56 ℃ for 10 minutes. High temperature and high humidity are the main conditions for the occurrence of the disease.

(1) Select disease resistant varieties. Such as Changfeng 1, Changfeng sweet pepper, Xiangyan 1, Xiangyan 2, Dafang wrinkled pepper, line pepper, etc. Choose disease-free land, disease-free single plants or disease-free fruits for seed saving.

(2) Crop rotation. Crop rotation with non-solanaceous crops for 2–3 years.

(3) Seed treatment. Soak the seeds in 55 ℃ hot water for 10 minutes, then move them to cold water to cool, and then germinate and sow. Or first soak the seeds in clean water for 10 hours, then soak the seeds in 1% copper sulfate solution for 5 minutes or in 1: 10 agricultural streptomycin solution for 30 minutes, rinse them clean with clean water, dry them and sow.

(4) Chemical prevention and control. Spray with 60% aluminium phosphide wettable powder 500 times liquid, 72% agricultural streptomycin 4 000 times liquid, 14% copper ammonia complex water agent 300 times liquid, 1: 1: 200 Bordeaux mixture, once every 7 days, continuous prevention and control 2–3 times.

040 What are the symptoms of pepper anthracnose?

Pepper anthracnose mainly affects the leaves and fruits. There are two types of symptoms:

(1) Red anthracnose symptoms. The infected fruits produce yellow-brown, sunken water-soaked spots, covered with orange-red granules arranged in a ring pattern, and pale red viscous matter oozes from the diseased part when it is damp.

(2) Black anthracnose symptoms. Both fruits and leaves can be affected, especially mature fruits and old leaves are prone to be invaded. When fruits are diseased, they initially appear as brown, water-soaked small spots, and quickly expand into circular or irregular black-brown sunken spots, with raised concentric rings on the surface of the spots, and many little black spots are born in rings on them. When leaves are diseased, it usually occurs on the old leaves, initially producing water-soaked spots that are discolored, which expand into circular or irregular shapes, with brown edges and gray-white centers, and small black spots are born in rings on them(Figure 26).

Figure 26 Pepper Anthracnose

041 What are the occurrence patterns and prevention methods for pepper anthracnose?

The pathogen of pepper anthracnose is a fungus from the genus Colletotrichum in the subphylum Ascomycotina. The pathogen can overwinter on the surface of seeds or hidden inside seeds, and can also overwinter in the soil with diseased residues, often invading through the wounds of the host and spreading through wind, rain, insects, and farming operations. The disease is prone to occur in high temperature and high humidity, overcrowding, poor ventilation, excessive use of nitrogen fertilizer. The disease is severe in plants with weak growth and many wounds, especially in the case of severe disease caused by sun burn fruits

(1) Select disease resistant varieties. Varieties with strong spiciness are more disease-resistant, such as B Early etc.

(2) Clean up the diseased residues. Timely remove the diseased fruits and leaves, clear the diseased residues after harvest, take them out of the field and bury them deep or burn them.

(3) Seed treatment. Soak the seeds in 55 ℃ warm water for 10 minutes, then immediately move them to cold water to cool them down, dry them and germinate them for sowing. Soak the seeds in 500 times liquid of 50% carbendazim, 75% chlorothalonil wettable powder for 30 minutes, wash them with clean water, dry them and sow them.

(4) Chemical control. Start spraying medicine at the beginning of the disease, you can choose 75% chlorothalonil wettable powder, 600 times liquid of 75% carbendazim wettable powder, 800-1 000 times liquid of 70% methotrexate wettable powder, 800 times liquid of

80% anthracnose fumei wettable powder.

042 What is the difference between cotton bollworm and tobacco budworm?

Both cotton bollworm and tobacco budworm primarily damage the buds, tender leaves, buds, flowers, and fruits as larvae. The damage to the buds and flowers causes a large amount of bud and flower drop. They start boring into the fruit from the third instar, causing rot and a large amount of fruit drop.

(1) Cotton Bollworm(Figure 27).

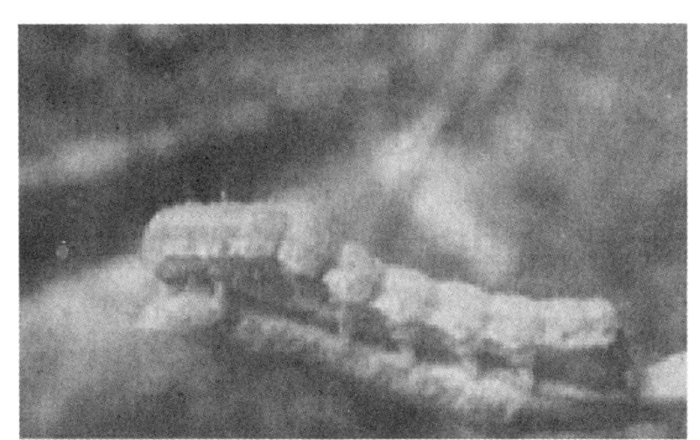
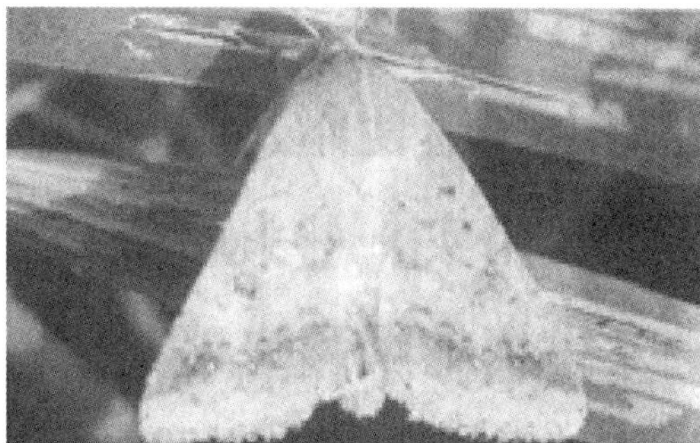

Figure 27 Cotton Bollworm Larvae, Adults

Adults: Body length 15-17 mm, wingspan 27-28 mm, body color generally reddish-brown in females, gray-green in males. The front of the forewing has a kidney-shaped pattern, ring-shaped pattern, and various transverse lines that are not very clear. The middle transverse line extends diagonally from the kidney-shaped pattern to the edge of the wing, ending directly below the ring-shaped pattern. The hind wings are gray-white.

Eggs: Hemispherical, taller than wide, diameter about 0. 5 mm, milky white.

Larvae: Mature larvae are 30-42 mm long, with colors including reddish-brown, green, yellow-white, light red, and black-purple. The head is yellow, the backline, subbackline, and stigmal line are darker than the body color, and the stigmata are white.

Pupa: Spindle-shaped, yellow-brown, 17-21 mm long, with sparse and coarse semicircular engravings on the back of the 5th to 7th abdominal segments. A pair of anal spines at the end of the abdomen are separated at the base.

(2) Tobacco Budworm(Figure 28).

Adults: Yellow-brown to gray-yellow, body length 13-15 mm, wingspan 28-37 mm.

Eggs: Flat circular, milky white.

Larvae: Newly hatched are rust-colored, about 2 mm long, 3-4 instars are 15-18 mm long, initially dark green, then gradually change to red to deep red.

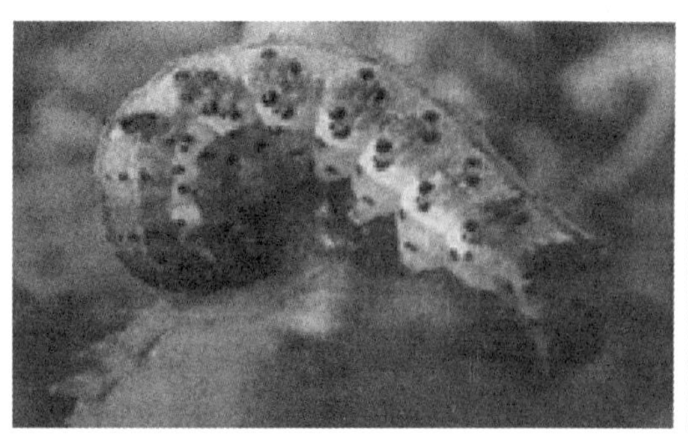

Figure 28　Tobacco Budworm Larvae, Adults

Pupa: The front part of the body appears short and thick, the spiracles are small and low, and rarely protruding.

043　What are the red dots crawling on the back of eggplant and chili leaves? How to prevent and control?

The red dots crawling on the back of eggplant and chili leaves are red spiders, also known as leaf mites. They feed mainly on eggplant, chili, melon, cowpea, etc. Both adult and young mites gather to cause damage, sucking the juice from leaves, flowers, and fruits. In the initial stage, very small yellow-white spots form on the leaves, and as the disease progresses, large yellow-white spots form. In severe cases, the leaves wither and fall off. High temperature and drought are conducive to the occurrence of red spiders(Figure 29).

Male mites are 0. 26 mm long, while female mites are 0. 42-0. 56 mm long, red in color, pear-shaped, with a dark brown spot on both sides of the body, and four pairs of legs. The eggs are spherical and range in color from light yellow to deep yellow.

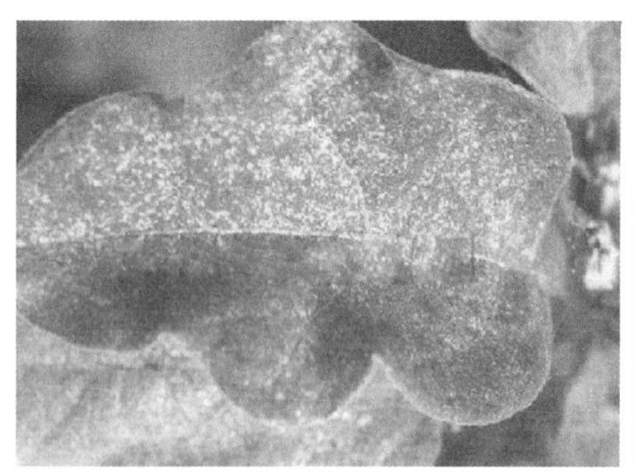
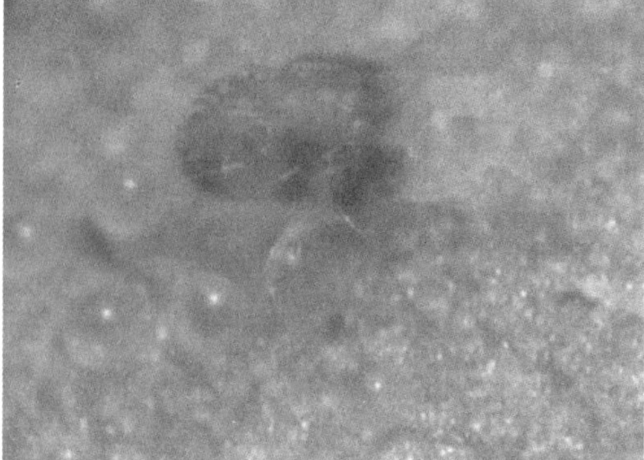

Figure 29　Red Spider(Mite)

There are 10-20 generations in a year(gradually increasing from north to south), with female adult mites overwintering in weeds, dead branches, fallen leaves, and soil crevices. When the temperature reaches above 10 ℃, they start to reproduce massively. From March to April, they first feed on weeds or other hosts, and gradually migrate upwards after ornamental plants sprout. Each female lays 50-110 eggs, mostly on the back of leaves. Reproduction can occur parthenogenetically, and its offspring are mostly males. Young mites and early-stage nymphs are not very active. Late-stage nymphs are lively and voracious, with a habit of climbing upwards. They first damage the lower leaves, then spread upwards. The development starting point temperature is 8 ℃, the optimum temperature is 25-30 ℃, the optimum relative humidity is 35% - 55%. Therefore, hot, low humidity environments, especially in dry years, are prone to large outbreaks. However, temperatures above 30 ℃ and relative humidity exceeding 70% are not conducive to their reproduction. Heavy rain has a suppressive effect.

(1) Clean up diseased residues. Timely remove the lower old leaves and burn them outside the shed.

(2) Adjust the temperature and humidity in the shed. Create a high temperature, high humidity environment, water in time to suppress its spread.

(3) Pesticide prevention and control. 35% acaricide oil emulsion can be used 1 200 times, or 20% acaricide oil emulsion 2 000 times, or 40% water amine phosphorus oil emulsion 2 000 times, 73% acaricide oil emulsion 2 000-3 000 times, 5% nisolone oil emulsion 1 500-2 500 times, sprayed once every 7-10 days, continuously sprayed 2-3 times. Focus on spraying the middle and upper leaves, tender parts, and fruits of the plant.

044 How to distinguish between potato ladybug and tomato twenty-eight star ladybug?

Both types of ladybugs mainly damage the host as adults and larvae. Newly hatched larvae group on the back of leaves, gnawing on the flesh of the leaves, leaving only the epidermis, forming many parallel irregular semi-transparent thin concave lines. As the larvae grow, they gradually disperse. Both adults and larvae can eat the leaves into holes, and when severe, only the veins are left, the leaves wither and turn brown, and the whole plant dies. In addition, it can also harm tender stems, petals, sepals, and fruits. The damaged plants not only decrease in yield, but also have poor fruit quality. The damaged parts of the fruit become stiff and bitter, and are not suitable for consumption(Figure 30).

Adult: semi-spherical, red-brown, densely covered with yellow-brown fine hair, each elytron has 14 black spots.

Eggs: bullet-shaped, initially light yellow, later turning yellow-brown.

Larvae: mature larvae are pale yellow, spindle-shaped, raised on the back, each segment on the back has neatly organized branch spines, the prothorax and the 8th-9th

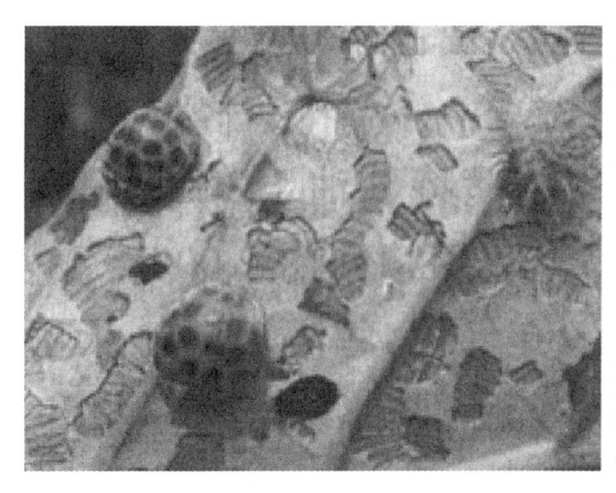

Figure 30 Tomato Twenty-Eight Star Ladybug

segments of the abdomen each have 4 branch spines, and the rest have 6.

Pupa: pale yellow, oval, with the molt of the final instar larvae wrapped around the tail end, with light black markings on the back.

Potato ladybugs overwinter in groups in windward sunny caves, stone cracks, tree holes, bark cracks, under rocks on slopes, in soil burrows, and among weeds. Adults are inactive in the morning and evening, and feed, fly, migrate, mate, and lay eggs during the day, the most active time being from 10: 00 am to 4: 00 pm. Eggs are mostly laid on the back of leaves, generally each egg mass has 20-30 eggs, standing on the back of the leaf, each female can lay nearly 400 eggs. Larvae have 4 instars, and after maturing, they pupate on the back of leaves, stems, or at the base of plants. Adults have a feigning death and oophagy habit, and larvae also have an oophagy habit.

The occurrence pattern of tomato twenty-eight star ladybugs is similar to that of potato ladybugs, but the overwintering aggregation phenomenon is not obvious, and adults feed day and night, with a habit of cannibalism and oophagy and pupophagy. Each egg mass usually has 15-40 eggs. Both adults and larvae have negative phototaxis, and often move on the back of leaves and other hidden places. The suitable temperature for reproduction of adult potato ladybugs is 22-28 ℃, they cannot lay eggs below 15 ℃ and die one after another above 35 ℃. The suitable temperature for reproduction of adult tomato twenty-eight star ladybugs is 25-28 ℃, with a relative humidity of 80%-85%, and they enter overwintering state when the temperature drops to 18 ℃.

045 What are the prevention and control methods for potato ladybugs and tomato twenty-eight star ladybugs?

Potato ladybugs and tomato twenty-eight star ladybugs are strongly regional and have an agglomeration distribution pattern in the field. The degree of occurrence is closely related to meteorological factors, ecological environment, and cultivation methods. Therefore, field surveys should be conducted in a timely manner for prevention and control. Prevent and control at the right time, actively adopt comprehensive treatment measures, and reduce the base number of initial pest sources.

(1) Clean the field. After the harvest of potatoes or eggplants, the field should be cleaned in time. Clean up the weeds, fallen leaves, and stones in the field and surrounding areas, as well as destroy the overwintering places of adult insects. Handle the leftover stems of potatoes or eggplants, eradicate the residual pests in the stems, and reduce the base number of overwintering pests.

(2) Manual killing. Use the overwintering habit of potato ladybugs and tomato twenty-eight star ladybugs to eliminate overwintering adults; during the peak period of egg laying, use their bright colored egg masses for easy recognition, and manually remove the egg masses on the back of leaves; you can also manually kill adults or remove egg masses on the back of leaves in the morning and evening; use the feigning death of adults, shake the plants, collect the fallen adults and larvae, and kill them.

(3) Pesticide prevention and control. During the occurrence period of overwintering adults and the hatching peak of the first generation of larvae, spray with 3 000 times dilution of Mieshaobi(21% cyanamide emulsion), 3 000 times dilution of 2. 5% bromocyanamide, 1 500 times dilution of 10% bromine horse emulsion, 1 000 times dilution of 10% cypermethrin emulsion, 1 000 times dilution of 50% xin sulphur emulsion, 2 000 times dilution of 2. 5% Kung Fu emulsion, 2 000 times dilution of abamectin, 1 500 times dilution of nicotine, 1 000 times dilution of 4% enovirin emulsion, etc.

046 What causes the water-stain like rot at the base of celery leaves?

The water-stain like rot at the base of celery leaves is often caused by celery soft rot. This disease often occurs during the celery transplanting recovery period or the early growth period after recovery. Generally, it starts from the tender and juicy base of the petiole. In the early stage of the disease, the lesion is pale brown, water-stain like, spindle-shaped or irregularly shaped, slightly sunken, and quickly expands to black-brown rot inside with a foul smell, and only the epidermis remains in the end(Figure 31).

The pathogen of celery soft rot is a bacterium of the genus Erwinia. The bacterium overwinters in the soil and invades through wounds, spreading through rainwater or irrigation water. The disease can occur at temperatures between 4 and 36 degrees Celsius, with the optimal temperature being 27−30 degrees Celsius. The disease is severe under high humidity conditions and is prone to occur when it is cold.

(1) Crop rotation. Implement crop rotation for more than 2 years, avoid rotation with

Figure 31 Celery Soft Rot

Solanaceae and cucurbits.

(2) Strengthen cultivation management. Reduce the occurrence of wounds, such as planting, loosening soil, weeding, and other farming operations to avoid root damage, timely pest prevention. Remove diseased plants in time and sprinkle with quicklime for disinfection. Plant reasonably, weed in time, drain water in time after rain to reduce field humidity, reduce watering or stop watering during the disease period. Planting should not be too deep, and soil should not be too high to avoid burying the petiole in the soil.

(3) Chemical prevention. At the beginning of the disease, you can use 72% agricultural streptomycin or new plant mycin 3 000-4 000 times liquid, 14% roxithromycin 350 times liquid, 40% bactericidal agent 8 000 times liquid spray, use one of the above drugs, or alternate use, spray once every 7-10 days, spray 2-3 times in a row, focus on spraying at the base of the petiole.

047 What is celery heart rot? What is the reason for its formation?

Celery heart rot mainly occurs in protected areas. After celery grows to 10 true leaves, there are chlorotic spots along the edge of the tender leaves, which soon turn brown and necrotic, and the heart leaves of severe diseased plants turn yellow and die. In a humid environment, the heart leaf part will rot due to infection by miscellaneous bacteria, commonly known as rotting heart(Figure 32).

The main reasons for the disease are as follows. Physiological disease caused by calcium deficiency. Due to high temperature, drought, improper fertilization or too high greenhouse temperature, lack of water, hindering root absorption of calcium, causing rotting heart; or too

Figure 32 Celery Heart Rot

much nitrogen, potassium, magnesium in the soil, due to antagonistic action hindering plant absorption of calcium; high salt concentration in the soil, causing difficulty in calcium absorption. Boron deficiency rotting heart. The available boron content in the soil is low or the plant cannot normally absorb boron due to antagonistic action. Excessive nitrogen fertilizer in dry conditions will affect the root's absorption of boron.

(1) Reasonable fertilization. Apply enough base fertilizer and increase the application of phosphorus and potassium fertilizer and boron fertilizer to cultivate strong plants and improve disease resistance. Do not use too much nitrogen and potassium fertilizer to avoid causing boron absorption obstruction, and appropriately increase the application of boron fertilizer.

(2) Reasonable irrigation. Do not make the bed surface too dry. Reasonable irrigation, not large water flooding. Keep warm and irrigate properly when the temperature is low.

(3) Strengthen cultivation management. timely removal of diseased residues, reasonable planting density, increase ventilation and light transmission.

(4) Foliar fertilization. In the early stage of the disease, foliar spraying with Meiling Efficient Calcium can be selected to supplement calcium, which has a very good prevention effect on celery heart rot. Generally, first dissolve 5 g of adjuvant in 15 kg of clean water, then add 50 g of Meiling Efficient Calcium to spray the heart leaves of celery until it drips. Spray twice. Be sure to spray on the heart leaves when spraying, spraying on other leaves is ineffective.

048 How are the black spots formed on the leaves of green onions?

The black spots on the leaves of green onions are called purple blotch of onion, also known as black spot disease, which occurs all over the country. This disease is serious in the northeastern region. In addition to harming onions, it also harms other lilies such as onions, leeks, and garlic. It mainly harms leaves and flower stalks. The initial disease spots are watery white spots, most of which are near the leaf tips or in the middle of the flower stalks, slightly concave. The disease spots are initially very small and gradually enlarge. The size is (1-3) cm × (2-4) cm, purple or brown, elliptical, with obvious concentric stripes, and black-brown mildew on the disease spots when wet. The disease spots can heal each other to form large spots(Figure 33).

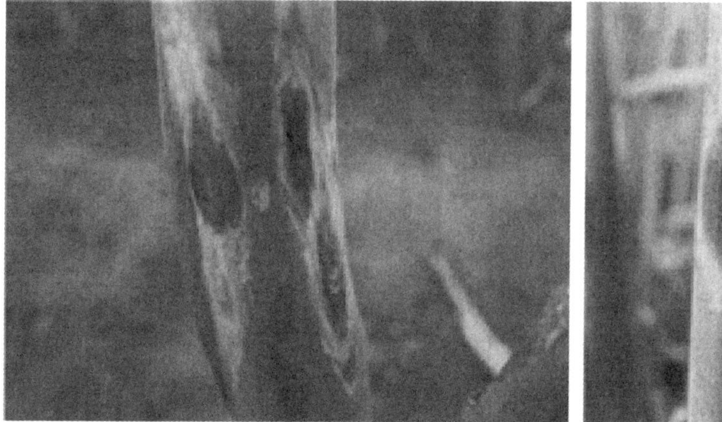
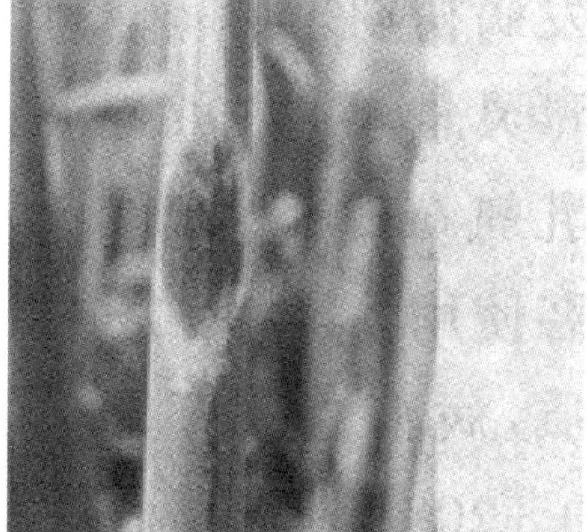

Figure 33 Purple Blotch of Onion

The pathogen of the purple blotch of onion is a chain spore fungus in the subphylum of Ascomycota. In the warm southern regions, the asexual spores of the pathogen harm all year round, and there is no overwintering phenomenon. But in the cold north, the mycelium overwinters in the host or with the diseased remains in the soil. In addition, the seeds can also carry the fungus. When conditions are suitable the next year, asexual spores are produced, which are transmitted by air flow or rainwater, invade from the stomata or wounds of the plant, or directly penetrate the host epidermis. After the disease occurs, asexual spores are produced in the diseased part for re-infection. The suitable temperature for the disease is 25-27 ℃, and it does not occur below 12 ℃. The germination and invasion of the pathogen require rainwater or dew, so heavy rain, long dew time, heavy fog, and large diurnal temperature difference cause heavy disease; poor land, continuous cropping, low-lying land, excessive watering, lack of nutrition, poor growth, and many wounds caused by pests also cause heavy disease; protected areas are heavier than open fields. There are differences between host varieties, and the red onion leaves have more waxy layers on the surface than the yellow and white onions, so they are more disease-resistant.

049 How to prevent and control the purple blotch of onion?

(1) Clean up the garden. After harvest, remove the diseased remains in the field, plow deeply and turn the soil to eliminate the overwintering fungus source. During the growing season, remove the diseased plants or pick off the diseased leaves and diseased flower stalks for centralized destruction.

(2) Crop rotation. In heavily diseased areas, practice crop rotation for more than 2 years with non-onion crops to reduce the fungus source in the field.

(3) Strengthen cultivation management. Choose flat, well-drained fertile loamy soil for planting. Apply more organic fertilizer, increase phosphorus and potassium fertilizer, cultivate soil fertility, and enhance the host's disease resistance. Harvest in time, store at low temperature, onion harvest should be after the top of the onion matures, harvest and dry, wait for the outer part of the bulb to dry before entering the cellar, the storage cellar should maintain low temperature(0−3 ℃) and low humidity(relative humidity less than 65%), and ventilate frequently. Timely eradication of onion thrips can reduce wounds and thus reduce the occurrence of diseases.

050 What are the harmful symptoms and prevention methods of onion downy mildew?

Onion downy mildew is a major disease of onions and garlic, and can also harm leeks and chives. Under suitable conditions, it can spread rapidly and cause severe damage.

Onion downy mildew infects the bulbs and often systemically infects the plant. Infected leaves turn greyish green, sometimes with white mildew. In severe cases, the leaves can become twisted and deformed and the plant will be stunted. When humid, a white mildew layer grows on the surface of leaves and stems; when dry, only white spots appear on the leaves. During the growing period, leaves and flower stalks show light yellow-green to yellow-white spots of varying sizes, elongated to oval, with white or purple mildew or dryness. Spots appear in the middle or lower part of the leaves, which wilt and dry up after hanging down. Early-onset plants twist, and late-onset infections cause ruptures and affect seed maturity (Figure 34).

Onion downy mildew often starts when the leaves reach 12−15 cm in length. The leaves turn light yellow-green or pale and are covered with white or light purple mildew before wilting and dying. In high humidity, the leaves rot and often break off from the infected part.

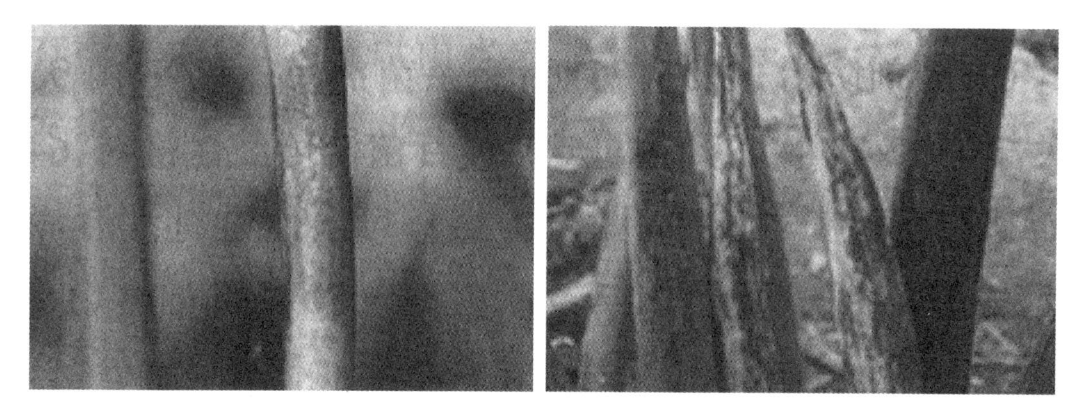

Figure 34 Onion Downy Mildew

The pathogen of onion downy mildew is a fungus from the genus Peronospora in the phylum Oomycota. The pathogen overwinters in the soil with oospores in the diseased residues, can survive for many years, or overwinters in the bulbs of the host plant with the mycelium, leading to systemic infection. In the spring of the second year, under suitable conditions, the oospores are spread to the leaves by rain or irrigation water, germinate to produce germ tubes, invade through the stomata, and cause disease. Sporangia are produced on the infected areas, mainly spread by air currents, but also by rain, insects, and agricultural operations. When the temperature is 15-20 ℃ and the relative air humidity is above 80%, a large number of sporangia will be produced, and the higher the humidity, the more sporangia will be produced. The germination and invasion of sporangia also require the presence of water droplets or a water film to complete. Therefore, the disease is prone to occur in rainy, foggy, and dewy weather, particularly during continuous heavy fog for more than 3 days, which is most likely to induce an outbreak of the disease. In addition, fields with low-lying terrain, sticky soil, heavy irrigation, excessive density, and poor growth are more likely to be severely infected.

Due to the fast spread of this disease and the difficulty in controlling it after infection, prevention should be the primary strategy, supplemented by treatment.

(1) Select disease resistant varieties. Choose disease-resistant varieties with pseudo-purple stems, thin leaf tubes, and a thick layer of wax powder.

(2) Crop rotation. Rotate with crops other than onions for 2-3 years.

(3) Clean field. Clean up the diseased residues during harvest, take them out of the field, and bury them deep.

(4) Seed treatment. Choose seeds from fields that have not been planted with onions, leeks, or garlic, which are high and dry. The seeds purchased should be soaked in water at 50 ℃ for 25 minutes, then cooled and sown.

(5) Strengthen cultivation management. Choose high and dry plots with good ventilation

and drainage, apply enough base fertilizer, appropriately increase the application of phosphorus and potassium fertilizers, and carry out high-ridge cultivation. Irrigate properly to avoid excessive moisture in the onion field. Increase the frequency of mid-cultivation to reduce field humidity and control disease occurrence. Diseased plants should be pulled out and properly disposed of in time. After harvest, deep plowing should be carried out to reduce the overwintering sources of the fungus.

051 What are the prevention and control measures for scallion rust disease?

Scallion rust disease occurs severely in autumn, and can also affect onions, garlic, leeks, etc. It mainly occurs on leaves and flower stems. At the beginning, spindle or oval orange-yellow raised spots (summer spore heaps) appear on the epidermis, and then the epidermis splits longitudinally to scatter orange-yellow powder. In late autumn, brown spots (winter spore heaps) appear on the orange-yellow lesions, which are not easy to rupture. When the disease is severe, the diseased leaves turn yellowish-white and die(Figure 35).

Figure 35 Scallion Rust Disease

The pathogen of scallion rust disease is a fungus of the subphylum Puccinia. It can occur all year round in warm areas. The summer spores of the pathogen are spread by airflow and rain splash, and they harm between onion and garlic vegetables. In northern areas, winter spores mainly overwinter with diseased residues or on overwintering garlic and seed onion plants. After the temperature rises in the spring, a new generation of summer spores is produced on the overwintering plants, which is spread by wind and rain and enters through the epidermis or stomata, infecting nearby plants and nearby fields, resulting in areas where diseased plants are concentrated in the field, that is "disease centers".

Scallion rust disease is prone to occur in areas with low temperature and heavy rain in summer and autumn, and it occurs severely in autumn. High plant density, excessive application of water and fertilizer, field shading, or low-lying terrain, prone to water accumulation, all contribute to the epidemic of rust disease. Insufficient fertilization and weak growth are prone to disease. There are obvious differences in disease resistance between varieties.

(1) Choose disease-resistant varieties. Eliminate highly susceptible varieties, avoid continuous or intercropping of onion and garlic vegetables. Strengthen water and fertilizer management to enhance plant growth and disease resistance. Pay attention to drainage and reduce field humidity. Severely diseased plots are harvested in time. At the beginning of the disease, diseased leaves are picked in time, deeply buried or burned.

(2) Chemical control. In early spring, find the disease center, spray medicine to block it, and then spray medicine in time according to the development of the disease and rainfall. Commonly used drugs are 25% triazole wettable powder 2 000-3 000 times liquid, 50% wilt rust spirit emulsion 800-1 000 times liquid, 65% manganese zinc wettable powder 400-500 times liquid, 70% new life wettable powder 600 times liquid, 25% enemy off emulsion 3 000 times liquid, etc., spray medicine once every 7 days, continuous prevention and control 2-3 times. The effect is better with alternating use of various drugs.

052 What are the causes and prevention methods of scallion soft rot disease?

Scallion soft rot disease mainly harms bulbs. Soft rot often occurs during the bulb enlargement period of onions. In the early stage of the disease, the bulb appears water-stained, semi-transparent gray-white patches appear on the lower part of the outer leaves of the plant, the base of the leaf sheath begins to soften and rot, which eventually leads to the side bending of the outer leaves and the expansion of the disease spots downward; the leaves turn yellow, thin, and the leaf tips dry up, the whole plant shrinks and does not grow; the root hairs turn yellow, and the root system turns brown and dies; there is a dark brown mold layer on the epidermis of the bulb (rotting), and the interior begins to rot after the onset of scallion soft rot disease, and will emit an unpleasant smell (Figure 36).

The pathogen of scallion soft rot disease is a bacterium of the genus Erwinia carotovora. The disease residues in the onion field, soil, and unripened fertilizer are the overwintering places of the pathogen. It is spread by fertilization, running water, and pests such as onion thrips and leafminer flies. It invades through wounds on the bulb. The suitable temperature for the growth of the pathogen is 25-30 ℃. High temperature and high humidity, low-lying continuous cropping, and plant elongation conditions are prone to disease.

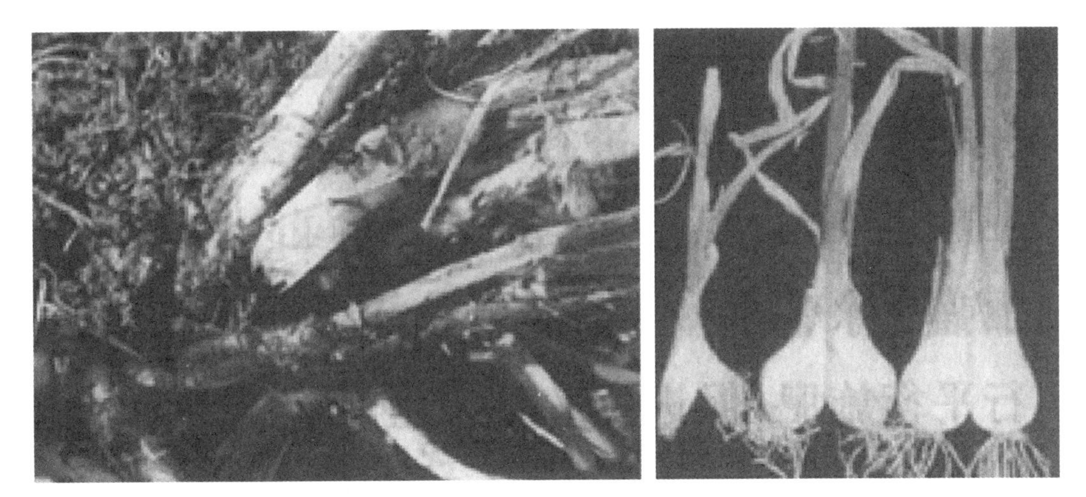

Figure 36 scallion Soft Rot Disease

(1) Seed treatment. The method is to first soak the seeds in 55 ℃ warm water for 20 minutes. Then soak in 72% agricultural streptomycin sulfate 2 000 times liquid for 30 minutes, shade dry and then sow.

(2) Rotation. Choose a high-lying, well-drained plot that has not been planted with onions, garlic, and cruciferous crops in recent years, which can effectively prevent the occurrence of onion soft rot. Farmers have a saying that "spicy against spicy, must be blind, onion garlic does not meet", which means that onion production is most afraid of continuous cropping. Onions and crops like corn can be rotated for more than 3 years, which can deteriorate the living environment of the pathogen and reduce the number of germ sources. Onions can also be rotated with cabbage and celery. In addition, since the onion is a shallow-rooted plant, it can also be rotated with deep-rooted root vegetables, solanaceae, legumes, and melons.

(3) Soil sterilization. For plots that have been severely diseased for many years, quicklime or 50% enemy sulfonate can be spread for soil sterilization before planting. The amount of quicklime is 25-40 kg per acre, and the amount of enemy sulfonate is 0. 5-1 kg per acre.

(4) Strengthen fertilizer and water management. Apply more organic fertilizer, rationally apply nitrogen fertilizer, heavily apply phosphorus and potassium fertilizer, implement balanced fertilization, and promote the disease resistance of onions. Timely control of pests such as onion thrips, leafminer flies and seed flies.

If the scallion encounters flooding during the growth period, you should take the "four early" management measures for remediation. "One early" is early drainage. Drain in time after flooding to minimize the time of field water accumulation. Because water accumulation can cause the soil to become compact and airtight, the root system of the scallion will suffocate due to insufficient air in the soil, causing root rot and root rot, creating an opportunity for the invasion of the pathogen. "Two early" is early hoeing. Due to water accumulation, a sticky skin will appear on the surface of the soil, making the soil airtight. Early hoeing can break the surface and disperse the soil, increase air permeability, and benefit the normal respiration of the scallion root system. "Three early" is early watering. Because the scallion field will become saline-alkali after water accumulation, it is often wet and not dry. At this time, watering can play the role of pressing alkali and promoting early drying in the field. The folk proverb "drought hoeing field, flooding pouring garden" means this. "Four early" is early topdressing. Because water accumulation will cause a large amount of nutrients in the soil to be lost, the scallion cannot get enough nutrient supply, the growth will be weak, and it will be easily infected by diseases. Therefore, it is necessary to topdress early with watering to supplement nutrients.

(5) Clean up the garden. Timely pull out the central diseased plants and take them out of the field for destruction.

(6) Chemical prevention and control. At the onset of the disease, spray with 77% Kocide wettable powder at 500 times dilution, 72% agricultural streptomycin sulfate wettable powder at 4 000 times dilution, 50% Daconil aqueous solution at 1 000 times dilution, 77% copper hydroxide wettable powder at 500 times dilution, 50% Diquat wettable powder at 200 times dilution, 50% Copper Octanoate wettable powder at 500 times dilution, 12% green copper lactate oil emulsion at 500 times dilution, 14% copper complex water solution at 300 times dilution, etc. These drugs can also be used to irrigate the roots. Depending on the condition of the disease, apply the drug once every 7–10 days. After 2–3 times, the disease can be effectively controlled.

Chapter 6 Q&A on Practical Techniques for Cut Flower Production

Manual of Horticultural Techniques
园艺技术手册

1. Basics of Cut Flower Production

001 What are cut flowers? What types of cut flowers are there?

Fresh cut flowers, or simply cut flowers, refer to the flowers, branches, leaves, fruits, etc., that are cut from cultivated or wild floral plants for decorative purposes. They include cut flowers, cut leaves, cut branches, cut fruits, etc. The cultivation method that uses modern cultivation techniques for protection or open-field cultivation, with high yield per unit area, short growth cycle, and capable of supplying fresh flowers all year round, is known as cut flower cultivation.

Roses, Iris, Chrysanthemums, Carnations, and Lilies are the top five cut flowers in the world with a large production volume. Emerging cut flowers include cut flower Gerbera, cut flower Tulips, cut flower Gypsophila, hybrid Euphorbia, cut flower Forget-me-not, etc.; cut branches include cut branch Willow, cut branch Gardenia, cut branch Amygdalus Davidiana, cut branch Acacia, etc.; cut leaf flowers mainly include cut leaf Ferns, cut leaf Asparagus, etc.

002 What are the characteristics of cut flower production? What are the cultivation methods for cut flowers?

Cut flower production is an important part of floral production, accounting for 60% of the total sales of flowers worldwide, and is the mainstay of floral production. Cut flowers have a high yield per unit area and generate large profits; they are suitable for large-scale production, and their cultivation in facilities can be artificially controlled. The harvesting and selling of cut flowers are highly intensive and can be supplied year-round. Cut flowers have a large consumption volume and a wide market. The international market's demand for roses, chrysanthemums, carnations, baby's breath, Iris, African daisy, lilies, and corresponding foliage plants is increasing year by year. Cut flower production has become a major direction for the adjustment of China's agricultural industry structure.

With the continuous improvement of people's living standards, fresh cut flowers have become a popular consumer product, especially for business openings and wedding occasions, where the decoration of fresh flowers adds a festive and romantic color. Bringing a bouquet of flowers when visiting relatives and friends can also reflect warmth and care.

Yunnan Province is China's largest cultivation base for fresh cut flowers in terms of production area and volume, accounting for 40% of China's annual cut flower production. Lingyuan in Liaoning Province is an important base for fresh cut flower production in northern China. In addition to meeting domestic market demand, most of the fresh cut flowers produced in China are exported to countries like Japan and Singapore. In recent years, Beijing, Shanghai, Shandong, Hainan, and other places have developed cut flower production bases, and the level of cut flower production is constantly improving.

From the perspective of cultivation facilities, the cultivation methods of cut flowers can be divided into open field cultivation and protected cultivation. Open field cultivation is highly seasonal, and the environment is not easy to control, making it difficult to ensure the quality of cut flowers. This cultivation method is commonly used in Fujian, Yunnan, and Hainan regions of China. In most parts of China, protected cultivation is the main production method.

003 What are the measures for preserving cut flowers?

The preservation of cut flowers refers to the practice of using low-temperature refrigeration or preservatives to extend the lifespan of fresh flowers after they have been harvested. It includes the overall measures for preserving fresh flowers during the post-harvest pre-treatment, storage, transportation, and retail sale processes, and is an important part of cut flower production and sales.

Cut flower preservation is a systematic project that involves every stage of cut flower production, including harvesting, packaging, transportation, sales, and application. Producing high-quality cut flowers free of pests and diseases is the foundation of preservation.

(1) Timely harvesting. This is very important for preservation. The appropriate time for harvesting depends on the season and the type of cut flower. For example, lilies can be harvested when the first bud shows color in winter, and can be harvested a little greener in summer.

(2) Proper processing after harvest. Cut flowers will wilt after losing 5% to 8% of their water. Insensitive flowers can return to their original state after rehydration, while sensitive flowers find it hard to recover even after rehydration. Therefore, the stems of most types of flowers should be immediately placed in cold water below 5 ℃ after being cut from the plant, with the ambient air humidity controlled at around 95% and the temperature maintained at 2–4 ℃, avoiding direct sunlight.

(3) Pre-treatment before boxing and transportation. It is best to pre-treat the cut flowers in a cold storage to dissipate the residual heat from the field. Necessary packaging should be used during transportation to reduce mechanical damage. The boxes used for transportation should not be too large. The transaction time should be minimized during wholesale sales.

(4) Good use of preservatives in retail. Silver Thiosulfate (STS) is a commonly used preservative.

Commonly used drug concentrations are: Indole Acetic Acid (IAA) 1-100 mg/L, Naphthalene Acetic Acid (NAA) 1-50 mg/L, Gibberellic Acid (GA) 1-400 mg/L, Silver Nitrate 10-100 mg/L, Silver Thiosulfate (STS) 20-400 mg/L, Boric Acid 100-1 000 mg/L, Sodium Hypochlorite 1 mg/L, Sugar 10-30 mg/L, Alcohol 5 mg/L.

In the specific preservation work, it should be noted that different flowers should have corresponding measures in different transportation and storage processes.

2. Q&A on Cut Rose Production Technology

004 What are the characteristics of cut roses?

Known as the "queen of flowers", the rose is one of China's top ten famous flowers. The cut rose is a shrub variety in the rose family that blooms all year round, generally referred to as a rose by consumers. It belongs to the prickly shrub of the rose family. Cut roses have strong upright branches, odd-pinnate compound leaves that are alternate, with 3-5 leaflets. Flowers are solitary or several cluster at the top of the stem, with numerous petals, and double-flowered types. There are many types of flower colors and shapes.

There are over 200 species in the rose genus, with more than 80 originating in China. The China rose, one of the main parent species of roses, originated in the central and southwestern regions of China, was introduced to Europe more than 200 years ago, and later hybridized with tea roses (produced in Yunnan), Turkistan roses (produced in Central Asia), French roses, etc., to breed various modern roses. There are now more than ten thousand varieties. Because there are many parents, they are referred to as cultivated varieties.

The basic characteristics of modern cut rose varieties include: beautiful flower shape, high heart curled edge or high heart upturned corner, especially when the flower opens 1/3-1/2, it is elegant and restrained, and the opening process is slow; the petals are hard, the flower vase life is long, the outer petals are neat, there are no broken petals, and the petals are not easily damaged by medicine; the flower color is bright, brisk, pure, and with a velvet sheen; the flower stem and flower stalk are stiff and straight, with strong support; the leaf size is moderate, the leaf surface is flat and glossy; the stem has fewer thorns, is resistant to pruning, has strong bud germination, and a high flowering rate; it has a high flower production, and the variety has a strong growth ability.

005 What kind of environment do cut roses prefer?

Cut roses prefer an environment with ample sunlight, a relative humidity of 70%-75%, and good air circulation. They require well-drained, fertile and moist loose soil, with a pH of 6-7. They need at least 5-6 hours of direct sunlight each day to grow well. The most suitable growth temperature during the day is 18-25 ℃, and 15.5-16.5 ℃ at night. Although they can withstand high temperatures of 35 ℃, and won't die at low temperatures of 5-15 ℃, diseases become severe in high temperatures above 30 ℃ and in humid low-temperature environments. If the environment is too dry and the temperature is below 5 ℃, the plants will enter a dormant or semi-dormant state. Dormant plants drop their leaves and do not flower.

006 What are the main varieties of cut roses on the Chinese cut flower market?

To produce high-quality cut roses, the variety should meet the following basic standards: generally, it should have a high heart with curled edges or upturned corners; strong double flowered, with many layers of petals arranged compactly; the petals should be thick and have a good texture, with neat outer petals and no fragmentation; the flower color should be bright, pure, and vibrant; the flower stem and stalk should be erect with strong support and a certain length; the flower opening process should be slow, with good vase life.

At present, the main varieties of cut roses produced in China are: the red series varieties "Carola" "Legend" "Bride" "Treasure" "Rhodes" "Glory" "Style" "Red Bishop" "Red Success", etc. (Figure 1).

Figure 1 Cut Roses

The pink series varieties include "Candy Snow Mountain" "Pink Snow Mountain" "Sweet Snow Mountain" "Gorgeous Pink" "Maria" "Awakening" "Pink Beauty", etc.

The yellow series varieties include "Golden Branches and Jade Leaves" "Golden Glow" "Intoxication" "Butterfly Yellow" "Asmer Gold" "Golden Badge" "Golden Medal" "Golden Age", etc.

The white series varieties include "Snow Mountain" "Titanic" "White Success" "Athena" "Wedding White" "Lychee".

Other color varieties include "Morika" "Holiday Princess" which are orange; "Peach Snow Mountain" is champagne color. There are also many bicolor varieties.

Multiflora varieties include "Multiflora Pink Lady" in pink, "Multiflora Violet" in deep red, "Multiflora Champagne Rose" in champagne, "Multiflora Carnival Bubble" in orange yellow, "Multiflora Juice Bubble" in orange red, "Multiflora Yellow Butterfly" in yellow, "Multiflora Leo" in white (Figure 2).

Figure 2 Multiflora Cut Roses

In production, the appropriate species and varieties should be selected according to the local natural conditions. The cultivation area and ratio of flowers with different colors and sizes should be properly adjusted to meet market demand. Generally, the ratio of red, scarlet, pink, yellow, and white is 40%: 15%: 15%: 20%: 10%. Medium-sized flowers have high yields and are suitable for greenhouse production in winter and spring. Greenhouse cultivated varieties should have strong resistance to powdery mildew, while open-field cultivated varieties should be resistant to black spot disease.

007 How are cut roses propagated?

The main propagation methods of cut roses are cutting and grafting, and grafting is commonly used in production. Grafted seedlings grow well, and the quality and yield of cut flowers are high. Cutting seedlings grow slowly and have low yield in the early stage, but grow steadily and have high yield in the later stage, and are often used in soilless cultivation.

Grafting is commonly used in nursery cultivation, with "T" budding being the most widely applied. The rootstocks can be seedlings or cuttings of "Thornless Dog Rose" "Wild Rose" "Pink Cluster Rose". The standard rootstock is a two-year-old seedling, with a diameter of 8−13mm, well-developed, and free from pests and diseases.

Budding is suitable to be carried out at 15−25 ℃ during the growth season. The current year's scions are selected, with the axillary buds on the scions being fully developed. The triangular buds at the base and tip of the branch are removed. The bud should be grafted onto the smooth bark of the rootstock 4 - 5 cm from the ground. If the rootstocks are planted densely in rows, it is best to have all the buds on the same side for easy management and inspection. The operation process is shown in Figure 3.

Figure 3 "T" Shape Budding

(1) Treatment of rootstocks. A vertical cut about 2.5 cm long is made from top to bottom on the smooth skin of the rootstock about 4–5 cm above the ground. A horizontal cut is made at the top end of the vertical cut, which accounts for about 1/3 of the circumference of the rootstock. Twist the budding knife slightly to peel off the two layers of skin.

(2) Preparation for budding. Start peeling from about 1.3 cm below the bud and continue until about 2.5 cm above the bud. Cut a horizontal slit about 2 cm above the bud, penetrating the cortex into the xylem, in order to take off the bud piece without the xylem.

(3) Insertion of the bud into the rootstock slit. Insert the bud piece into the peeled-off two layers of skin. Until the bud piece and the horizontal cut of the rootstock are aligned. Tie the budding joint tightly with a plastic strip.

Precautions for grafting: The grafting knife should be sharp and disinfected to prevent infection at the grafting site. The length and thickness of the bud should be appropriate, and the rootstock and bud should be closely fitted, firmly tied, and sealed. Water the field before and after grafting to keep it moist, but do not wet the bud. The grafting operation should be quick and accurate. After survival, promptly wipe off the buds sprouting on the rootstock. When the bud grows to more than 15 cm, cut off the scion on the rootstock. The nursery of grafted seedlings should timely loosen the soil, weed and control pests and diseases.

Cutting propagation can be carried out throughout the growth period, generally suitable from April to October. Select semi-mature and robust branches that have been flowering for about a week, and take care not to choose "blind branches" that have not flowered. Cut the branches into small sections with 2–3 buds, retain 1 compound leaf to reduce water evaporation, dip the base of the cutting in some growth regulators, such as IBA or NAA, to promote rooting. The cuttings are spaced 3–5 cm apart, inserted 2.5–3.0 cm deep into a river sand substrate, sprayed with water after insertion, and properly shaded to retain moisture. Using full-light mist propagation can improve the survival rate.

008 How to prepare the soil before planting cut roses?

Cut roses can maintain a production capacity for 4–5 years after successful planting, with a root system that goes deep and wide into the soil. Keeping a good growing environment for the root system is key to the steady and high yield of cut roses.

If the soil is heavy and sticky, soil amendments (a mixture of bark, sawdust, rice bran, and other organic matter) should be filled into the planting trench. The purpose of using such soil amendments is to make the soil loose and well-aerated, with good drainage. The amount of amendments is about 20% of the total soil volume, and it should be thoroughly mixed with the original soil.

Before planting, apply fully decomposed organic fertilizer, deeply turn the soil to 40–50 cm, and then prepare the planting beds. In the rainy and humid south, it is suitable to make high beds, while in the arid north, it is suitable to make low beds. For greenhouse production, small high beds can be made and used after soil sterilization. The reference amount of base fertilizer per 100 m^2 is as follows: compost or pig manure 500 kg, cow manure 300 kg, fish waste 20 kg, sheep manure 300 kg, oil residue 10 kg, bone meal 35 kg, superphosphate 20 kg, wood ash 25 kg. It should be applied during soil preparation, serving as the basic fertilizer for 2–3 years or more after rose planting.

The planting bed should be 15–20 cm above the ground, the width of the double-row planting bed should be 50 cm, the distance between beds should be 40–45 cm, and the planting bed should run from north to south, with the length depending on specific conditions.

The entire preparation work before planting should be completed within one month before planting.

009 What should be noted when planting cut roses?

The planting time is generally between March and September, with the best time being from May to June. After planting, flowers usually start to appear after about 6 months. Cutting seedlings should be replaced after 2–3 years, and grafted seedlings should be replaced after 4–5 years. Single-row double-row cultivation is recommended, with a plant spacing of 20–30 cm and row spacing of 40–50 cm. Planting should be done in cloudy, low-temperature weather, preferably in the early morning or evening.

During planting, the root system should be spread out in all directions. For grafted seedlings, the grafting site should be 1−2 cm above the soil surface, and the soil should be slightly mounded at the grafting site so that the soil around the plant is higher than the ground. After watering thoroughly to establish the roots and the soil has settled, loosen the soil and level it, keeping the grafting interface above the soil surface to prevent the scion from producing its roots. For seedlings grown in nutritional pots, the original soil clump should be broken up during planting, allowing curled roots to spread out. If the roots of the bare-root seedlings are found to be dry, they should be soaked in water for 24−48 hours before planting to ensure they are fully hydrated.

After planting, ensure that the soil is moist, spray water on the leaves during the day, and provide some shade. After 3−5 days, it is possible to check whether white new roots have been produced. If a large number of white new roots have been produced, it indicates that the planting has been successful.

010 How to shape and prune young rose seedlings?

There are two basic types of buds on rose branches. The buds at the base of the 3-leaflet leaf and the first 5-leaflet leaf on the flower branch are sharp buds. The buds below the second 5-leaflet leaf are round buds. The flower branches that sprout from the round buds are long, and the flowers at the top are also large, suitable for use as flowering branches. (Figure 4)

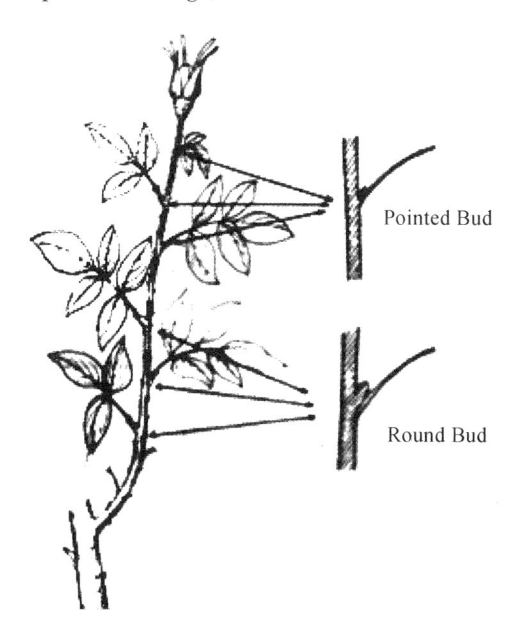

Figure 4 Distribution of Bud Types on Rose Branches

In the seedling stage, it is necessary to remove the flower buds in time, allowing the plant to remain in vigorous nutritional growth, promoting the development of the root system, and good leaf development. This will allow more substantial branches to form at the base of the plant, cultivating them into blooming mother branches. When operating, remove the flower buds about 0.5 cm at the top of the branch, retain the branches and leaves, and wait for the buds at the base to form branches with a diameter of 0.6–0.8 cm, which can be kept as the main flowering branch. If the thickness is not enough, continue to remove the buds and remove the side buds sprouting from the upper part.

011 How to manage the branches of roses for blooming?

When the flower buds on the branches reserved for blooming mother branches start to color, cut the branches from the second 5-leaflet leaf slightly above the base. When the round buds in the leaf axils below develop into sufficiently thick branches or when the flower bud diameter grows to 0.5 cm, cut off the tip above the second 5-leaflet leaf and leave it as the main branch. The shoots of the main flowering branches should not be cut too early. The branch length should be more than 50–60 cm, with more than 8–10 leaves, to ensure that enough flowering branches can sprout from it.

012 What daily pruning should be done for cut roses?

When the flower buds on the branches are medium-sized, remove the side buds and side buds on the branches in time to protect the robust growth of the branches and the blooming of the top buds (remove the top buds for multi-headed roses and leave multiple side buds).

Timely cut off the sprouts emitted from the rootstock, check whether there are self-rooting at the base of the scion and remove them in time.

Timely remove the flower buds of weak branches and retain the leaves as auxiliary branches. The weak branches that grow on weak branches should be cut off; the weak branches that grow on robust branches should be cut off at the joint; the robust branches that sprout on robust branches should be cut off above the first healthy 5-leaflet leaf (Figure 5).

To maintain a uniform tree vigor and upward growth of flowering branches, pruning usually involves leaving outward buds to allow the plant to spread. Diseased branches and leaves should be removed promptly and taken out of the nursery for destruction.

In July and August, do not prune, only pick flower buds, retain leaves, and after the beginning of autumn, cut off the upper part, leaving 2–3 leaves. By late September, the plant can enter the peak blooming period.

In mid to late December, carry out a low position heavy pruning, that is, cut back the plant to a height of about 60 cm from the ground. This can produce early spring flowers around the Qingming Festival when the flower price is high, and reach the peak blooming period around "May 1st".

Figure 5 Handling of Weak and Strong Branches Sprouting From Branches of Different Diameters

1—Treatment of weak branches growing on weak branches; 2—Handling of weak branches growing from robust branches; 3—Strong branches sprouting on robust branches should be cut off above the first healthy 5-leaflet leaf

013 How to control the flowering period of cut roses?

(1) Treatment with plant growth regulators. Spraying 100 mg/L of gibberellin in late autumn and early winter can allow the harvest of cut roses to be advanced by 7-15 days, but the use of hormones is easily affected by temperature, concentration, rose variety and growth period, so the variables are quite large. Therefore, attention should be paid to the timing, concentration, variety and other issues of use. It is advisable to conduct small-scale experiments before large-scale treatments.

(2) Pruning regulation method. Generally, roses need 40-45 days in summer and 60-65 days in winter from pruning to flowering. If pruned in late September, it can bloom in mid-November, and if pruned in mid-October, it can bloom in December. You can choose the pruning time according to this rule to control the flowering period.

If flowers are supplied for Christmas (the supply date is from December 10-23), it is necessary to prune and trim the branches from October 25 - 30, and maintain the night temperature in the greenhouse at 8-10 ℃, and daytime at 18-25 ℃. From late November to early December, pay attention to nighttime insulation, and adjust the day and night temperatures according to the growth of rose buds.

If the peak period of flower harvesting is advanced, the temperature can be lowered by opening windows during the day, maintaining the greenhouse temperature at 20-22 ℃ and the night temperature at 8-9 ℃ to delay the harvest time; when the peak period of flower harvesting is delayed, the greenhouse temperature can be maintained at 24-26 ℃ during the day and 12-13 ℃ at night. After entering the harvest period in mid-December, pay attention to nighttime insulation and heating to reduce the temperature difference between day and night.

If flowers are supplied for the Spring Festival and Valentine's Day, the external temperature is relatively low at this time, and it is necessary to prune and trim the branches from December 1–10 (for the Carolina variety, it is from November 10–15), and maintain the greenhouse temperature at 8–10 ℃. During this period, there is a large difference in temperature and humidity between day and night. On sunny days, pay attention to ventilation and cooling at noon, close doors and windows at night to keep warm, heat up at night to reduce humidity, and reduce the temperature and humidity difference between day and night. Adjust the day and night temperatures 15–20 days before harvest according to the growth of rose plants and buds. When the flower harvest time is advanced, open windows during the day to lower the temperature, maintain the greenhouse temperature at 20–22 ℃, and heat up at night to maintain at 8–9 ℃; when the flower harvest time is delayed, only open small windows at noon for ventilation, and maintain the greenhouse temperature at 24–26 ℃; pay attention to snow removal on snowy days to protect the greenhouse facilities and maintain the growth of roses, and maintain the night temperature at 12–13 ℃.

(3) Staggered planting period. Planting can be staggered from March to October each year to avoid concentrated flowering periods. Generally, flowers are harvested about 100 days after planting, and this method can be used to control the flowering period for early production.

014 What are the main pests and diseases of cut rose flowers? How to prevent and control them?

Common diseases of cut rose flowers include downy mildew, powdery mildew, rust, and black spot diseases. The pests mainly include red spider mites and aphids.

(1) Powdery Mildew. Symptoms of the disease include the infection of leaves, stems, and petals. In the early stages of leaf infection, bright yellow spots form along the main veins, side veins, or edge of the leaf. When the humidity is high, a sparse gray-white mold layer can be faintly seen on the back of the spots. The spots can connect to form large leaf spots, and the leaflets are prone to falling off, leading to the falling of leaf stems, and the upper part of the flower branch sheds leaves from bottom to top, resulting in bare stems. When the pathogen infects the flower branch, the infected part's epidermis turns pale purple-red early on, the color gradually deepens, and it slightly sinks. As the disease progresses, the sunken part's epidermis cracks longitudinally, and the upper part of the branch gradually wilts and dies. When the pathogen infects the flower buds, dark red dot-like spots form on the petals in the early stages. The spots gradually expand into round to oval slightly raised spots, with the center not changing color and the edge of the spots being dark red. Later, the

petals turn brown and gradually rot. Dark red spots of the pathogen that are slightly raised often form on the petals of red varieties, while only dark red spots form on the petals of light-colored varieties.

Occurrence pattern. For greenhouse-grown roses, the initial infestation period begins in early December and continues to cause damage until May of the following year. Greenhouses with poor insulation, large temperature differences, and high humidity are severely affected by the disease. Continuous cloudy weather and foggy conditions are inducing factors for this disease, and different varieties have different resistance levels.

Prevention and Control. After watering, ventilation should be ensured and leaves should be kept as dry as possible to prevent dew formation on the leaf surface.

Preventive spraying should begin immediately after the roses are planted in the autumn. Towards the end of autumn, a fungicide specifically for powdery mildew disease should be used. If there are cloudy and rainy days, protective spraying should be done immediately when the weather clears up.

Prevention period drug selection. Spray with 1% Bordeaux mixture or 600 times diluted Antai life, every 10–15 days. If there are cloudy and rainy days, spray every 5–7 days.

Timely medication when disease occurs: Spray thiazoline-300 times solution + 1 500 times solution of Rou Shuitong; Fluorine mazone 600 times solution + 1500 times solution of Rou Shuitong, on both sides of the leaf. Spray every 3 days when the disease is severe, and once every 5 – 7 days after the disease is under control. Appropriate concentrations of Carbendazim and Metalaxyl are also effective.

(2) Powdery Mildew. Symptoms of the disease mainly affect the leaves, stems, and flowers. The mycelium grows on both sides of the leaves, forming white circular powdery spots, which expand and merge to form larger patches as the disease progresses. The spots are covered with a white powdery layer, significantly impacting the photosynthesis of the leaves and leading to premature senescence.

Occurrence pattern. Dry conditions favor the disease. In greenhouse cultivation, there is varying degrees of harm throughout the year.

Prevention and Control. Prune and remove severely diseased leaves and branches, take them out of the greenhouse and burn them in a concentrated manner. Strengthen fertilizer and water management, avoid applying nitrogen fertilizer excessively.

Preventive spraying should begin immediately after the roses are planted in the autumn. A fungicide specifically for powdery mildew disease should be used at the end of summer and beginning of autumn.

Prevention period. Spray with 40% Dakning suspension 600 times solution, 50% Methylthiophanate 1 000 times solution. Spray every 10–15 days.

Treatment. 20% Triazole emulsion 2 000 times solution, 12.5% Cyproconazole emulsion 3 000 times solution; 40% Fuxing emulsion 7 000 times solution. To improve the prevention and control effect, add 1 500 times solution of Rou Shuitong to each sprayer. Spray every 3 days when the disease is severe in spring, and once every 5-7 days after the disease is under control. Alternate medications should be used. In winter, sulphur fumigation in a closed greenhouse environment yields good results.

(3) Rust. The disease is severe under high humidity conditions in winter. Small orange-red spots occur on the stems and leaves. Soon after, the epidermis ruptures, forming powdery spots.

In the early stages of the disease, spray with 15% Rustning wettable powder 1 000-1 500 times solution, once every 10 days, for a total of 3-4 times; or use 12.5% Alkenazole wettable powder 3 000-6 000 times solution, 10% Segao water dispersible granules diluted 6 000-8 000 times, 40% Fuxing emulsion 8 000-10 000 times solution for spray prevention and control.

(4) Black Spot Disease. This disease not only affects roses but also threatens nearly a hundred varieties of plants in the Rosaceae family and their hybrids, including roses, yellow thorns, mountain roses, golden cherries, and white begonias. The disease mainly affects the leaves. In the early stages of infection, brown spots appear on the leaves, which gradually expand into circular or nearly circular spots. The edges are irregularly radiated, the tissues around the disease turn yellow, and there are black spots on the disease spots, which are the conidia of the pathogen. In severe cases, the disease spots become continuous, and even all the leaves of the whole plant fall off, leaving only bare stems. The spots on the young branches are long oval, dark purple-red, and slightly sunken.

This disease is a common and severe disease of roses. It often causes yellow leaves, dead leaves, and leaf fall in summer and autumn, affecting the blooming and growth of roses. The disease is prone to occur in conditions such as water accumulation in low-lying areas, poor ventilation, insufficient light, improper fertilization and watering, and poor sanitation. Different varieties of roses have different resistance to the disease, with light-colored yellow flower varieties generally more susceptible.

Prevention and Control Measures.

Strengthen maintenance and management to enhance the plant's disease resistance; use disease-free plants for cultivation; apply fertilizer and rotate crops reasonably. The planting density should be appropriate to facilitate ventilation and light transmission and reduce humidity; pay attention to watering methods to avoid sprinkler irrigation; pot soil should be updated or disinfected in time.

Eliminate the initial source of infection, thoroughly remove diseased fallen leaves and dead plants, and burn them in a centralized manner. Spray 3-5 Baume degree lime sulfur mixture during the dormant period.

Use fungicides during the disease period, especially in the early stages of the disease, such as 47% Jia Rui Nong wettable powder 600-800 times solution, 40% Fuxing emulsion

8 000–10 000 times solution, 10% Segao water dispersible granules 6 000–8 000 times solution, 10% Multi-antimycotic wettable powder 1 000–2 000 times solution, 6% Le Bi Geng wettable powder 1 500–2 000 times solution.

(5) Aphids. Aphids, both adults and nymphs, gather on the young shoots, flower buds, petals and leaves of rose plants, sucking sap and causing the leaves to shrivel, curl, or form galls, affecting blooming and inducing sooty mold disease.

Adult aphids come in two types: winged and wingless. The same applies to nymphs. Both adult and nymph aphids have a pair of abdominal tubes capable of secreting waxy substances and informational alarm hormones.

Prevention and Control: Pay attention to quarantine and pest situation, and emphasize early prevention and control. When aphids are significantly present, spray with 50% malathion emulsion diluted 800–1 200 times.

Use yellow boards to lure and kill aphids. In flower cultivation areas or greenhouses, place yellow sticky boards to attract and trap winged aphids. You can also use silver-white tin foil to reflect light, discouraging aphids from settling or migrating.

(6) Spider Mites. They generally feed on the leaves of plants, directly damaging the leaf tissue, hence the name leaf mites. They mainly harm the spring shoots, young leaves, flower buds, young fruits, etc. The affected leaves lose their green color, showing pale spots or patches, or the leaves curl, shrivel, deform, and mite galls appear. In severe cases, the entire leaf becomes scorched and falls off.

Prevention and Control. Thorough garden cleaning is essential, including the complete removal of weeds, dead branches, fallen leaves, and the excavation and incineration of dying old trees and pest-infected remains. Deep plowing and soil amendment in winter can kill pests and improve the environment. In dry winters and springs, frequent watering is necessary to increase the humidity in the garden, promote robust plant growth, and enhance the ability to resist diseases and pests.

In the early stages of infection, spray with 40% trichlorfon emulsion diluted 800–1 200 times or 50% trichlorfon wettable powder diluted 800–1 200 times, which are effective against adult mites, nymphs, young mites, and eggs. It is best to spray again 10–15 days later.

3. Q&A on Cut Chrysanthemum Production Technology

015 What are the characteristics of cut chrysanthemums?

Chrysanthemum is a perennial herbaceous flower in the Asteraceae family, native to China, and is one of the traditional famous flowers with the longest cultivation history in our country. Cut chrysanthemums are a group of chrysanthemums suitable for being used as cut flowers. They are one of the four major types of cut flowers in the world, and their production ranks first among the four major types of cut flowers. They have advantages such as diverse flower types, rich colors, wide uses, resistance to transportation and storage, long vase life, easy reproduction and cultivation, year-round supply, low cost, and high output. Considering the domestic and foreign development environment and its own characteristics, the cut chrysanthemum industry is bound to have a broad development prospect.

Japan is a major producer of cut chrysanthemums in the world, accounting for more than 50% of the world's cut chrysanthemum production. It is also a major consumer of cut chrysanthemums, consuming 3–4 million stems on average every day. The domestic production of chrysanthemums in Japan cannot meet the demand, and the production of chrysanthemums is gradually shifting to developing countries with suitable climates and lower production costs. The steady growth of the world's flowers and the shift in the industry have created a favorable external environment for the further development of China's flower industry.

016 What kind of environmental conditions do cut chrysanthemums prefer?

Chrysanthemums favor abundant sunlight, a cool climate, high and dry terrain, and a well-ventilated environment. For the production of cut chrysanthemums, it is advisable to choose a sunny, well-drained site. The soil should have good drainage and aeration, with a pH of 6.5–7.2, and sandy loam soil rich in organic matter is preferred. If two or three crops of chrysanthemums are continuously planted, crop rotation is necessary, and soil disinfection can also be carried out.

Chrysanthemums have a certain cold resistance, with the optimal growth temperature being 15 – 25 ℃, and different varieties have different degrees of cold resistance. It is recorded that the roots of some chrysanthemums can withstand low temperatures of -30 ℃.

After a certain period of vegetative growth, chrysanthemums enter the reproductive growth stage, which is generally when there are about 10 leaves and the plant height is above 25 cm. Most chrysanthemums are short-day plants, and flower buds will only differentiate and bloom when the daylight length is less than 12 hours. Autumn chrysanthemums and cold chrysanthemums belong to this category. Another part of the varieties are sensitive to temperature, summer chrysanthemums and summer-autumn chrysanthemums belong to this category, with flower bud differentiation at 10–15 ℃. If the plants encounter low temperatures or insufficient daylight hours during bud differentiation, a bunch of willow-leaf-shaped leaves will grow at the top of the plant, called "willow buds" or "willow bud heads".

017 What types and varieties of cut chrysanthemums are available for production selection?

The varieties of chrysanthemums that are suitable for cut-flower production are often those with round and pure-colored flowers, tall stems and short, thick necks, fresh green and thick, flat leaves that are not large, suitable for long-distance transportation, and have a long vase life. According to the whole branch method, cut chrysanthemums are divided into standard chrysanthemums and spray chrysanthemums. Standard chrysanthemums have one flower at the top of each stem, usually large and medium-sized varieties; spray chrysanthemums are prolific varieties, with multiple flowers per stem, usually small-flowered varieties. In the northern regions, varieties that are resistant to low temperatures should be chosen for planting cut chrysanthemums that are on the market around Qingming Festival.

Common types and varieties of cut chrysanthemums in China are shown in Table 1.

Table 1 Different types of cut chrysanthemum varieties and their flower bud differentiation and development responses to day length and temperature.

Type	Flowering Period	Temperature Response	Daylight Response	Varieties
Summer Chrysanthemums	Mid-China region from April to June; cold northern regions from June to August	The flowering bud differentiation temperature is around 10 ℃, for some it's 5 ℃. High temperature inhibits bud development	Insensitive	Xin Jingxing, Bai Jingxing, Summer Red, Golden Splendor, Gandaiki, and Battle of Chibi, etc.

Table 1(Continued)

Type	Flowering Period	Temperature Response	Daylight Response	Varieties
Summer-Autumn Chrysanthemums	July to September	The flowering bud differentiation temperature is around 15 ℃. Low temperature inhibits bud differentiation	Quantitative short-day	Jingyun, Jingjun, Bai Tianhui, Baozhi Mountain, and Summer Peony, etc.
Autumn Chrysanthemums	October to November	The flowering bud differentiation temperature is around 15 ℃. The critical low temperature is 10 ℃. High temperature inhibits bud development	Short-day	Xiufang's Force, Autumn Sakura, Huge Treasure, Sun Orange, Japanese Snow Green, and Light of Four Seasons, etc.
Cold Chrysanthemums	December to the following January	The flowering bud differentiation temperature is 6 - 12 ℃. High temperature inhibits bud development	Short-day	Gold Imperial Garden, Silver New Year, Cold Lady Red, Cold White Plum, Cold Sun, Divine Horse, etc

Note: Quantitative short-day refers to the type that can bloom normally under short-day conditions, and short-day treatment can promote blooming.

018 How to Obtain Seedlings of Cut Chrysanthemums?

The production of cut chrysanthemums is often propagated by softwood cuttings.

(1) Training of Cutting Mother Plants. In September and October, after the field is fully fertilized, the virus-free tissue cultured seedlings are planted at a density of 25–30 plants/m^2, and water and fertilizer management is strengthened. When the plant grows to 12–15 cm, it is topped once, removing 1–2 cm of the top bud to promote lateral branches. 20 days later, the plant is topped again to promote more lateral branches for cuttings. Generally, the number of cutting mother plants can be determined according to the ratio of 1∶10 between mother plants and nursery plants. Each mother plant can be cut 3–4 times.

(2) Making of Cutting Beds. The cutting bed is generally 90 cm wide, and the length is

determined according to the actual situation of the shed. A fixed nursery bed can be built with bricks, or directly with soil, with a layer of old plastic at the bottom. The nursery bed with a depth of 15-20 cm is filled with 10-15 cm of cutting substrate, and the aisle is 24 cm wide.

The cutting substrate can be made of sand, perlite, vermiculite, or peat moss, and can be used in a 1: 1: 1: 1 mixture of the four materials, or sand, vermiculite, or peat moss can be used alone. The cutting substrate should not be reused.

(3) Cutting and Management. In the northern regions, cut chrysanthemum cuttings can be taken from November to May of the following year. The time of cutting can be determined according to the planting period and whether topping is carried out.

Cut off a tender shoot of 5-8 cm from the mother plant, with 5-7 leaves, remove the lower leaves, and retain the upper 2-3 leaves. On the cutting bed, use a wooden stick that is close to the thickness of the cutting to make holes of 3 cm × 4 cm, insert the cutting into the substrate for 2-3 cm, compact it, and water it. Maintain a temperature of 15-20 ℃, a relative humidity of 80%-90%, and the cutting can take root in about 10 days, and can be transplanted into a seedling for planting after 20 days.

019 How to plant cut chrysanthemums?

The timing of planting cut chrysanthemums varies depending on the cultivation season. Generally, the planting is done in the afternoon in summer, and in the morning on sunny days in winter. If there is a large amount of planting, it can be done all day under shading conditions.

Before planting, a nylon support net with a mesh size of 10 cm × 10 cm is laid on the cultivation ridge. The surface of the net should be tightened to make each mesh square. The two ends of the net are fixed with baffles and iron pipes. If conditions permit, using an iron wire net would yield the best results.

For single chrysanthemums, the plant row spacing is 10 cm×10 cm, and one cut flower seedling is planted in each mesh. Each ridge is planted with 4-5 rows, with 27 000 plants per acre. When planting, the stem of the seedling is buried 2 cm into the soil to allow the root system to expand. After planting, water thoroughly immediately, and then immediately proceed with seedling support and supplementation. Chrysanthemum seedlings with exposed root systems should be re-covered with soil. Use a shade net for shading.

020 How to manage light exposure after planting standard cut chrysanthemums?

(1) Supplemental lighting cultivation. If the cultivated variety is a typical short-day autumn chrysanthemum variety, supplemental lighting should be carried out when natural

sunlight is shorter than 13.5 hours. High-pressure sodium lamps or incandescent lamps can be used for supplemental lighting. The arrangement of supplemental lighting lamps should be determined according to the actual power of the lamp. Generally, every 100 W can cover 9 m^2. The supplemental lighting lamp should be installed at a position 1.7–1.8 m away from the top of the plant. This height is the most reasonable combination of light-irradiated area and light intensity. The supplemental lighting time can be gradually lengthened as the day length shortens, generally from the beginning 2 hours to the later 4 hours. Supplemental lighting is generally from 23:00 at night to 2:00 in the morning of the next day, and the light intensity should be above 50 lux. When the plant height reaches 60 cm, the electric lighting should be stopped to allow the plant to enter reproductive growth. At this time, water can be appropriately controlled. According to the growth of the plant, Jiulai can be used to control the plant type. After stopping the electric lighting, the night temperature must reach above 18 ℃. If the temperature is too low, the flower buds cannot differentiate. The temperature is best not to exceed 25 ℃. If the temperature is too high, the flowers will deform.

(2) Shading cultivation. Shading cultivation and electric light supplementation are corresponding. When the natural day length is far longer than the critical day length of the cultivated variety, the cultivated variety should be shaded. The shading treatment must make the light intensity in the shed less than 5 lux, and people cannot read newspapers smoothly in the shed under this illuminance. The height of the plants during shading varies with different varieties and different cultivation times.

The shading material is an important link that affects the success of shading. Shading materials generally choose materials that are stretchable, opaque, and lightweight. There are two ways of shading: external shading and internal shading. External shading means directly covering the shading material on the outer film of the greenhouse; internal shading involves setting up a steel wire to form a house-like structure inside, and then covering it with shading material. If the shading effect is not good (the material has a high light transmittance or is missing), it can cause phenomena such as double-layer calyxes, empty buds, and too few petals. If the temperature is high during shading, the shading material should be opened at night and forced ventilation should be carried out to reduce the temperature in the shed, and then it should be shaded before dawn the next day. If the temperature during the shading period is higher than the suitable temperature for flower bud differentiation for a long time, it will cause phenomena such as deformed flowers, thick calyxes, twisted petals, and too few petals.

After the chrysanthemums used in early spring are planted, the sunlight duration is less than 14 hours, which can cause the chrysanthemums to differentiate their flower buds prematurely without reaching the specified height, so artificial supplementary lighting treatment is needed.

A week after planting the chrysanthemums, every day before sunset in the afternoon, lower the straw curtain for artificial supplementary lighting treatment. To save electricity, you can also use the intermittent lighting method, using high-intensity light to irradiate for 5 minutes from 22: 00 at night to 2: 00 in the early morning of the next day, to achieve the effect of long daylight. Such treatment may cause degradation in chrysanthemums, which can be improved by alternating between long daylight and short daylight treatments. Due to the low temperature in winter, the growth of chrysanthemums is relatively slow, and generally about 90 days of supplementary lighting is needed, which should stop in the middle and late January.

After the middle and late January, the natural daylight time is less than 12 hours. Chrysanthemums can complete flower bud differentiation and display buds within a month under the condition of less than 12 hours of sunlight, and can bloom from the end of March to the beginning of April.

021 How to manage cut chrysanthemum plants on a daily basis?

(1) Raising the net. When the plant height on the net reaches about 25 cm, the net should be raised in time to keep the length of the upper part of the net around 15 cm. If the upper part of the net is too long, the plants tend to bend; conversely, if the upper part of the net is too short, the plants can also easily bend because they have not completely lignified. It is best to raise the net in the afternoon on sunny days, because the leaves are relatively soft at this time and are not easily damaged when the net is raised. When raising the net, tighten the flower net to the outside and lift it up at the same time. The work of raising the net must be timely.

(2) Pinching buds. When the axillary buds grow to 2 cm, they should be pinched off in time. Bud pinching should be timely, carried out with both hands at the same time, and the buds should be removed when they are no longer than 2 cm. If the buds are pinched off too late, a large wound will be left; also, bud pinching should not be done when the buds are very small, as this can easily remove the leaves and fail to reach the standard height. Method one: When pinching buds, hold the petiole with your middle finger and index finger, and use your thumb to remove the axillary bud; Method two: Use your thumb to support the flower stem, your index finger on the inner side of the petiole, and knock off the axillary bud along the petiole downwards; Method three: Use your index finger to support the flower stem, your thumb on the inner side of the petiole, and knock off the axillary bud along the petiole downwards. When pinching buds, make sure not to drop leaves, leave stubs, or miss any buds.

(3) Bud stripping. 29 days after the flower buds begin to differentiate, the side buds next to the main bud have grown to the size of green beans, at this time the side buds should

be promptly removed. If the side buds are removed too late, it will affect the growth of the main bud, causing large wounds and small main buds; if the side buds are very small when they are removed, it's easy to damage the main bud. Depending on individual proficiency, the principle should be to remove the side buds without harming the main bud. Be careful not to leave stubs when removing buds (the production of cut chrysanthemums with multiple small heads requires the removal of the main bud to promote the development of side buds).

022 What are the main pests and diseases of cut chrysanthemums? How to prevent and control them?

The main diseases of cut chrysanthemums include rust, powdery mildew, and grey mold; the main pests include aphids and leaf miner flies.

(1) Rust. It occurs more severely in cold, rainy, warm during the day and cold at night, and humid weather. The pathogen of rust disease is a type of fungus. When the temperature is between 6 and 20 ℃ and the humidity conditions are suitable for the occurrence of summer spores, the spores spreading in the air will produce on live plants, and rust disease will become more serious. After the disease, the back of the leaves is densely covered with small reddish-brown or orange-yellow spots, which gradually expand. After the epidermis ruptures, orange-yellow powder is scattered and blown away by the wind, eventually causing the leaves to turn yellow and fall off.

Prevention and treatment methods. Improve ventilation and reduce humidity. Clean up and destroy diseased branches and leaves. During the disease period, you can spray 500 times liquid of 80% mancozeb, 1 500 times liquid of 25% triadimefon wettable powder, 400 times liquid of 20% validamycin emulsion, 800 times liquid of 25% carbendazim wettable powder, or 500 times liquid of 75% chlorothalonil wettable powder every 7 – 10 days, alternate use, spray 3 to 4 times in succession, can achieve good prevention and control effect.

(2) Powdery mildew. It often occurs between August and October. The spores that spread through the air remain on living plants and prefer cool, high humidity climates, and often occur in high-density plant bodies. It is most susceptible to infection and disease when the humidity is high, light is low, ventilation is poor, and the temperature difference between day and night is around 10 ℃. After the disease, white and gray powdery spots often appear on the leaves and stems and quickly merge into patches. Sometimes it can cause leaf deformities, and chrysanthemum leaves wilt and die. Chrysanthemums grown in greenhouses are more prone to powdery mildew.

Before the disease, spray 800–1 000 times the liquid of 50% tobujin wettable powder, 1 000 times the liquid of 50% mancozeb wettable powder once a week, and continue to spray 4–5 times. Remove and destroy the diseased leaves at the beginning of the disease, and then use medicine for prevention and treatment.

(3) Gray mold. The pathogen is Botrytis cinerea, which prefers low temperatures (10–16 ℃) and relatively high air humidity, and can produce spores on dead plant tissues. Infected plants change color and appear water-soaked spots on the flowers, and the leaves at the bottom begin to rot. Rainy weather, excessive application of nitrogen fertilizer, over-dense planting, and sticky soil texture all contribute to the occurrence of gray mold.

Before planting new chrysanthemums, the roots can be soaked in 300 times the liquid of 65% mancozeb wettable powder for 10–15 minutes; in the early stage of the disease, 0. 3–0. 5 Bordeaux mixture or specific fungicides such as succinic acid, puchain can be sprayed.

(4) Aphids. They can harm chrysanthemums all year round. They are very small in size, visible to the naked eye in black, and are particularly active. They generally gather at the top of the plant to suck sap, causing the plant's stems and leaves to turn yellow and affecting flowering. The higher the temperature, the heavier the pest damage.

Prevention and control. Use natural enemies such as ladybugs, green lacewings, and predatory mites to play a role in controlling pests.

When pests occur, spray 2 000–2 500 times liquid of 10% imidacloprid wettable powder or 2 500–4 000 times liquid of 20% imidacloprid, 1 000–1 500 times liquid of 50% imidacloprid emulsion, 1 000 times liquid of 40% imidacloprid emulsion, 1 000–1 500 times liquid of 50% aphid-resistant wettable powder.

(5) Leaf miner flies. Both adults and larvae can cause damage. The female stings the host leaves with its ovipositor, forming small white spots, and feeds on the sap and lays eggs in it. The larvae burrow into leaf tissues, forming snake-like white spots with wet black and dry brown areas; adult oviposition and feeding also cause injury spots. The severely damaged leaves are covered with white snake-like tunnels and points, and finally the leaves wilt.

The prevention and control of aphids and leaf miner flies: 1. 8% abamectin emulsion can be used at 4 000–6 000 times the liquid, spray continuously 3 times, or use yellow boards to trap and kill adults, 22% dimethoate smoke agent can be used for greenhouse fumigation.

4. Q&A on Lily Cut Flower Production Technology

023 What are the characteristics of lilies?

Lilies are perennial herbaceous flowers of the lily genus in the lily family, with spherical or flat bulbous underground, made up of many fleshy scales, hence the name lily. Lilies are mainly distributed in the temperate and cold regions of the Northern Hemisphere. China is one of the main production areas of lily plants and also the origin center of lilies worldwide. According to the classification of horticultural cultivars, lily horticultural species are divided into 9 species systems, among which the commonly cultivated three species systems are the Asian lily hybrid system (Figure 6), Oriental lily hybrid system (Figure 7), and trumpet lily hybrid system (Figure 8).

Figure 6 Asian Lily Hybrid System

Figure 7 Oriental Lily Hybrid System

Figure 8 Trumpet Lily Hybrid System

Lilies are autumn-planted bulbs, the root primordia at the bottom of the bulb plate usually germinate and form roots after the autumn coolness, and germinate new buds, but the new buds often do not emerge from the soil. The basic roots are fleshy and branched, can maintain a lifespan of 2 years, and their main function is to absorb nutrients for stem and leaf growth and flowering. After overwintering in natural low temperatures, they germinate and grow above-ground stems in the following spring, and grow and bloom. The natural flowering period is from late spring to summer, with above-ground wilting in autumn and winter, and overwintering in the soil as dormant bulbs.

After lilies bloom, the bulbs enter a dormant period. After enduring the high temperatures of summer, dormancy can be broken. Then, after being induced by low temperature vernalization, flower buds are formed at suitable temperatures. The temperature and time to break low temperature vernalization vary with different varieties. Generally, breaking dormancy requires conditions of 20–30 ℃ for 3–4 weeks. The Asian lily system needs to be refrigerated at -2 ℃, and Oriental lilies and trumpet (musk) lilies usually need to undergo the low temperature vernalization period at -1–1.5 ℃, and start forming flower buds when the stem and leaves reach 8–10 cm.

024 What types and varieties of cut flower lilies are there, and what are their characteristics?

The main types of cut flower lily production are as follows:

(1) Asian Lily Hybrid System. The flowers are small, without fragrance, the plants are tall, and have strong cold resistance. The flower colors are rich, mainly yellow, white, pink, and dark red. It takes about 80–100 days from planting to flowering. Commonly produced varieties are the yellow "Alaska" and "London", the orange-yellow "Goodnight"; the red series "Red Lobo" and "Nile"; and the white "Tribea".

(2) Oriental Lily Hybrid System. The flowers are larger and fragrant, also known as perfume lilies, they have a long growth period and flower later, it takes about 100–120 days from planting to flowering. Varieties commonly used for production cultivation include the mixed-color series "Soborg" "Robin" "Tongjing" "Fairy Ning Fu"; and the pure color series "Corridor".

(3) Trumpet Lily Hybrid System. It belongs to the high-temperature type of lily. The flowers are trumpet-shaped, with a special light musk scent, also known as musk lilies, and the color is mostly white. It takes about 100 days from planting to flowering. Varieties commonly used for production cultivation include "White Heaven".

025 What kind of environmental conditions do cut lilies prefer?

Cut lilies prefer conditions with abundant light, cool and humid climates. They require fertile, humus-rich, deep soil with excellent drainage, such as sandy soil. Most varieties grow well in slightly acidic to neutral soil.

Lilies are particularly averse to continuous cropping and heavy clay soils. Therefore, it is necessary to thoroughly sterilize the soil before planting. However, different hybrid lines have slightly different requirements for their growing environment.

Before planting, apply sufficient base fertilizer, use 5 000 kg of mature cow dung per acre, 10 kg of diammonium phosphate, 15 kg of potassium sulphate, and deep tillage to 30 cm. Use a high ridge planting method, with a ridge height of about 20 cm, a ridge width of 1–1.2 m, and a walkway width of 0.4–0.5 m, which is convenient for daily production operations.

026 What are the sources and treatment methods of bulbs in cut lily production?

The bulbs of lilies can be divided into first-generation bulbs and second-generation bulbs. Currently, the first-generation bulbs are all imported from the Netherlands. After the first-generation bulbs are planted and harvested, the bulbs can be planted for the second time after dormancy and disinfection treatment, which are called second-generation bulbs. The flower quality produced by the second-generation bulbs is lower than that of the first-generation bulbs.

Before planting, you should choose good bulbs, select robust growth, disease and pest-free new bulbs. The circumference of Asian lilies is 12–14 cm, the circumference of Oriental lilies is 14–16 cm, and the circumference of musk lilies is 12–14 cm suitable for production.

Oriental lilies planted in the northern regions mainly use imported Dutch bulbs. The Dutch bulbs are generally transported in a frozen state in boxes. Before planting, the frozen bulbs need to be slowly thawed. First, open the box lid and the plastic packaging, and place them under 10–15 ℃, and avoid placing them under high temperature or direct sunlight. After the bulbs are thawed, they need to be germinated in a cold storage of 8–12 ℃ for one week. When the buds grow to 5–6 cm, they can be planted (this process is generally completed at the bulb dealer's place).

027 What issues should be paid attention to when planting cut lilies?

Cut lilies are generally planted on high ridges, and the planting depth and density depend on the variety and bulb size. The soil covering depth should be about 8 cm. Generally, smaller bulbs are planted slightly shallower, and larger bulbs are slightly deeper; clay can be shallower, while loose and poor water-retaining soil should be deeper.

In winter, off-season cultivation has insufficient light and low temperature, so the planting should not be too dense. The suitable planting density for different specifications of bulbs can refer to Table 2.

Table 2 Planting density for bulbs of different specifications

Specification (circumference/cm)	12–14	14–16	16–18	18–20
Planting Density (pieces/m^2)	35–45	30–40	25–35	25–35

028 How to manage Oriental lilies after planting for cut flower production?

(1) Watering. The first three waterings after planting are crucial. Immediately after planting the lily bulb in moist soil, water it thoroughly. Water it again after 2–3 days, and a third time after 5–7 days, depending on the soil's moisture level and the weather conditions. After this, water it every 7 – 15 days, depending on the soil's drainage and weather conditions. Spray or drip irrigation can be used.

(2) Fertilizing. Begin fertilizing when the lily grows to around 25 cm and starts to leaf out. Apply thin but frequent layers of fertilizer. Increase phosphorus and potassium levels during the middle phase and decrease the amount of nitrogen until 3 weeks before harvest. If you plan to cultivate seed balls after the lily flowers, apply a quick-acting fertilizer rich in phosphorus and potassium to promote the enlargement and fullness of the bulb scales.

(3) Netting. Start netting at the beginning of planting. Use plastic or nylon nets that are as wide as the ridges for support. The net's grid should be 15–20 cm wide. Fix it with posts, usually one every 2 m. The height above ground should be 80–100 cm so that the support net can be raised as the lily grows.

(4) Bud thinning. When the buds grow to about 0. 5 cm, keep only four. Remove any deformed buds early to allow the remaining buds to fully absorb nutrients, further improving the quality of the cut flowers.

029 How to harvest and handle cut lilies after harvesting?

Cut lilies are usually harvested in the early morning, and can be harvested when there is at least one colored bud in the 5 buds.

The best time to harvest is in the morning. The cutting position of the flower stem depends on whether the bulb is to be reused. If the bulb needs to be reused, 20 cm should be left at the base of the plant, and care should be taken not to damage the remaining leaves. If the bulb is not to be reused, it can generally be cut 15 cm above the ground, or the stem can be cut according to the market's requirements for the length of the flower stem.

After harvesting, the flowers should be graded according to the number of buds on each stem, the length and firmness of the flower stem. Remove the leaves below 10 cm from the base of the flower stem, then tie every 10 lilies into a bundle, put on packaging, and stick a label on the bag. The label should indicate the variety name and the number of buds.

After bundling, place the lilies in a plastic bucket filled with clean water. The water in the bucket should be sufficient to keep the lilies hydrated. The temperature in the storage room should be kept between 2-15 ℃.

When shipping, put the lilies in a box with holes to ventilate and disperse the high concentration of ethylene produced by blooming, preventing wilting and bud drop.

030 How to control the flowering period of cut lilies?

(1) Cold storage treatment of bulbs.

To make lilies bloom earlier, sufficient cold storage treatment of the bulbs should be carried out before planting. Generally, larger bulbs of the species are selected and treated at 13-15 ℃ for 6 weeks, then at 8 ℃ for 4-5 weeks. During the treatment, the bulbs should be buried in a substrate such as peat or fresh sawdust, and placed in a plastic box with a plastic film wrap for moisture retention. Long-term cold storage or long-distance transportation of lily bulbs should be carried out at -2--1 ℃. Defrosting treatment should be done when planting.

For inhibition cultivation, long-term cold storage is needed, pre-cooling at 1 ℃ for 6-8 weeks. Asian lily varieties can be stored for over a year at −2 ℃, while Oriental lilies and musk lilies can be stored for 6-8 months at −1.5 ℃.

After being treated with cold storage, lily bulbs can be planted at any time under the temperature conditions that satisfy their growth, in order to achieve the purpose of adjusting the flowering period.

(2) Adjustment of temperature and light.

Lily bulbs that have been treated with cold storage generally only need 60-80 days from sowing to blooming. When low temperatures arrive, it is necessary to increase the temperature to above 15 ℃, maintain a daytime temperature of 20-25 ℃, and a nighttime temperature of 10- 15 ℃, avoiding continuous high temperatures above 25 ℃ during the day and low temperatures below 5 ℃ at night.

Lilies are long-day plants. The Asian lily species are sensitive to the length of light exposure. For winter production, to prevent blind flowers, bud drop, and bud abscission, artificial light supplementation is needed. Artificial light supplementation should start when the first bud on the inflorescence is 0.5 – 1 cm, and continue until the cut flowers are harvested. Under a temperature condition of around 16 ℃, artificial light can be supplemented for 5 weeks. The light supplementation time is from 20:00 in the evening to 4:00 in the early morning, adding 8 hours of light, which has a significant effect on preventing lily bud abscission, promoting early flowering, and improving the quality of cut flowers.

5. Q&A on Carnation Cut Flower Production

031 What are the characteristics of carnations?

Carnations, also known as Dianthus, are perennial herbaceous flowers of the Caryophyllaceae family. They have erect and highly branched stems, with enlarged nodes. The flowers come in a variety of colors, including pink, red, goose yellow, white, dark red, purple-red as well as variegated and edged varieties. Each flower has a long blooming period and good decorative effect, making it one of the most popular cut flowers in the world.

Carnations are famously known as the flower of Mother's Day, symbolizing maternal love that is kind, warm, sincere, and unrequited.

032 What kind of environmental conditions do carnations prefer?

Carnations prefer a warm and cool climate, avoiding extreme cold and heat, and they need ample sunlight to grow well; The suitable growth temperature is 18 – 24 ℃ during the day, and 12 – 18 ℃ at night. The flowering period is 10 – 20 ℃, and if the day-night temperature difference does not exceed 12 ℃, it can prevent the occurrence of calyx splitting. They like fertile soil with good ventilation and drainage, and rich in humus. If conditions are suitable, they can bloom all year round.

033 What are the varieties of carnations for cut flower cultivation that can be chosen for production?

Cut flower carnation varieties can be divided into two groups. Large-flowered carnations, which have large flowers and each main stem ends in a single flower; this is a type that is popular in the market (Figure 9); Spray carnations, also known as multiflora carnations, have smaller flowers, with several branches on each main stem, and flowers growing on the branches (Figure 10). These have been trendy in the market in recent years.

According to their adaptation to the environment and trait performance, they can be divided into summer and winter types. The main varieties of the summer type include "Tangading" "Harris" "Roma", etc., while the main varieties of the winter type include "Nora" "Dallas" "Chic Bride", etc.

Figure 9 Large-flowered Carnation

Figure 10 Multiflora Carnation

In greenhouse production, varieties can be chosen based on their robust growth, upright flower stems, bright colors, few split calyces, strong disease resistance, cold resistance, and high yield. The "William Sim" "Master" "Nora" "Kale" and others that have been introduced to China over the years all have their own characteristics and good marketability, and can be chosen for energy-saving daylight greenhouse cultivation.

Additionally, in production, the appropriate proportion of flower colors can be matched according to market consumption needs, such as 30% each for red, white, and pink, and a total of 10% for yellow, orange, and mixed colors.

034 How are the seedlings of cut flower carnations obtained?

Seedlings for cut flower production are obtained through a combination of tissue culture and cutting methods. The establishment of excellent variety gardens and cutting gardens is the first step in cutting propagation. The variety gardens and cutting gardens should be covered with insect-proof nets for de-virus tissue culture seedlings.

(1) Cultivation of cutting parent plants. The soil for the cutting garden should be fertile. The plant spacing for parent plant cultivation is 15 cm×20 cm. After planting, when the seedling is about 15 cm tall, the first heart picking is carried out to promote the development of lateral branches. 15–20 days later, the second heart picking is performed to promote more lateral branches for cutting. The ratio of cutting parent plants to seedlings for production is 1:25. To ensure the quality of cut flowers and prevent virus infection, the parent plants need to be replaced every year.

(2) Cutting. The cuttings for production seedlings are best taken from the 2–3 lateral branches of the parent plant. High-quality cuttings are 8–10 cm long and have 4–5 pairs of intact leaves and complete stem tips, thick stems, tight internodes, and thick leaves. The cuttings retain 3 to 4 pairs of leaves, and the rest are all removed. They are bundled in 20–30 branches, soaked in clean water for 0.5 hours to absorb enough water, and quickly dipped

in a 50 mg/L naphthaleneacetic acid solution to promote rooting. The cutting substrate can use perlite. The cutting plant spacing is 2-3 cm, and it is inserted about 3 cm into the substrate. After cutting, water is immediately applied, and rooting can occur in 15-21 days under suitable conditions.

Producers can purchase seedlings produced by professional seedling companies, which is a major production expense.

035 What kind of soil conditions are required before planting cut flower carnations?

The selected soil should have good drainage and aeration. In the north, it should be carried out in a protected facility. If the soil does not meet the requirements, try to replace the soil. A large amount of farmyard manure must be used to make the soil loose and fertile. Deeply plow the soil until it' s smooth. If the soil is at risk of nematode damage, soil disinfection treatment should be carried out before planting.

After leveling the soil, make ridges. The ridge should be 15-20 cm high and 80-100 cm wide. The length depends on the conditions of the facility.

036 What should be noted when planting cut flower carnations?

(1) Planting time. The time from planting to flowering of cut flower carnations varies according to light intensity, temperature and photoperiod, with the shortest being 100-110 days and the longest about 150 days. Adjust the planting time appropriately according to market supply.

(2) Planting density. The plant spacing is 10 cm × 10 cm. Varieties with strong branching should have a smaller density, while those with weak branching can be planted more densely. Medium and small flower varieties can have a larger density, while large flower varieties can have a smaller density. The general density is 33-40 plants/m^2, requiring about 14 000 seedlings per 667 m^2. For short-term cultivation where flowers are harvested only once, the density can be increased to 60-80 plants/m^2.

(3) Planting operation. Carnations are most sensitive to wilt disease and are prone to stem rot. Plant shallowly and keep the seedlings upright. The reasonable depth is that the upper part of the cutting seedling in the original medium is slightly exposed to the soil surface, and the general planting depth is 2-5 cm. The soil should be moist when planting, and do not remove the heart-protecting soil.

After planting, shallow furrows can be drawn on the soil surface between the rows with your fingers, and an appropriate amount of water can be poured into the furrow. Try not to let the water splash on the leaves and cause the plants to fall over.

037 What kind of management is required after planting carnations?

(1) Fertilizer and water management. It is necessary to ensure sufficient long-acting base fertilizer, frequent application of thin top dressing, and focus on comprehensive nutrition.

Nitrogen fertilizer is best in nitrate form. Potassium fertilizer and calcium fertilizer are beneficial for neat flowering and improve cut flower quality. Boron is easily deficient, which can cause plants to be short, internodes to be short, stems to crack, the base of the stem to become fat, and the bud not to form buds. Symptoms such as petal browning will occur during the bud period. Boron deficiency is prone to occur when the pH is too high, and boron will also be deficient when the soil is too dry. Commonly used boron fertilizers are borax, boric acid or boron magnesium fertilizer. The most suitable soil pH is 6.0–6.5.

Carnations require a lot of water during the growing period, but they cannot be watered too much at once. It is necessary to ensure good root aeration and even water absorption. Sudden changes in water volume during the harvest period will cause the phenomenon of calyx cracking. After planting, moderate water control "squatting seedlings" should be carried out to promote the robustness of the plant root system.

(2) Temperature and light management in facility cultivation. Carnations are suitable for cool environments, with the optimal average growth temperature being 15–20 ℃. In summer, shading nets and misting measures are used to cool down, while in winter, insulation and heating are equally important, especially strengthening insulation at night, so that the night temperature is within the range of 5–12 ℃, which can ensure cut flower production.

Carnations are originally long-day plants, and most cultivated varieties are day-neutral. If the daylight can be extended to 16 hours, it is conducive to nutrient growth and flower bud differentiation, early flowering, and improving yield and quality. In production, artificial light sources are often added during the flower bud differentiation stage, each time for 25–50 days.

(3) Topping. There are 3 types of topping for carnations, including 1-time topping, 2-time topping, and 1.5-time topping. The 1-time topping method only tops the planted plants once, generally when there are 6–7 pairs of leaves. After topping, 3–4 side branches are sprouted per plant to form flowering branches. This method flowers the earliest, in a short time, and there are two peak harvests.

The 2-time topping method is to top all the side branches again when there are about 5 nodes on the side branches after the main stem is topped, so that the number of flowers formed per plant reaches 6–8 branches. This method can form more flower branches at the same time, the first batch of harvest is more concentrated, and the second batch of flowers is weak.

The 1.5-time topping method is an improvement on the basis of the previous two methods. After topping once, only half of the side branches are topped for the second time, and the other half is not topped, so that flowering is carried out in two stages. This solves the contradiction of wanting to bloom early and balance flower supply.

(4) Netting, bud stripping and shoot picking. During the growth of carnations, in order to keep the stems upright and prevent lodging, netting should be started when the plant height is 15 cm. The size of the net used is equal to the row spacing of the planted seedlings, so that there is one seedling per grid. As the plants grow, the grid should gradually rise, and the stems should often be disturbed into the grid, and the number of net layers should be increased according to the layer spacing of 20 cm.

038 How to prevent calyx cracking, flower head bending, and blind flowers?

Large flower varieties are prone to calyx cracking when blooming, losing their commercial value, and seriously affecting economic benefits. The main reasons are large temperature differences between day and night during the flowering stage; too much watering and fertilizing during the low-temperature period; and an unbalanced supply of nitrogen, phosphorus, and potassium, which causes the petals to grow rapidly beyond the growth of the calyx. Too many petals squeeze and break the calyx, causing the calyx to crack.

The phenomenon of flower head bending will occur when too much fertilizer is applied during the flower bud differentiation period, resulting in excess nutrients, or when the daylight hours are too short.

Both low temperatures and poor nutrition can lead to blind flowers. If the temperature drops below 0 ℃ during the bud stage and then quickly rises, it is easy to cause petal deformity or flower failure; boron deficiency during the bud differentiation period will produce petal-less deformed flowers, and calcium deficiency will cause bud death.

039 How to regulate the flowering period of large-flowered Dianthus caryophyllus?

(1) Adjusting the planting period. The time required for large-flowered Dianthus caryophyllus to grow to flowering varies due to different temperatures, light conditions, and management levels. The shortest is 100–110 days, and the longest is 150 days. The planting period can be calculated backwards according to the cultivation season, conditions, and the desired flowering period.

(2) Pinching to control the flowering period. The growth status and peak flowering period of Dianthus caryophyllus can be regulated by pinching, achieving a balanced supply to the market throughout the year. For example, pinching in July can lead to marketable cut flowers from December to January of the following year. Pinching in early January can result in marketable cut flowers for the "May Day" and Mother's Day.

6. Q&A on Off-season Cultivation Technology of Tulip Cut Flowers

040 What are the characteristics of tulips?

Tulips, also known as foreign lotus flowers or grass musk, belong to the lily family and are perennial bulbous herbaceous flowers. They have a variety of colors and flower types, including red, yellow, white, orange, purple, pink, and bicolor; flower types include bowl-shaped, cup-shaped, lily-shaped, etc. Tulip flowers symbolize "sacredness and happiness", and the most popular ones on the market are red and yellow flowers (Figure 11).

Figure 11 Cut Tulip Flowers

Tulips prefer a warm and humid environment in winter and a cool and slightly dry environment in summer. They thrive in sandy loam soil that is rich in humus and well-drained.

Tulips are autumn-planted bulbs, germinating in late autumn, flowering in early spring, and entering a bulb dormancy period in early summer. After being planted in the open field in autumn, the root system begins to grow, with the optimal temperature being 9-13 ℃, and growth stops below 5 ℃. After planting, the first vigorous growth occurs, and the second growth peak is at the beginning of the following year, with 60%-80% of the total root length occurring in the year of planting. The three weeks before flowering is a period of vigorous stem and leaf growth, with the optimal temperature being 15-18 ℃, and the stem and leaf growth stops by the flowering period. During the summer dormancy period, flower bud differentiation occurs, with the optimal temperature being 20−23 ℃.

041 What is meant by 5 ℃ bulbs and 9 ℃ bulbs in tulips?

Tulips are autumn-planted bulbous flowers and are low-temperature vernalization plants. The bulbs must undergo a certain low-temperature treatment to be able to root, germinate, and flower. The bulbs imported from the Netherlands to China have already completed this process. After the bulbs are harvested in the place of origin, they need to be treated at 18−25 ℃ for 2−3 weeks. Then, it takes about 2 weeks to lower the temperature from 18 ℃ to 9 ℃, from 9 ℃ to 7 ℃, and from 7 ℃ to 5 ℃. The time for 5 ℃ treatment depends on the characteristics of the variety, and they can be planted when the bud is 2 cm long. Bulbs that are treated at low temperature only with 9 ℃ are called 9 ℃ bulbs.

042 What are the considerations for tulip planting?

(1) Determination of planting time. In the production of sunlight greenhouses, the time from sowing to harvesting is about 45 days, and the sowing period can be determined according to the harvesting period. Cut flower tulips are generally planned to be on the market from the Spring Festival to Valentine's Day. According to the full growth period of 55−65 days for tulip cut flowers, the sowing period is arranged 7−10 days in advance.

(2) Treatment of bulbs before sowing. Before sowing, the membranous false scales on the surface of the bulb should be removed, commonly known as peeling. Remove the accessory buds on the bulb and leave a robust main bud. The bulbs are disinfected with a mixture of 70% methyl tobutin 1 000 times solution, 50% carbendazim 800 times disinfectant solution, and acaricide special 1 000 times solution. Pay attention to the following when disinfecting: the disinfectant can only be used for half a day at most; the bulbs to be disinfected must be completely immersed in the disinfectant; the disinfection soaking time is 5−20 minutes; the bulbs after disinfection should be planted after draining the water; the bulbs disinfected on the same day should be planted on the same day.

(3) Planting method. The development degree of tulips is 10−15 cm, so the planting density is very flexible, and the plant row spacing can be (10−15) cm×(15−25) cm. High planting density is conducive to the heightening of cut flower stems, but it is prone to diseases.

(4) Planting depth. For bulbs with the skin removed, it is best to slightly expose the top of the bulb. For bulbs that have not removed the skin, the planting depth can be 3−4 cm below the soil surface.

After sowing, water thoroughly to allow the bulbs to fully contact the soil for rooting.

043 How to adjust management measures according to the growth characteristics of tulips after planting?

(1) The key period for rooting and germination of tulips is 15 days after sowing. During this stage, temperature management should be done well. The most suitable temperature is 9–10 ℃. If the environmental temperature is high during this period, shading methods can be used to lower the temperature.

(2) Management 15 – 30 days after planting. After the seedlings grow to 5 cm, the environmental temperature should be increased, raising it by 2 – 3 ℃ each day until the temperature rises to 15 ℃ and is maintained at 15–17 ℃ to promote vigorous plant growth.

Start to water heavily 15 days after planting, generally watering once every 7 days, ensuring that the soil humidity is between 60% and 80%. Ventilate and dehumidify after watering.

Tulips have already been fully fertilized before planting, so they only need to supplement calcium fertilizer after germination. Spray the leaves with a 0. 1% calcium nitrate solution and spray twice every 7 days.

(3) Management 30–45 days after planting. 30 days after planting, the plants grow to about 25 cm. The temperature is maintained at 15 – 17 ℃, and light management is strengthened to ensure more than 8 hours of natural light each day. Water every 3–4 days to keep the soil moisture at 85%–90%.

044 What is the standard for harvesting tulip cut flowers?

When the tulip plant grows to about 40 days, the color of the flower bud is still green, and in a few days, it will start to show color, at which point it can be harvested. During harvesting, the bulb can be pulled out together with the plant, and care should be taken not to break the flower stem. After harvesting, place the tulips on a newspaper, align the flower heads, tie ten stems lightly at the leaf part, then wrap the flower heads with the newspaper and tie lightly.

045 How to prevent and control tulip virus disease?

Tulip breaking virus disease is the main viral disease.

(1) Symptom identification. This disease mainly affects the flowers and leaves, causing color changes, but different varieties have different reactions to this disease. On light or white varieties, the symptom of petal breaking color is not obvious; on red and purple varieties, the flowers change color greatly, producing broken color flowers, with varying sizes of mottled

spots or stripes on the petals; black flowers turn to light black. When the leaves are affected, they appear light green or gray-white stripes, sometimes causing flower leaves.

(2) Pathogen. Tulip breaking virus, the virus particles are bent linear, with a size of 700 nm × 12 nm × 13 nm. The lethal temperature of the tulip virus is 65–70 ℃, the dilution end point is 10 - 5, and the survival period outside the body is 4 - 6 days. The virus overwinters in diseased bulbs, becoming the initial infection source for the next year. The tulip breaking virus is transmitted by aphids and sap. In addition, bulbs formed on diseased plants from the previous year will also be infected with the virus, and they will often get sick even if they are planted in disinfected soil. Under general cultivation conditions, double flowers are prone to disease.

(3) Prevention and control measures.

Strengthen quarantine to prevent diseased seedlings and other propagation materials from entering disease-free areas. Use healthy and disease-free cuttings, bulbs, etc. as propagation materials, establish a virus-free mother garden, and avoid artificial transmission. Viral bulbs can be soaked in 45 ℃ warm water for 1.5–3.0 hours.

Obtain virus-free seedlings by using the meristem tissue culture method to reduce the occurrence of viral diseases.

In daily field management, such as pinching, breaking buds, pruning, etc., disinfect hands and tools with 3%–5% trisodium phosphate or hot soapy water.

Regularly spray insecticides to prevent insects from transmitting the virus. Diseased plants should be removed and thoroughly destroyed as soon as they are discovered.

Drug prevention and control: In recent years, with the development of science and technology, several drugs effective against viral diseases have been developed, such as Virus A, Virus Special, Virus Spirit, 83 Antagonist, Antidote No. 1, etc. You can choose to use them according to the actual situation.

7. Q&A on Baby's Breath Cut Flower Production Technology

046 What kind of environmental conditions does Baby's Breath prefer?

Baby's Breath, also known as Gypsophila or Silk Carnation, is a plant of the Carnation family Gypsophila genus. It prefers a sunny, dry, and cool environment. It is very cold-resistant, can withstand temperatures below 0 ℃, and is also heat-resistant. However, the growth of the flower stem is inhibited below 10 ℃ and above 30 ℃, easily forming a rosette shape. The suitable temperature for cut flower production is 10–25 ℃. It prefers calcareous, slightly alkaline soil. It is a long-day flower, and flower bud differentiation requires more than 13 hours of daylight.

047 What are the cultivated varieties of Baby's Breath?

Currently, the main varieties of Baby's Breath on the world cut flower market are the following five, including three white flower varieties, "Fairy" "Perfect" "Diamond"; and two red flower varieties, "Flamingo" "Red Sea".

In China, the two white flower varieties, "Fairy" and "Perfect" (Figure 12), are widely used in large-scale production. "Fairy" is a small flower type, with strong adaptability and high yield. It can be produced throughout the year and is a representative variety. "Perfect" belongs to the large flower type, sensitive to light and temperature, and is difficult to cultivate, but it has a good market and high price, suitable for corsages and bouquets. "Diamond" is a medium flower variety selected from "Fairy", between "Fairy" and "Perfect", combining the advantages of both. The plant is shorter than "Fairy", with shorter internodes. Flowering will be delayed under low temperature conditions. The cut flower shape is beautiful, making it a popular variety. "Flamingo" is light pink, the flower color fades easily, the stem is slender, and the flower is like "Fairy". "Red Sea" has peach-red flowers, large and hard stems, blooms in autumn, the flower color is very bright, not easy to fade, and it is a variety welcomed by the market.

Figure 12 Baby's Breath as Cut Flowers

048 How to cultivate Baby's Breath seedlings?

For small and medium-scale production, cuttings can be used for seedling cultivation, while tissue culture seedlings are commonly used for large-scale production.

For the cutting method, it is important to ensure that the quality of the mother plant is pure, robust, and free of pests and diseases, and that the flower stem has not unfolded. The top of the plant is first pinched to promote the growth of side branches, and then these side branches are used for cuttings. Side buds about 5 cm long with 4–5 nodes are cut from the mother plant to make cuttings. The lower leaves are removed, and 10 – 30 cuttings are bundled together. After treating with rooting hormone, they are inserted into perlite or zeolite substrate. The plant row spacing is 3 cm × 3 cm. Under moist and shaded management, roots can be grown in 15–20 days.

049 What needs to be prepared before planting Baby's Breath?

Baby's Breath has fleshy roots and is extremely intolerant of waterlogging, so the cultivation site should be high, with loose soil and good drainage. Before planting, sufficient organic fertilizer should be applied and deep plowing should be carried out. A suitable amount of phosphorus and potassium fertilizer, such as wood ash and superphosphate, should be added to the organic fertilizer.

After leveling the land, high ridges should be made. The ridge surface should be about 30 cm above the ground. The ridge width can be 0. 9–1 m, and the distance between the ridges should be 45–50 cm.

050 What should be considered when planting Baby's Breath?

The planting time depends on the variety and harvest time, and seedlings can be planted when they grow to about 6 nodes. Planting in summer should avoid rain, and it is best to plant on cloudy days. If planting on sunny days, a sunshade net should be used. Before planting, the seedlings should be thoroughly watered. When planting, each ridge should be double-rowed, with a plant spacing of 30–40 cm and a row spacing of 50–60 cm for cross-planting. It is suitable to plant shallowly, with a planting depth of 3 – 5 cm, and after planting, water thoroughly to make the root system closely connected with the soil.

051 What should be done for field management of Baby's Breath?

(1) Pinching and bud removal. About one month after planting, when the seedlings have grown about 8 nodes, the top 4–5 nodes are pinched off to promote the germination of side buds. Two weeks after pinching, a large number of side branches will grow. For standard large-flowered cut flowers, only 3 – 4 branches are left. After the growth of side branches, the inferior ones should be removed and the weak buds should be wiped off to ensure the quality of cut flowers.

(2) Netting. When the plants grow vigorously and the plant clumps are large and prone to lodging, netting can ensure the upright growth of cut flowers. When the plant height is 15 cm, the net is pulled, the net grid is 20 cm × 20 cm, the net height is 40–50 cm, and the net grid does not need to be moved up afterwards.

(3) Fertilizer and water management. The applied fertilizer is mainly nitrogen fertilizer in the early stage to promote plant growth. After 40 days of planting, phosphorus and potassium fertilizers are applied. Fertilization is stopped half a month before flowering. Foliar spraying of 0. 2% potassium dihydrogen phosphate and 5mg/L boric acid can promote flowering.

(4) Lotus seat transformation or treatment method for no spike in mid-October. 100–200 mg/L of gibberellin can be sprayed on the top of the plant. After 5 days, the treatment is repeated until the spike is emitted, which can promote the flowering branch.

052 How to harvest Baby's Breath cut flowers?

(1) The best time to harvest flowers is when 50% of the flowers have bloomed, and the

small flower buds have fully differentiated, before the first blooming flowers have changed color. Generally, each branch of flowers should weigh no less than 25 g, and each branch should have 3 branches.

(2) Post-harvest processing. Immediately after harvest, immerse in clean water. After immersion, bundle 10 branches together, and cover the cut with a small plastic cup and cotton ball containing preservative. It can be preserved or transported under conditions of 2–3 ℃.

053 How to control the flowering period of Gypsophila?

The flowering period of Gypsophila can be controlled by manipulating the photoperiod. Gypsophila is a long-day plant, short-day conditions promote stem and leaf growth, and flower bud differentiation and blooming require long-day conditions. Large-scale cut flower production relies on artificial light adjustment to meet the needs of growth and development.

The best effect is achieved by adding light starting from the 3rd week after seedling topping, and continuous treatment for 6–9 weeks can achieve 60%–80% flower stem growth and blooming. The methods of long-day treatment include extending the bright period, night interruption, and gap lighting. The extension of the bright period is to turn on the lighting facilities after sunset and supplement light until 16 h, and it is also very effective to supplement light for 4 h in the early morning. The night interruption method is to add light from 22:00 at night until 2:00 in the early morning, but the effect is not ideal. The gap lighting method begins at sunset, using an automatic control device to add light every 1 h for 10 min until dawn, which gives the best result, allowing flowers to bloom 30–50 days earlier.

The light intensity is best controlled at 50–100 lux. Generally, a wire is pulled above the cultivation bed, and the light bulb is 1.0 m away from the top of the plant to ensure even light distribution.

8. Q&A on Anthurium Cut Flower Production Technology

054 What types of Anthurium cut flowers are there?

Anthurium, also known as Flamingo Flower, belongs to the Araceae family and is a perennial root flower. The variety cultivated as a cut flower is also called large-leaf Flamingo Flower. Depending on the color of the spathes, they can be divided into bright red varieties such as "Gloria" and "Alexia"; the scarlet variety "Lydia"; the white variety "Margare-tha"; and there are also pink and greenish-white spathe varieties (Figure 13). The most commonly used in production are the varieties with bright red spathes.

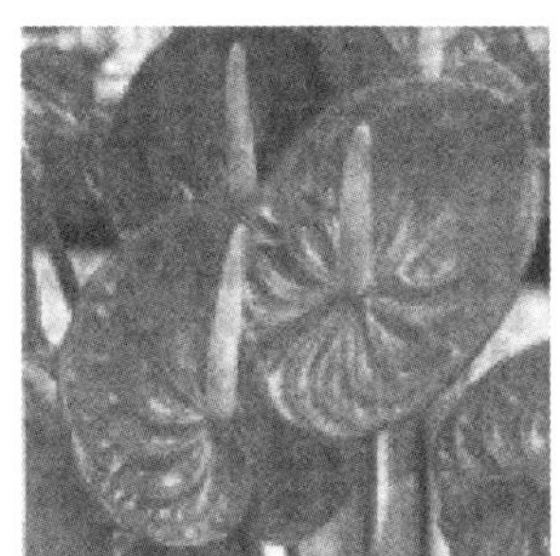

Figure 13 Various Colors of Anthurium Cut Flowers

Different varieties have different levels of cold tolerance, heat tolerance, stem hardness, and yield.

055 What kind of environmental conditions does Anthurium prefer?

Anthurium, native to the humid, semi-shaded valleys of tropical rainforests in South America, prefers a warm, high-temperature, well-drained environment. It is not cold-tolerant, and the optimal growth temperature is a daytime temperature of 25 – 28 ℃ and a nighttime temperature of 20 ℃. At temperatures above 35 ℃ in summer, plant growth is slow, and growth stops at temperatures below 18 ℃. It is prone to frost damage at temperatures below 13 ℃. Anthurium prefers humid conditions but does not tolerate waterlogged soil; the suitable relative air humidity is 80%–85%. It prefers semi-shade, with an appropriate light intensity of 15 000–20 000 lux. However, it requires adequate sunlight in winter for the root system to develop well and for the plant to be robust. It requires loose, well-drained humus soil.

With appropriate environmental conditions, it can bloom year-round.

056 What kind of cultivation soil is used for Anthurium cut flower production?

The commonly used cultivation substrate is made by mixing coarse peat, perlite, and coconut coir in a 3∶1∶1 ratio. Depending on the resources available, materials like bark, planer shavings, sawdust, and pine needles can be mixed in the appropriate proportions to make the cultivation substrate. In northern regions, pine needle soil can be directly used as the cultivation substrate. It can be shaped into beds that are 30–50 cm high and 30–50 cm wide, depending on whether it is a single or double row, ensuring a 25 cm thick cultivation substrate. A 10 cm thick layer of crushed stone can be laid at the bottom of the bed as a drainage layer.

057 What are the key points of cut flower production management for Anthurium?

For the production of cut flowers, tissue-cultured seedlings should be used.

When cultivating cut flowers, it is necessary to deeply turn the cultivation substrate 20–30 cm and apply ripe base fertilizer to keep the soil moderately moist. In order to supply flowers all year round, planting can be done in batches, usually flowering one year after planting.

Seedlings should be planted from January to May. The seedlings should have 6–7 leaves and be about 30 cm high. The plant row spacing should be 40 cm × 50 cm, and they should be planted in a staggered triangular pattern, with about 30 000 seedlings per hectare. After 7–8 years of individual plant cultivation, the growth vigor declines and needs to be updated in time.

During the growth period, attention should be paid to the regulation of temperature, humidity, and light. During the high-temperature period in summer, temperature can be lowered through shading, spraying, and ventilation, and the night temperature should be maintained at around 18 ℃ in winter. The shading rate in summer should be 75%–80%, and it should be 60% – 65% in winter. Fertilize 2 – 3 times a year, and Anthurium does not tolerate heavy fertilizer, so topdressing is generally liquid fertilizer. More nitrogen fertilizer can be applied during the seedling stage, and nitrogen, phosphorus, and potassium should be balanced during the flowering stage.

058 How are Anthurium cut flowers harvested and how are they treated after harvesting?

The best time to harvest Anthurium is when half of the small flowers on the spadix are open, at which time the color of the spathe has been fully displayed and the flower stalk has hardened. Harvesting should be done before 10 a. m. on sunny days. Select the plants that are ready for harvest and cut them 2–3 cm from the base of the flower stalk and place them in a bucket. During the harvesting process, attention should be paid to separate the cut flowers to prevent squeezing and damage to the spathe.

After harvesting, immerse the cut flowers in a solution of 170 mg/L silver nitrate for 10 minutes to extend the preservation time. When packaging, sort by color and size. Each flower is packaged in a plastic bag and placed in a specific carton.

059 How to control the flowering period of Anthurium?

Anthurium has strict environmental requirements during the growth period. Generally, it can bloom 120–150 days after planting. If the environmental management is proper, it can bloom all year round and continue to bloom for 4–6 years. After 5–6 years, it can be propagated by dividing the plant.

In production, plant growth regulators can be used to control the flowering period of Anthurium. Gibberellin and cytokinin can both make Anthurium bloom earlier. Spraying 50–150 mg/L gibberellin during the seedling stage of Anthurium can make the flowering period earlier, and the effect is best at 100 mg/L. Spraying it four times every 7 days can advance the flowering period by about 30 days, and the quality of the cut flowers is good. The effect of 50 mg/L of zeatin is the best, but it can only advance the flowering period by about 7 days.

Shading treatment can advance the flowering and extend the flowering period.

Chapter 7 Q&A on Practical Techniques for Edible Mushroom

1. Basis of Strain Production

001 What is a fungal seed?

In a broad sense, a fungal seed refers to a reproductive material with the ability to propagate and relatively stable genetic characteristics, including spores, tissues, or mycelium. In practical production, the term fungal seed typically refers to mycelium that is cultured in suitable nutrient media. It is a mixture of mycelium and the media it grows on.

002 What are the different categories of edible fungal seeds?

Edible fungal seeds can be classified into solid-state seeds and liquid-state seeds based on their physical characteristics. In practical production, the most widely used classification is based on the source of the seed, the number of propagation generations, and the purpose of production, dividing the fungal seeds into three levels: parent strains, original strains, and cultivated strains.

003 What is a parent strain and what is its role?

The parent strain refers to the pure mycelium obtained through the initial isolation from the natural environment. It includes the mycelium that has undergone subculturing. The parent strain is the first step in fungal seed production. The metabolic capacity of the mycelium is relatively weak, and it has higher requirements for the nutrient media. Therefore, it is typically cultured in test tube media. Hence, it is also known as a primary strain or test tube strain. The parent strain is mainly used for propagating original strains and preserving fungal seeds.

004 What is an original strain?

The fungal seed obtained by transferring the parent strain onto solid culture media is called an original strain. Typically, the parent strain is inoculated into a fungal bottle or bag containing substrates such as sawdust, cottonseed hulls, grains, or straw for cultivation. Therefore, the original strain is also known as a secondary strain or bottled seed. During the cultivation process on solid culture media, the mycelium of the parent strain adapts to the media and exhibits robust growth. As a result, the original strain can also be directly used as a cultivated strain for field production.

005 What is a cultivated strain?

The fungal seed obtained by scaling up the cultivation of the original strain is called a cultivated strain, commonly referred to as a production strain or tertiary strain. Cultivated strains are typically used solely for production purposes and should not be used for further propagation. Otherwise, it may result in decreased vitality, strain degeneration, and subsequent losses.

006 How are excellent varieties of edible mushrooms obtained?

Natural selection, also known as conventional breeding, is a simple and effective method for obtaining excellent varieties. In the natural survival and evolution of edible mushrooms, different genetic traits have been developed due to various environmental conditions. It is possible to isolate and screen edible mushroom varieties with excellent agronomic traits suitable for cultivation. Many of the currently used varieties have been isolated, screened, domesticated, and cultivated from wild edible mushrooms. Both wild and cultivated edible mushrooms undergo variations during their growth and development. Selecting superior varieties from mutated individuals is also a method of natural selection. For example, the white button mushroom was bred from a mutated individual of the grayish-brown Agaricus bisporus, and the low-temperature variety of the blackish-gray oyster mushroom was bred from a mutated individual of the grayish-white Pleurotus ostreatus.

007 What is fungal seed isolation? What are the methods for isolating edible mushroom seeds?

The process of obtaining pure cultured mycelium by transferring valuable tissues or spores of fruiting bodies onto slanted test tube media is called fungal seed isolation.

The isolation of edible mushroom seeds can be achieved through three methods: spore isolation, tissue isolation, and intra-strain mycelium isolation. Among them, spore isolation is a method of sexual reproduction, while tissue isolation and intra-strain mycelium isolation are methods of asexual reproduction.

008 What is tissue isolation? What are the advantages of tissue isolation in edible mushroom cultivation?

Tissue isolation is a method of cultivating pure mycelium by using any part of the fruiting body, mycelial strand, or mycelial core of edible mushrooms.

Tissue isolation in edible mushroom cultivation offers several advantages, including ease of operation, high success rate of isolation, and the ability to preserve the genetic characteristics of the original strain. Therefore, it is the most commonly used method for fungal seed isolation in production.

009 What is grain spawn?

Grain spawn refers to fungal seeds cultivated on the grains of crops. It includes various types such as wheat spawn, millet spawn, corn spawn, and grain spawn in general. Some grain spawn may also contain additional materials or even chemical components. Grain spawn is characterized by its clean, robust, and vigorous mycelium, convenient handling, quick colonization, and early fruiting. It helps improve the survival rate and quality of the fungal seeds.

010 What is branch spawn?

Branch spawn refers to fungal seeds primarily produced using branches of broad-leaved trees as the main substrate. Branch spawn is commonly used to produce tertiary spawn, also known as stick spawn or small stick spawn. It offers several advantages, including fast inoculation speed, quick recovery after inoculation, three-dimensional radial spread within the cultivation bag, shorter cultivation time, and improved mushroom synchronization. It finds application in the production of various fungal species.

011 What is liquid spawn for edible mushrooms?

Liquid spawn refers to the liquid form of edible mushroom spawn produced through deep cultivation (liquid fermentation) techniques in bioreactors using liquid culture media.

The term "liquid" refers to the physical state of the culture media, and "deep cultivation in liquid" refers to the techniques used in fermentation engineering. "Liquid spawn production" essentially involves using bioreactor technology to produce liquid spawn, replacing traditional solid spawn methods. By utilizing the principles of bioreactor fermentation, it provides the optimal conditions of nutrition, pH, temperature, and oxygen supply for mycelial growth, resulting in rapid mycelial growth and proliferation. This method allows for the production of a large quantity of mycelial biomass and metabolic products within a short period. Liquid spawn can be used as mother spawn or primary spawn and can also be directly used for cultivation purposes.

012 What are the advantages of liquid spawn?

Liquid spawn has six advantages, including fast spawn production, strong vitality, and high purity.

(1) Fast spawn production. In the bioreactor, the fungal cells in liquid spawn are cultivated under optimal conditions of temperature, pH, oxygen, carbon-nitrogen ratio, etc. The mycelium divides rapidly and can generate a large quantity of mycelial biomass within a short period of time. In contrast, solid spawn cultivation involves mycelial growth at a uniform speed from top to bottom, while liquid spawn cultivation accelerates mycelial proliferation at a geometric rate. Generally, a cultivation cycle can be completed in four days, whereas the cultivation of mother spawn or primary spawn typically takes more than 22 days.

(2) Strong vitality. By controlling the nutrition, temperature, oxygen, pH, and other environmental factors in liquid spawn cultivation, the growth requirements of the fungal cells are maximally met. Metabolic waste gases produced during respiration are promptly eliminated, resulting in vigorous metabolism. As a result, all the mycelial biomass in liquid spawn has a similar age, leading to strong mycelial vitality. In contrast, in solid spawn cultivation, metabolic byproducts can accumulate and mix within the nutrient medium, naturally affecting the quality of the fungal cells. Furthermore, solid spawn often exhibits inconsistent mycelial ages between the top and bottom of the bottle (bag), with the mycelium at the top potentially aging and losing vitality while the mycelium at the bottom has just grown.

(3) High purity. The mother spawn used in liquid spawn cultivation is generally prepared through selection and purification, resulting in high purity of the genetic source. In contrast, the slant mother spawn used in solid spawn cultivation often experiences a decrease in purity due to unlimited subculturing by some vendors. Liquid spawn cultivation takes place in sealed containers, and the oxygen supply system is filtered, ensuring a purity level of 99.99% and guaranteeing the cultivation of pure liquid spawn.

(4) Low contamination rate. Liquid spawn has fluidity, making it easy to disperse after inoculation. It has multiple germination points, and the mycelial biomass quickly germinates upon inoculation. Under suitable conditions, the mycelium covers the inoculation surface within approximately 24 hours, effectively controlling cultivation contamination. In contrast, solid spawn requires 5-7 days from mycelial colonization to sealing the inoculation surface, involving detachment, tearing, and sterilization with chemicals, resulting in a higher

probability of contamination by unwanted microorganisms.

(5) Fast mushroom colonization. Liquid spawn accelerates the mushroom colonization process by approximately double the speed. This results in concentrated fruiting and shortened cultivation cycles, leading to reduced costs in labor, energy consumption, and facility requirements.

(6) Low cost. The shortened colonization time reduces electricity and labor costs during the colonization phase. Additionally, it eliminates the need for materials and costs associated with producing mother spawns and third-generation spawns.

013 What is the liquid spawn shake flask cultivation method?

The liquid spawn shake flask cultivation method involves inoculating the mother spawn of edible mushrooms into sterilized triangular or infusion bottles containing liquid culture media. These bottles are then placed on a shaker table for agitation during the cultivation process. The shake flask cultivation method requires minimal investment, simple equipment, and is suitable for general mushroom spawn production facilities.

014 What is deep-layer cultivation in edible mushroom fermentation tanks?

Deep-layer cultivation in edible mushroom fermentation tanks refers to the production of mushroom spawns using liquid deep fermentation. In this method, pure and superior mushroom strains are inoculated into a liquid culture medium in fermentation tanks, allowing the mycelium to proliferate and form numerous small fungal balls. The liquid mushroom spawn is then transferred into substrate bags to cultivate the growth of fruiting bodies.

Deep-layer cultivation offers advantages such as shorter production cycles, higher yields, and increased profitability, making it an important method for large-scale production of edible mushrooms. Based on the sterilization method of the fermentation tanks, there are two types: mobile fermentation tanks and fermentation tanks with control cabinets.

015 What are the requirements for a sterile inoculation room?

A sterile inoculation room is typically suitable at a size of 4-6 m^2 and should include a buffer room and an inoculation room. Both rooms should have movable doors to prevent air circulation and should be equipped with a 30 W ultraviolet lamp and a 40 W fluorescent lamp. The ultraviolet lamp should be positioned at a height not exceeding 2.2 m for disinfection and sterilization purposes, while the fluorescent lamp is used for illumination during inoculation. The floor should be made of cement, and the walls should be smooth. Avoid painting the floor as the paint may absorb ultraviolet light and reduce sterilization effectiveness.

016 How to check the sterilization effectiveness of a sterile room or inoculation box?

The sterilization effectiveness and air contamination level of the sterile room or inoculation box should be regularly inspected. Two commonly used methods for inspection are the agar plate method and the slope method.

(1) Agar Plate Method. Prepare meat broth agar and potato glucose agar mediums. Following aseptic procedures, pour the mediums into petri dishes to create agar plates. Prepare six plates for each medium and place them inside the sterile room or inoculation box. During the inspection, open three plates of each medium according to aseptic procedures, expose them for a certain period, and then cover them again. Keep the other three plates closed as controls. Incubate both sets of plates at 30 ℃ and observe them after 48 hours. Based on the presence or absence of microbial colonies, the number of colonies, and their morphology, the contamination level and types of contaminants in the sterile room or inoculation box can be determined.

(2) Slope Method. Take six test tubes each of meat broth agar and potato glucose agar mediums prepared in standard procedures. Place them inside the sterile room or inoculation box. Following aseptic procedures, open three of the test tubes' cotton plugs (the removed cotton plugs should be placed in sterile culture dishes) for a certain period and then reseal them with the cotton plugs. Keep the other three test tubes with unopened cotton plugs as controls. Incubate both sets of test tubes at 30 ℃ and observe them after 48 hours. Based on the presence or absence of microbial colonies, the number of colonies, and their morphology, the contamination level and types of contaminants in the sterile room or inoculation box can be determined.

2. Mother Seed Production

017 What should be considered when introducing cultivated edible mushrooms?

Before introducing cultivated edible mushrooms, it is important to conduct market research, seek expert guidance, and ideally conduct on-site visits to avoid unnecessary complications.

(1) Understand if the supplier of the mushroom spawn has obtained the "Edible Mushroom Spawn Production and Operation License".

(2) Assess the professional knowledge and practical experience of the producer.

(3) Understand the characteristics and cultivation techniques of the intended mushroom variety. Pay attention to differences between varieties from different regions and be aware of situations such as synonymous names for different strains or different strains with the same name. Inquire about the generation of the mushroom spawn from the supplier.

(4) Conduct small-scale cultivation experiments before large-scale cultivation. Regardless of the variety being introduced, it is important to conduct small-scale mushroom cultivation experiments to ensure that it is suitable for local conditions before proceeding with large-scale cultivation. This allows observation of its growth habits, adaptability, and resistance to stress and contaminants, reducing unnecessary losses.

018 How to choose edible mushroom spawn?

When purchasing edible mushroom spawn, in addition to selecting reputable universities, research institutions, or formal edible mushroom spawn manufacturers as purchasing options, careful selection should be made for on-site purchases or bulk shipments, with a focus on the following four aspects.

(1) Appearance. For the same variety, when cultured under the same conditions using the same culture medium, the appearance should be similar. While different edible mushroom strains may have slight color variations, when grown on a natural woody fiber medium, the mycelium should be predominantly white.

(2) Growth Rate. For the same variety, when cultured under the same conditions using the same culture medium, the growth rate should be similar. Whether in the growth phase or

fully colonized, the mycelium should appear full, dense, robust, and uniform.

(3) Odor. Normal mushroom spawn should emit a strong mushroom fragrance when the bottle (bag) is opened. If the odor is weak or absent, it indicates an issue with the spawn and it should not be used.

(4) Identification. Yellowing mycelium, shrinkage of the culture medium, and detachment from the walls indicate aging and should not be used or used sparingly. Varieties without production dates and detailed biological characteristics should be avoided or used sparingly. Varieties that have not been tested in the production base should be used sparingly or avoided altogether. Companies without production qualifications should not be used.

019 How to prepare mother spawn culture medium?

(1) Formula Selection. Choose a suitable culture medium formula based on the edible mushroom strain to be cultivated. Consider local resource availability and prioritize materials that are widely sourced and affordable. The formula for Potato Dextrose Agar (PDA) culture medium is as follows: 200 g potatoes, 20 g glucose (or sucrose), 20 g agar, and 1 000 mL water.

(2) Material Pre-treatment. Clean the potatoes thoroughly, remove the skin and sprouts. Accurately measure the required quantities of each ingredient according to the formula. Prepare experimental equipment such as an electric stove, aluminum pot, beaker, test tubes, funnel, and cheesecloth.

(3) Preparation. Cut the peeled potatoes into small pieces and place them in an aluminum pot with 800 mL of water. Bring it to a boil and let it boil for about 20 minutes. Stop heating when the potatoes are tender but not mushy. Filter the potato solution using a double-layered cheesecloth into a beaker, discarding any impurities in the pot. Rinse the pot, then pour the filtered solution back into the pot. Heat it over low heat and add glucose (or sucrose) and agar, stirring continuously until fully dissolved. Adjust the water content and finally make up the volume to 1 000 mL.

(4) Packaging. The culture medium should be packaged while still hot. For slanted agar tubes, glass tubes with dimensions of approximately 18 mm × 18 mm are commonly used. Use a packaging barrel for distribution. Fill each tube with the culture medium, amounting to 1/5 to 1/4 of the tube's length, taking care to avoid any medium on the tube's mouth or walls. If the tube mouth gets dirty, it can be wiped clean with a clean cloth to prevent contamination of the cotton plug.

(5) Cotton Plugging. Ordinary cotton is typically used to make cotton plugs, with a length of about 5 cm. When inserting the cotton plug into the tube, it should fit tightly against the tube wall, with smooth ends and no loose fibers. The cotton plug should have moderate tightness, ensuring it does not easily fall out when lifted by hand. The cotton plug should be

inserted about 2/3 into the tube mouth, leaving 1/3 outside the tube.

(6) Sterilization by bundling. Bundle 5 to 10 test tubes together, wrap them with parchment paper, secure with a tying rope, label them, and place them in a high-pressure autoclave for sterilization. Generally, sterilization at a temperature of 121 ℃ for 30 minutes is sufficient to ensure complete sterilization.

(7) Slanting the agar tubes. After sterilization, remove the test tubes while still hot. Place a support under the cotton plug end of the tubes, and lower the bottom end to create a slanted surface for the agar medium. The length of the slant is typically half the length of the tube. Allow the tubes to cool, and the agar medium will solidify naturally into a slanted surface.

(8) Checking sterilization effectiveness. After preparing the agar tubes, incubate them at a temperature above 20 ℃ or at room temperature for several days. If no fungal mycelium or bacterial colonies appear on the surface of the agar medium, it can be used for inoculation.

020 Which part should be selected for mushroom tissue isolation?

For the isolation of fruiting body tissues of edible mushrooms, it is recommended to choose young and tender tissues. The specific procedures are as follows:

(1) For mushrooms with a universal veil, such as Agaricus bisporus and Volvariella volvacea, select small and immature mushroom buds with intact veils. The size can be adjusted flexibly based on the convenience of the isolation procedure. During isolation, take the young and tender flesh of the gills without spores as the isolation material.

(2) For larger fruiting bodies without a universal veil, choose young and small fruiting bodies. The flesh at the junction of the cap and stem can be taken as the isolation material.

(3) If only mature and larger fruiting bodies are available, take the flesh above the gills at the outer edge of the cap as the isolation material.

(4) For smaller fruiting bodies, such as Flammulina velutipes, take the active meristematic tissue at the junction of the stem and cap. The specific method is to hold the stem with one hand in a sterile environment, and with the other hand, separate the middle finger and index finger, quickly remove the cap along the stem, exposing a hemisphere-shaped junction at the top of the stem. Use a sharp inoculation loop to take a small piece from it.

021 How to perform tissue isolation from large fruiting bodies?

Select fresh, clean, and well-developed fruiting bodies in a sterile inoculation box. Immerse them in 0.1% mercuric chloride solution for 0.5 to 1 minute. Then rinse them with sterile water and pat dry, or wipe the cap and stem twice with a 75% alcohol cotton ball for surface disinfection. Tear open the fruiting body at the junction of the cap and stem, and pick a small piece of tissue. Transfer the tissue onto a mother spawn culture medium and incubate it at around 25 ℃.

When the mycelium grows to about halfway up the slant, transfer the tip of the mycelium to a new culture tube to cultivate as regenerated mother spawn. Expand the regenerated mother spawn to create original strains and cultivate them to induce fruiting. Select high-yielding and high-quality mother spawn for further cultivation.

022 How to perform tissue isolation for jelly fungi?

Tissue isolation for jelly fungi can be challenging due to their thin and somewhat tough fruiting bodies. Taking Auricularia auricula-judae (black fungus) as an example, there are two methods that can be used:

(1) In the first method, rinse the fruiting bodies with sterile water and gently pat them dry with sterile gauze. Then, disinfect them by wiping with an alcohol-soaked cotton ball. Finally, use a sterile scalpel to cut the fruiting bodies into small pieces of about 0.5 cm^2. Transfer these pieces onto a culture medium and incubate at 28 ℃.

(2) In the second method, dissect the tissue blocks from the undeveloped fruiting bodies within the earing mass.

023 What is the mycelium isolation method? When is it applied?

The mycelium isolation method is a technique used to obtain pure mycelium by utilizing the substrate on which edible mushrooms grow. It is typically employed when it is difficult to obtain fruiting bodies or when the fruiting bodies are too small and thin for tissue or spore isolation. This method is also commonly used for mushrooms that have associated companion fungi, such as Tremella fuciformis (silver ear fungus).

024 How to perform mycelium isolation from the substrate?

In edible mushroom production, mycelium isolation is mainly conducted using cultivation bags (bottles), substrate beds, or wood logs as the source of mycelium. If using materials from bags (bottles), it is common to select the young and tender mycelium from the bottom of the bag (bottle) for isolation. During isolation, a clean piece of mycelium is taken, and under sterile conditions, the mycelium block is broken apart. Small mycelium blocks of about the size of a green bean with vigorous mycelial growth are then transferred to a test tube.

025 What is mother spawn expansion?

The expansion cultivation of the slant mother spawn of edible mushrooms is referred to as subculturing or transfer. Regardless of whether the mother spawn is obtained from an outside source or isolated internally, it needs to be appropriately subcultured to produce a large quantity of regenerated mother spawn for continuous production. The vitality of the regenerated mother spawn often decreases with each subsequent subculture, and it is generally recommended to limit the number of subcultures to no more than 5.

026 How to expand mother spawn for edible mushrooms?

(1) Preparation of inoculation materials. Place the test tube containing the mother spawn (wrapped in newspaper to avoid exposure to UV light), blank slant culture medium, alcohol lamp, 75% alcohol cotton balls, label paper, marker pen, and inoculation tools in the inoculation box or on a laminar flow cabinet.

(2) Pre-treatment of inoculation equipment. Prior to inoculation, fumigate the inoculation box with a disinfectant spray for 30 minutes. In the laminar flow cabinet, turn on the UV light 30 minutes before the inoculation process (the UV light should be turned off 10 minutes before starting the operation) and start the fan 20 minutes before to ensure a sterile environment.

(3) Surface disinfection and sterilization before inoculation. After wiping hands with a 75% alcohol cotton ball, reach into the inoculation box or laminar flow cabinet. Then, wipe the outer surface of the test tube containing the mother spawn and the inoculation tools with the alcohol cotton ball for surface disinfection. Ignite the alcohol lamp to create a sterile area around the flame, and perform the inoculation procedure near the flame to avoid contamination by airborne microbes.

Hold the inoculation loop with the right hand and sterilize the tip by heating it until it turns red in the flame. Any part that may come into contact with the test tube during the inoculation process should be thoroughly burned in the flame for sterilization.

(4) Transfer of the spawn. Gently pick up a small amount of mycelium with the inoculation loop, and quickly transfer it to the center of the blank slant culture medium with the aerial mycelium facing upwards. Withdraw the inoculation loop, burn the mouth of the test tube, and singe the cotton plug until slightly charred, then place the cotton plug near the flame. Repeat the above steps until the original spawn is exhausted. Generally, one test tube of mother spawn can be used to inoculate 30–40 test tubes.

(5) Labeling. The label should be affixed to the front end of the test tube, not obstructing the cotton plug or the slope of the culture medium. The name of the mushroom

species should be written in larger font on the first line of the label. The strain number is typically written in smaller font after the species name. The name of the inoculator or the inoculating unit can be written at the beginning of the second line, using a smaller font. The inoculation date should be written after the name of the inoculator or unit on the second line, also using a smaller font.

(6) Cultivation. Wrap the upper part of the test tubes with paper, grouping them in bundles of 10, and place them in a light-avoiding incubator at 25 ℃ for cultivation.

3. Production of Original and Cultivated Seeds

027 What should be considered when purchasing mushroom spawn?

(1) Firstly, choose to purchase spawn from a reputable supplier with appropriate qualifications. Secondly, before purchasing, it is important to have a detailed understanding of the variety characteristics, cultivation conditions, and key cultivation points.

(2) The buyer should inquire about the type, variety, grade, type of culture medium, inoculation date, storage conditions, shelf life, and other relevant information concerning the purchased mushroom spawn to ensure a clear understanding.

(3) Examine the appearance of the spawn to assess its quality. Look for spawn that is clean, plump, and robust, without signs of aging such as wrinkling or the presence of yellow liquid. It is recommended to inspect each bottle (bag) individually.

028 What materials are needed for producing branch spawn?

(1) Preparation of branches. Wood materials that can be used for cultivating edible mushrooms, such as poplar, mulberry, oak, basswood, and willow, can be used for making branch spawn. Typically, choose loose-textured broadleaf branches, cut them to a length of 12–15 cm, with one end pointed and the other end flat, and then dry or oven-dry them for later use. Alternatively, popsicle sticks, fruit tree branches, or convenient chopsticks can also be used. However, branches that have been fumigated with sulfur or bleached with hydrogen peroxide should not be used, and fruit tree branches with excessive pesticide residues should also be avoided.

The length of the branches is typically 12–15 cm, with a width of 0.5–0.7 cm and a thickness of 0.5–0.7 cm. The specifications of the branches should be selected based on the size of the cultivation bags.

(2) Preparation of auxiliary materials. Wood shavings (50%), wheat bran (48%), gypsum (1%), lime (1%), with a moisture content of 60%–65%, are used to fill the gaps between the branches. Sugar, potassium dihydrogen phosphate, and magnesium sulfate are used to prepare the nutrient solution for soaking the branches.

(3) Preparation of cultivation bags. Polypropylene bags with dimensions of 17 cm × 30 cm

or 15 cm × 28 cm and a thickness of 0.048–0.052 mm are commonly used.

029 How to produce branch spawn?

(1) Soaking the branches. Immerse the entire bundle of branches in a lime water tank for a minimum of 24 hours. Prepare a nutrient solution by dissolving 1% white sugar, 2% potassium dihydrogen phosphate, and 0.1% magnesium sulfate in water, in appropriate proportions. Place the soaked branches into the nutrient solution and let them soak for about 15 hours. Adjust the soaking time based on the size of the branches and the temperature, ensuring that the branches are thoroughly soaked.

(2) Bagging. Prepare the auxiliary materials by mixing them in appropriate proportions until the moisture content reaches 60%–65%. Mix the soaked branches with the prepared auxiliary materials, ensuring that each branch is coated with the materials. Start by adding a small amount of auxiliary materials to the bottom of the bag, enough to cover the bottom of the bag with the growing medium. Place the prepared branches neatly and vertically into the plastic bag, sprinkle with the growing medium, and vibrate the bag to ensure that the wood shavings of the growing medium fill the gaps between the branches. Add another layer of growing medium to cover the branches, known as "bridging". Then tighten the plastic collar and seal it with a plastic plug.

Bagging requires a moderate tension, not too loose. Each gap between the branches should be filled with the growing medium. After bagging, it is important to ensure that the spawn bag adheres tightly to the branches inside the bag, without any separation between the bag and the branches. Otherwise, the aerobic fungi between the branches and the spawn bag may age, die, and lead to the appearance of clear, yellow, or red liquid after the spawn has grown, affecting the quality of the spawn. However, the bag should not be packed too tightly, as it may cause the spawn bag to burst and prevent the mycelium from germinating and penetrating into the wood of the branches, resulting in poor survival upon inoculation.

Follow the standard procedure of high-pressure sterilization, inoculation, and cultivation.

030 How to cultivate pure culture and spawn?

The cultivation room should be cleaned thoroughly and strictly disinfected. During the initial stage of cultivation, the temperature should be maintained at around 25 ℃. Then, reduce the temperature by 1 ℃ every 10 days until the bags (or bottles) are fully colonized. To make full use of space, spawn bottles or bags can be placed on cultivation shelves. Initially, the spawn bottles should be placed upright. Once the mycelium has germinated and established, they can be placed horizontally and stacked. Placing the bottles upright can result in the accumulation of dust and contaminants in the bottle caps, as well as water sinking to the bottom of the culture medium, leading to dryness in the upper part and water accumulation in the lower part, making it difficult for the mycelium to fully colonize the medium. Placing the spawn bottles horizontally allows for regular rotation, ensuring even distribution of moisture inside the bottles. As for spawn bags, the number of layers and placement can be adjusted based on the room temperature, with more layers in colder seasons. It is necessary to reverse the orientation of the spawn bags (top to bottom and inside to outside) every week to maintain a consistent temperature and uniform growth. In hot seasons, the spawn bags should be arranged in a "well" shape or placed in a single layer to facilitate ventilation and cooling, avoiding heat damage.

The relative humidity in the cultivation room should be maintained at 60%-70%, with light avoidance and regular ventilation. The cleanliness of the cultivation room should be maintained to prevent contamination from other microorganisms. The growth of spawn is faster than that of pure culture. When the mycelium reaches about one-third of the culture medium in the spawn bottles (or bags), the temperature in the cultivation room can be reduced by 2-3 ℃ to prevent high-temperature inhibition caused by enhanced metabolic activity during mycelial growth.

During the cultivation of spawn, regular inspections should be conducted to promptly eliminate inferior or contaminated individuals. Typically, the first inspection is performed when the mycelium has grown to a depth of about 1 cm in the culture medium, the second inspection when it has grown to about one-half to one-third, and the third inspection before full colonization. The inspections mainly focus on the normal germination of the mycelium, presence of contamination, vitality, and growth vigor. Slow germination or weak mycelium should be removed promptly, and each individual should be checked individually to avoid overlooking any. Contaminated bottles should be immediately eliminated, and the contamination source should be isolated.

4. Liquid Strain Production

031 How to produce shaken bottle liquid seed?

The manufacturing process of liquid culture medium: calculation→ weighing → boiling → volume adjustment → packaging → sterilization

(1) Preparation of culture medium. Calculate the quantity of each ingredient according to the selected culture medium formula. Weigh the various substances according to the formula. Add the measured soybean powder to 2/3 of the total amount of water, boil it for about 30 minutes, and filter it three times through 4 layers of gauze. Dissolve the soluble starch in a small amount of cold water to form a paste, add it to the boiled soybean powder filtrate, and stir well. Continue to heat over low heat, accurately weigh the remaining drugs, add them to the filtrate, and stir until completely dissolved, and make up the volume with water. Pay attention to stirring to prevent overflow or scorching.

(2) Packaging. When packing in triangular bottles, avoid the culture liquid sticking to the bottle wall. The filling coefficient for triangular bottles is 60%. Add 10–13 small glass beads or glass fragments with a diameter of less than 0. 8 cm to each culture medium bottle. Insert the cotton plug into the bottle mouth for 3–4 cm, and wrap it with cowhide paper or seal it with 8 layers of gauze measuring 8 cm × 8 cm.

(3) Sterilization. Sterilize at a pressure of 0. 12–0. 15 MPa for 50 minutes, remove and cool, and place it in a sterile room for later use.

(4) Inoculation. Put the test tube mother culture (wrapped in newspaper to avoid UV exposure), liquid culture medium, bottle rack, inoculation tools, alcohol lamp, lighter, and other supplies into the inoculation box or super clean workbench. Before inoculation, the inoculation room, inoculation box, or super clean workbench must be disinfected to ensure that the inoculation operation is performed under strict aseptic conditions.

When the temperature cools to below 28 ℃, inoculate under aseptic conditions. The method is to select the test tube mother culture, loosen the cotton plug of the triangular bottle, use an inoculation hook to pick up 5 mm × 5 mm × 0. 2 mm of the culture 5–6 pieces, and quickly place them in the triangular bottle. Each mother culture can inoculate 2–3 bottles. After firing the cotton plug twice, seal it with double-layered newspaper, paste a label on the triangular bottle, indicating the name of the strain, culture medium formula, and inoculation date.

(5) Cultivation. After inoculation, place the triangular bottles in a light-avoiding environment at 25–26 ℃ for 48–72 hours (the cultivation time may vary depending on the variety). When the inoculation blocks grow to 1 cm above the liquid surface (for individual culture blocks), transfer them to a shaking incubator. For reciprocating shaking incubators, the oscillation frequency is 80–100 times/min with an amplitude of 6–10 cm. If a rotary shaking incubator is used, it is generally controlled at 160–200 rpm. The temperature in the incubator is maintained at 24–25 ℃, usually for about 7 days. Only when the mycelium sedimentation reaches more than 80% after 20 minutes of settling, it can be inoculated into the fermentation tank.

032 What are the characteristics of high-quality liquid spawn?

The diameter of the fungal balls is around 1 mm, with uniform size. Even after a period of rest, the fungal balls remain suspended in the culture medium without sedimentation. They are viscous and emit various distinctive mushroom aromas. Different strains exhibit different colors in the culture medium, varied morphologies of fungal balls, and different smells. The culture medium of Flammulina velutipes (enoki mushroom) appears light yellow, clear and transparent, with numerous small mycelial balls suspended in the liquid, accompanied by aromatic fragrance. The culture medium of Agaricus bisporus (button mushroom) is brown, clear and transparent, with a pleasant aroma. The culture medium of Auricularia auricula-judae (wood ear mushroom) is greenish-brown, viscous, and has a sweet aroma, indicating the completion of cultivation.

033 How to produce liquid spawn in a fermentation tank with a control cabinet?

The production process of liquid spawn in a fermentation tank with a control cabinet includes: tank cleaning and inspection→ sterilization of the tank → preparation of ingredients → sterilization of filter, filter cap, air inlet pipe, inoculation gun, and tubing → loading of ingredients → sterilization → cooling → inoculation → cultivation → inoculation of bags (bottles).

(1) Tank cleaning and inspection. Before production, the fermentation tank must be thoroughly cleaned to remove any adhering fungal balls, blocks, and residue. Once the tank walls are free of suspended particles and residual fungal balls, fill the tank with water above the heating pipes. Start the equipment and check if the control cabinet and heating pipes are functioning properly, valves are not leaking, pressure gauge and safety valve are working correctly, and ensure proper water and electricity supply.

(2) Tank sterilization. Close the inoculation valve and air inlet valve at the bottom of the

tank. Fill the tank with water through the feed port until it reaches the middle line of the sight glass. Close the feed port cover tightly and close the exhaust port. Start the heater. When the temperature reaches 100 ℃, release the cold air and slightly open the exhaust valve. When the temperature reaches 123 ℃ and the pressure is 0. 12 MPa, maintain it for 35 minutes. Then close the exhaust valve and let it sit for 20 minutes. Open the exhaust valve, inoculation port, and air inlet port to drain the water from the tank.

(3) Preparation of ingredients and loading. Prepare the culture medium according to the mother culture. Close the air inlet valve and inoculation valve at the bottom of the fermentation tank. Pour the culture medium into the fermentation tank through the inoculation port and add about 12 mL of Paodi. Rinse the cooking pot with water, and the loading amount should be 60%-80% of the total volume of the tank, with a height above the top edge of the sight glass by 10 cm. Tighten the feed port cover.

(4) Sterilization. Close all valves. Press the sterilization button on the control cabinet. Before the temperature of the culture medium reaches 100 ℃, introduce unfiltered air directly into the medium and stir to prevent the medium from sticking to the heating rods at the bottom of the tank. When the temperature reaches 100 ℃, slightly open the exhaust valve until sterilization is complete.

During the first, middle, and final stages (0, 17, 30 minutes) of the 35-minute sterilization period, drain the medium three times. Open the air inlet valve and the inoculation port valve slightly to let out a small amount of gas and medium. Each draining session should last 3-5 minutes, with a total drainage of 3-5 liters, to remove the raw medium at the valve and sterilize the valve and pipelines.

(5) Cooling. Drain the hot water from the jacket, and blow air through the first-stage and second-stage filters using an air pump. At the same time, inject cold water into the jacket to lower the temperature inside the tank. When the pressure on the pressure gauge of the tank cover drops below 0. 05 MPa and it is confirmed that the first and second-stage filters are dry, open the air inlet pipe that passes through the filters into the tank cover and blow air onto the filters to agitate the culture medium and rapidly cool it down.

(6) Inoculation in the tank. Prepare the spawn, sterilized gauze soaked in 75% alcohol, a lighter, 95% alcohol, wet gloves, wet towels, and other supplies. Gradually open the exhaust valve until the pressure in the fermentation tank drops to "0". Close the exhaust valve and turn on the air pump. Place the sterilized gauze ring soaked in alcohol above the feed port and ignite it. Quickly open the feed port cover, remove the stopper from the spawn bottle in the middle of the flame, rotate the bottle to sterilize the upper part and the mouth of the bottle on the flame, and quickly pour the spawn into the tank. Close the feed port cover, tighten it, remove the gauze ring, and complete the inoculation.

(7) Cultivation. Start the control cabinet and enter the cultivation phase. Slightly open the exhaust valve to maintain the tank pressure at 0. 02-0. 04 MPa. Check the cultivation

temperature and ensure an airflow rate of 1.2 m^3/h or above. The cultivation temperature is set differently for different varieties, with a range of 24–26 ℃ for standard varieties and 28–31 ℃ for high-temperature varieties. The airflow rate should be adjusted to 1.2 m^3/h or above, the tank pressure should be maintained at 0.02–0.04 MPa, and the cultivation time is 72–96 hours.

034 How to check the quality of liquid spawn in a fermentation tank?

After 24 hours of inoculation, a sample can be taken every 12 hours to observe and record the growth and germination of the spawn. Sampling method: Open the valve and drain the condensate water from the steam pipe. Adjust the exhaust valve to release a small amount of steam and sterilize the sampling port. When sampling, discard the initial portion of the culture medium that comes out, close the sampling valve, flame-seal it, and then reopen the valve to collect the desired amount of spawn in a sterilized triangular bottle. Close the sampling valve and seal it with the bottle cap.

The quality of the liquid spawn is generally assessed by three visual observations and one smell test. Firstly, observe the color of the spawn. Normal spawn has a pure color, although it can be light yellow, orange-yellow, light brown, etc., but it should not be turbid, and most of the time, the color gradually becomes lighter (except for Armillaria, Flammulina, and Auricularia, which may darken). Secondly, observe the clarity of the liquid spawn, which is usually clear and transparent (although Auricularia and Ganoderma may have some mycelium debris). In the early stages of cultivation, the liquid may appear slightly cloudy, but as the cultivation progresses, there should be no small particles or flocculent substances in the liquid, and the liquid should become clearer and more transparent. Otherwise, it is considered abnormal. Thirdly, examine the presence of noticeable spikes around the spawn and the growth of the spawn. Edible mushroom spawn usually has small spikes around it, which can be long or short, soft or hard (e.g. Hypsizygus marmoreus has hard spikes). The smell test involves smelling the liquid spawn. The sweet smell of the medium will become lighter as the cultivation time increases, and in the later stages, there should be a faint fragrance of the liquid spawn. Auricularia and Flammulina have a mild fragrance, Pleurotus eryngii has a strong fragrance, Ganoderma has a medicinal smell, and Agaricus bisporus has a light sake lees aroma.

After 48 – 72 hours, the concentration of the spawn increases rapidly, and when the volume percentage concentration is ≥80%, it can be transferred to bags.

After sampling, place the triangular bottle on a tabletop and let it stand for 5 minutes. If the spawn neither floats nor precipitates, the spawn accounts for more than 80% (volume percentage) of the liquid, the color of the liquid becomes lighter, the liquid becomes clear and transparent, and there is a clear boundary between the spawn and the liquid, indicating that the mycelium is mature and can be used. If the liquid becomes turbid, has a moldy or alcoholic smell, or the color darkens, it indicates spoilage.

035 How to inoculate spawn bags with liquid spawn?

After the liquid spawn has been cultured, prepare a sterilized inoculation tube and inoculation gun under the protection of a flame. Keep the pump running and close the exhaust valve slightly. When the pressure stabilizes at 0.03–0.05 MPa, open the inoculation valve. Sterilize the tip of the inoculation gun in the flame of an alcohol lamp and inoculate under sterile conditions. The inoculation amount should be sufficient to cover the surface of the bottle or bag, typically around 15–18 mL. It is important to avoid using too little spawn, as this may result in localized germination and contamination of the cover.

5. Quality Identification and Preservation of Spawn

036 How to use a microscope to examine the purity and condition of liquid spawn?

Take 2–3 loops of fermentation liquid with an inoculation loop and prepare temporary slides. Observe them under a microscope or stain the slides for microscopic examination. Check for the presence of contaminants and assess the vitality of the mycelium based on the depth of color. Normal mycelium appears clumped together, relatively dense, with branching root-like structures at the edges. Mycelium contaminated with mold tends to be thicker and does not clump together. Mycelium contaminated with bacteria is shorter and finer.

Microscopic examination using a microscope is a simple and fast method that allows for the timely detection of contaminants. However, due to the small sample size and limited observation area, it may not be able to detect small amounts of contaminants easily.

037 How to use meat broth to culture and examine liquid bacterial strains?

Take a loopful of fermentation broth and inoculate it into a test tube containing phenol red meat broth. Incubate at 28–30 ℃. If the solution turns from red to yellow, it indicates the presence of bacteria in the fermentation broth. If the solution remains red, no bacterial growth is present.

Phenol red meat broth medium formula: Peptone 1%, glucose 0.3%, beef extract 0.30%, sodium chloride 0.5%, phenol red 0.003% (prepare a 1% alcohol solution of phenol red in advance), pH 7.4–7.5; sterilize at 120 ℃ for 30 minutes.

038 How to use streak plate method to culture and examine liquid bacterial strains?

Generally, the sample is streaked directly on the agar medium and incubated at 28–30 ℃. Alternatively, the sample can be first cultured in meat broth medium and then streaked on the agar medium. The streak plate method has a slower response time and is more complex to perform, but it can detect a small amount of contaminating bacteria.

039 What are the characteristics of excellent spawn for edible mushrooms?

(1) Uniform Growth. Spawn of the same variety and from the same source, when propagated, should show no significant differences in growth appearance between tubes, regardless of the amount of propagation. This includes growth rate, color, thickness of the mycelium, and the abundance of aerial mycelium.

(2) Normal Growth Rate. Different species and varieties of edible mushrooms have their inherent growth rates under specific nutritional and cultivation conditions. For example, Agaricus bisporus strain 10, when cultured on comprehensive PDA medium at 25 ℃ in the dark, should fully colonize the slant in 6 days. If the growth rate deviates from this range, it indicates insufficient vitality, aging, or degeneration of the spawn, making it unsuitable for further use.

(3) Morphological Characteristics. Each species and variety of edible mushrooms have their unique morphological features, such as colony morphology, abundance of aerial mycelium, color of mycelium, presence or absence of pigments in the medium, and characteristics of the growth edge. For instance, Auricularia polytricha strain ACCC50249, when cultured at 26–28 ℃, secretes yellow pigment by the aerial mycelium within 2–3 days after fully colonizing the slant, while strains of Flammulina velutipes do not produce such pigments. Most varieties of Auricularia auricula and Flammulina velutipes do not secrete brown pigments into the medium before fully colonizing the slant. If the spawn has a brownish color in the medium or a darker color on the back of the inoculum block during growth, it is considered abnormal. Such strains often fail to produce mushrooms or have low yields even if they do.

(4) Colony Edge Appearance. The edge of the colony should be full, neat, and vigorous in growth.

040 How to identify the quality of spawn for edible mushrooms?

The spawn for edible mushrooms serves as the original source of the strain in the production process, and its quality is crucial for mushroom cultivation. Every excellent strain should possess the characteristics of being "pure, true, robust, moist, and fragrant". The standards for identifying the quality of spawn are as follows:

(1) High purity. The strain should be free from contamination by other microorganisms.

(2) Pure and vibrant mycelium color. The mycelium of most species should be pure white, glossy, and without signs of aging or discoloration.

(3) Robust mycelium growth. The mycelium should be thick, with abundant branching

and dense growth. It should readily consume the substrate when inoculated onto the culture medium and exhibit vigorous growth.

(4) Moist culture. The spawn should be moist and firmly adhered to the walls of the test tube.

(5) Possession of characteristic aroma. It should have a distinct fragrance specific to the particular strain, without any moldy or putrid odors.

041 How to distinguish the quality of spawn for edible mushrooms?

To determine the quality of spawn for edible mushrooms, it should exhibit the basic characteristics of the specific mushroom species. The mycelium should penetrate the bottom of the spawn bottle (bag), showing uniform and consistent growth with elasticity. The mother spawn block should be present in the original spawn bottle (bag), while the original spawn block should be present in the cultivation spawn bottle (bag). These indicate that the spawn is of good quality. However, it is not recommended to use spawn in the following situations:

(1) The mycelium in the spawn bottle (bag) gradually disappears, showing consumed patches or signs of mite activity, indicating contamination by mites.

(2) The spawn bottle (bag) contains various colored contaminating fungal spores such as red, yellow, black, green, and there are clear antagonistic lines between two or more types of fungi on the bottle (bag) walls. The presence of foul odors such as sourness or putrefaction indicates contamination by molds, bacteria, yeast, or other contaminants. In such cases, the spawn should be eliminated and promptly disposed of by deep burial or incineration.

(3) The surface of the spawn bottle (bag) shows a thick, dense, and tough fungal skin, the mushroom columns shrink and detach from the walls, mycelium exhibits autolysis, the bottom of the spawn bottle (bag) accumulates a large amount of yellow-brown liquid, and excessive primordia or ear-like protrusions appear on the surface and surroundings of the spawn bottle (bag). These signs indicate aging or degeneration of the spawn and it is not suitable for use.

042 How to preserve spawn for edible mushrooms? How long can it be stored?

Mother spawn should be stored at temperatures between 0–4 ℃ and should not be stored for more than 3 months.

Original spawn and cultivation spawn can be stored indoors, under clean, ventilated, dry (with relative humidity around 60%), and light-shielded conditions at 24 ℃ for up to 10 days. They can also be stored at temperatures between 0–4 ℃ for a maximum of 40 days.

043 How to prevent spawn degeneration for edible mushrooms?

(1) Preventing cross-contamination of spawn. Strengthen "variety" isolation during spawn transfer and fruiting trials to minimize cross-contamination between different varieties and ensure the stable genetic composition of superior strains over a longer period of time.

(2) Controlling the number of spawn generations. Strictly control the number of spawn generations during production.

(3) Using effective spawn preservation methods. Spawn preservation should combine short-term, medium-term, and long-term methods. Different preservation methods should be applied based on the requirements of different edible mushroom species to ensure the long-term preservation of their original superior traits.

(4) Creating favorable nutritional and environmental conditions for spawn growth. Optimal nutrient conditions in the spawn medium are essential for robust spawn growth and to minimize degeneration.

(5) Being vigilant against possible viral infections in spawn. Timely testing and identification should be conducted for suspicious spawn to detect viral infections. Strains with high levels of viral particles and severe impacts on the morphology of mycelium and fruiting bodies should be promptly eliminated.

044 How to rejuvenate spawn for edible mushrooms?

(1) Revitalization through transplantation. Utilize low-temperature preservation to keep the spawn at a lower level of metabolic activity. However, if the spawn has been stored for too long, the mycelium may age. For spawn stored for 3 – 4 months, it is necessary to re-transplant it. After transplantation, the spawn should be cultured at suitable temperatures for 7–15 days until the mycelium covers the substrate surface. It can then be further preserved at low temperatures or used for cultivation. The preservation medium should contain potassium dihydrogen phosphate, dipotassium phosphate, and peptone to prevent changes in the pH of the medium.

(2) Nutrient component replacement. Most edible mushrooms have a preference for changing nutrient components in the culture medium. Repeatedly using the same proportion of the medium can lead to poor nutrition for the mycelium and spawn degeneration. Therefore, it is necessary to vary the nutrient components of the culture medium according to the characteristics of different mushroom species to improve the vitality of the spawn and obtain superior strains.

(3) Suitable environmental conditions. Create suitable environmental conditions during spawn cultivation, such as temperature, light intensity, and air quality. This ensures robust

mycelium growth and stable characteristics.

(4) Selection of superior strains. When preserving or using spawn, it is important to carefully select the strains. Strains showing significant changes in vitality or mycelial characteristics, or those contaminated with other microorganisms, should be discarded rather than used. This is an essential guarantee for obtaining superior strains.

045 What are the manifestations of spawn degeneration in edible mushrooms? What are the underlying causes?

Under normal cultivation conditions, the manifestations of spawn degeneration in edible mushrooms include decreased yield, deterioration of variety, decrease in uniformity, decline in vitality, and reduced resistance to stress.

There are many reasons for spawn degeneration. External factors include excessive generations and unfavorable conditions. Internal factors involve unfavorable genetic variations in the spawn's genetic material.

046 What are the manifestations of spawn degeneration in the mother culture stage of edible mushrooms?

(1) Abnormal colony morphology. For example, a variety that originally had flat colonies may exhibit tightly wrinkled colonies. Aerial mycelium may change from filamentous to snowflake-like, or it may appear in varying amounts or disappear altogether. The color of the mycelium may change, with degraded strains showing a slight yellow, light brown, or other colors, or transitioning from bright to dull.

(2) Mycelial lodging. If the mycelium appears to be lodged or flattened, it indicates a weakening growth vigor.

(3) Slow mycelial growth. Under certain culture media and conditions, both excessively fast or slow mycelial growth rates are abnormal, particularly slow growth.

(4) Inconsistent mycelial growth rate and appearance. In a superior strain, there should be no difference in mycelial growth rate and appearance during propagation.

(5) Pigmentation of mycelium. Most edible mushroom varieties do not secrete pigments during the mother culture stage, except under unfavorable high-temperature conditions or when aging.

To evaluate the quality of spawn and detect issues such as strain degeneration, aging, pathogen invasion, contamination, and variety mixing, it is necessary to compare the typical biological characteristics (including morphological features, physiological and ecological characteristics, and cultivation habits) of the strain against reference standards. Evaluating spawn quality requires considering the aspects to be assessed and determining the appropriate standardized methods. Results are influenced by the methods and conditions used, and without comparability, it becomes impossible to differentiate. Therefore, the establishment of standards is necessary. These standards include the culture media, culture conditions (temperature, humidity, pH, light exposure, etc.), and the age of the strain.

6. Basic Knowledge of Edible Mushroom

047 What is edible mushroom disease?

Edible mushroom disease refers to the phenomenon where the mycelium and fruiting body of the edible mushroom cannot grow and develop normally during their growth process due to unsuitable environmental conditions (such as too high CO_2 concentration, excessive moisture, high temperature, inappropriate culture medium) or being infected or contaminated by other fungi, bacteria, viruses, nematodes, etc. This results in the death of the mycelium and the shrinkage, rot, or death of the fruiting body. This is referred to as edible mushroom disease.

048 What are the types of edible mushroom diseases?

Edible mushroom diseases can be divided into two major categories according to the cause of the disease: infectious diseases (pathogenic diseases) and non-infectious diseases (physiological diseases). Infectious diseases are further divided into parasitic diseases and competitive diseases.

Parasitic diseases refer to diseases caused by the direct infection of the mycelium and fruiting body of edible mushrooms by pathogenic organisms such as fungi, bacteria, viruses, and slime molds.

049 What harm do competitive fungi cause to edible mushrooms?

Competitive diseases refer to diseases that pose a threat to the production of edible mushrooms by harmful fungi that infect the culture medium and can even secrete toxic substances. Although competitive fungi do not directly infect the mycelium and fruit bodies of edible mushrooms, they can cause seed waste if they occur during the seeding stage, and yield reduction or even crop failure if they occur during the cultivation stage. The pathogens include fungi and bacteria, etc.

050 What is an infectious disease?

Infectious diseases are diseases caused by various pathogenic microorganisms invading edible mushrooms, leading to metabolic disorders. The main pathogenic microorganisms are fungi, bacteria, viruses, and nematodes, which are infectious and possess invasive characteristics. The feature of these diseases is that the pathogenic microorganisms directly absorb nutrients from the mycelium or fruiting body of edible mushrooms to build themselves, which hinders the normal physiological activities of edible mushrooms, leading to symptoms and a decrease in both the yield and quality of edible mushrooms.

051 What is the pattern of fungal diseases in edible mushrooms?

The majority of fungi that cause diseases in edible mushrooms belong to the mold category and have filamentous mycelium. After infecting for a specific period, these pathogenic fungi form lesions and reproductive bodies-spores, on the surface of the infected edible mushrooms, with airflow and sprinkling water being the main modes of transmission. Most of these fungal pathogens prefer high temperature, high humidity, and acidic environments. Common fungal diseases include brown rot, soft rot, Trichoderma disease, and Penicillium disease, etc.

052 What is the pattern of bacterial diseases in edible mushrooms?

The majority of bacteria that cause diseases in edible mushrooms are various kinds of Pseudomonas. These bacteria mostly prefer high temperature, high humidity, low oxygen pressure, and nearly neutral substrate environment. They can be spread by airflow, substrate, water flow, tools, operations, and insects, etc. Common bacterial diseases include pleurotus ostreatus disease and bacterial blotch, etc.

053 What is a viral disease in edible mushrooms?

Viral diseases in edible mushrooms are caused by a group of organisms similar to molds. They compete with edible mushrooms for space and nutrients, and can also harm the mycelium and spores of the mushrooms, significantly affecting their production. Viral diseases are a common and important disease in the production of edible mushrooms with strong infectivity. The symptoms vary depending on the virus concentration and infection conditions, such as numerous small old mushrooms, pinhead mushrooms, mummy-like mushrooms, etc., which are easily confused with the symptoms of physiological diseases. This virus mainly affects mushrooms like Agaricus bisporus, shiitake, oyster mushroom, straw mushroom, Poria, and Tremella, etc.

054 What is the pattern of nematode diseases in edible mushrooms?

Nematodes are a type of microscopic protozoa that harm the mycelium and often damage the fruiting bodies. Due to their tiny size and being invisible to the naked eye, they are often mistaken for harm caused by miscellaneous fungi or damage from high temperatures "burning" the mushrooms. Different edible mushrooms show different symptoms when harmed by nematodes. Nematodes are widely distributed in soil and culture media, and their main modes of transmission are soil, substrate, and water flow.

055 What is a physiological disease in edible mushrooms?

Physiological diseases refer to diseases caused by unsuitable living conditions for edible mushrooms. When the ecological environment conditions cannot meet the minimum requirements for the development of edible mushrooms, the mushrooms will undergo metabolic disorders, leading to deformities. These diseases are not communicable, but they can cause varying degrees of yield reduction and product quality decline, even leading to a total loss of the crop.

056 What are the causes of drug damage in edible mushrooms?

During the growth of mycelium and fruiting bodies of edible mushrooms, they can be sensitive to certain drugs, which can easily lead to malformation or death. For example, fumigation with sulfur can easily cause the death of mushroom species; pesticides such as methotrexate and carbendazim have strong fungicidal properties and are used to suppress mycelial growth in the substrate; spraying dichlorvos during the mushrooming period can lead to the death of primordia, malformed mushrooms, or long-term failure to produce mushrooms; some oyster mushroom varieties will deform or fail to produce mushrooms when using carbendazim at concentrations exceeding 800 times; using dichlorvos during the fruiting period of oyster mushrooms can cause dead or malformed mushrooms.

057 How to prevent and treat drug damage?

(1) Choose safe, specialized pesticides for targeted treatment, and strictly control the timing and dosage of medication.

(2) Once drug damage occurs, rinse several times with clean water, promptly remove the affected mushrooms, and allow them to regrow.

058 What are the reasons for fertilizer damage in edible mushrooms?

When different varieties of edible mushrooms are in the mycelial growth stage or the fruiting body growth stage, if the carbon-to-nitrogen ratio in the formula is out of balance, especially when there is too much nitrogen fertilizer, the mycelium and fruiting bodies are prone to pathological phenomena. For example, the mycelium of oyster mushrooms grows vigorously in grain culture media, but no fruiting bodies are produced; in the case of shiitake mushrooms, oyster mushrooms and other varieties, when there is too much bran in the culture medium, the mycelium becomes thicker and it is difficult for mushrooms to develop; in the culture medium with too much nitrogen fertilizer and a strong smell of ammonia during the composting of Agaricus bisporus, the mushroom species cannot germinate and eat the culture medium, and may even be fumigated to death by ammonia, resulting in the entire culture medium being discarded.

059 How to prevent fertilizer damage in edible mushrooms?

(1) Reasonably adjust the formula and strictly control the use of nitrogen fertilizer. Reduce or do not use nitrogen-containing fertilizers such as urea and ammonium sulfate, and use more organic nitrogen fertilizers, such as cake fertilizer, bran or soybean meal, etc.

(2) When it is found that excessive nitrogen fertilizer inhibits the growth of mycelium, the culture medium should be taken out in time and mixed with new material for re-fermentation and re-seeding.

060 How to deal with diseased fruiting bodies and mushroom bags?

(1) Fungal diseases. If detected early, you can remove the fruiting bodies and be careful not to dispose of them randomly. At the same time, strengthen ventilation, and use drugs such as carbendazim to kill the surface bacteria of the mushroom bags in the disease area.

(2) Bacterial diseases. Timely removal of diseased fruiting bodies, spray an appropriate amount of fungicide on the mushroom bags in the disease area, and continue the management of the next mushroom crop. Mushroom bags with serious diseases should be discarded.

061 How to deal with contaminated culture medium?

Mushroom substrate that is slightly diseased or has a short mushrooming time can be made into organic fertilizer after proper medicinal treatment; heavily diseased mushroom substrate must be treated with biological fermentation using specific microorganisms before it can be made into organic fertilizer.

062 How to deal with the cultivation site after disease occurrence?

(1) Treatment inside the mushroom shed.

Medicinal treatment. Before the mushroom bags are brought into the shed, use fungicides to disinfect the shed depending on the length of time the shed has been in use and the disease situation from the previous production cycle. Especially the entrances and exits, ventilation ports, wooden pillars and roof of the shed must be strictly sprayed with no dead corners lcft.

High-temperature fumigation. Use the high temperature of the sun and the fumigation of medicine to disinfect the shed. This method has low cost, less pollution, simple operation, and good effect.

Sun exposure method: Choose a sunny day to remove the shed film for sun exposure.

(2) Treatment outside the mushroom shed. Keep the environment around the mushroom shed clean. The raw material warehouse, mushroom room, and ingredient field should maintain a certain distance from the mushroom shed, and try to remove the sources of pollution.

063 How to use pesticides in edible fungi?

(1) During the mushrooming period, the use of pesticides should be cautious, as they can contaminate the fruiting bodies and their residues can affect product quality.

(2) It is forbidden to directly use highly toxic organomercury, organophosphorus, and other drugs for mixing materials, stacking materials, and bed pest control; pesticides with a long residual period, difficult to decompose, and a pungent smell cannot be used for edible fungus cultivation. Especially during the mushrooming period, it is forbidden to use drugs with strong toxicity, a long residual period, or a pungent smell.

(3) To prevent and control diseases and pests of edible fungi, highly effective, low-toxicity, and low-residue drugs should be used, and the type and concentration of drugs should be selected according to the prevention and control object. For example, dichlorvos has fumigation and contact killing effects and has special effects on adult and larval flies and

springtails, but its killing effect on mites is poor. Fenitrothion is a new type of high-efficiency, low-toxicity, and low-residue organophosphorus insecticide. In addition to its special effects on flies and springtails, it also has a good contact killing effect on mites, and its efficacy is better than dichlorvos. If fenitrothion and acaricide are used to control mites, it is better than other pesticides. The appropriate concentration should be chosen when applying pesticides. Fenitrothion and acaricide should be used at a 500-fold dilution for mite prevention from the composting stage to before the mushrooming stage, but the concentration should be reduced to 1 000-fold during the fruiting body stage.

(4) When using pesticides, you should first familiarize yourself with the properties of the pesticides. Misuse of pesticides can sometimes form a layer of toxic substances on the surface of the cover soil or culture medium, affecting mycelial growth and causing a decrease in yield.

(5) As far as possible, use plant-based insecticides and microbial preparations, such as pyrethrum, rotenone, and yield-increasing bacteria preparations.

(6) Protect natural enemies and do not abuse pesticides. When predatory mites and certain leather mites appear on the bed surface, they should not be killed but protected instead.

064 What are the agricultural prevention and control measures for pests and diseases of edible fungi?

In production, we should adhere to the prevention and control policy of focusing on agricultural prevention and control and supplementing with chemical prevention and control, and carry out comprehensive control of the occurring pests and diseases.

(1) Maintain good environmental hygiene and eliminate insect and fungal sources. Maintain good environmental hygiene in the inoculation room, culture room, and cultivation site. Check regularly and immediately deal with any contaminated mushroom bags found. Mushroom rooms should be thoroughly cleaned and disinfected before use.

(2) Choose excellent strains. In addition to requiring high yield, high quality, and strong resistance, excellent strains should also have high purity, no pest infection, and suitable mycelial age. Different temperature-type varieties are matched according to different production seasons. The source of the strain must be correct, and the number of generations of propagation should be strictly controlled.

(3) Choose the right raw and auxiliary materials. Use fresh raw and auxiliary materials without mold and pests, adjust the nutritional components of the culture medium, do a good job in the pretreatment of the culture medium such as fermentation and sterilization, to form a substrate environment that is suitable for the growth of edible fungus mycelium and can effectively inhibit the occurrence of miscellaneous bacteria.

(4) Create suitable environmental conditions for growth. During the cultivation management process, measures such as ventilation, water control, and shading can be taken to create environmental conditions that are conducive to the growth of edible fungi and not conducive to the growth of miscellaneous bacteria, in order to achieve the purpose of controlling disease pollution.

(5) Pay attention to reasonable rotation and tide change. Turning over and sunning the site, combined with spreading lime or other drugs, mixed in the soil, can kill various pests in the soil; after the site is rotated, it can greatly reduce the number of pests.

(6) Shorten the time for mushrooms (ears) to come out. Shorten the growth cycle of the fruiting bodies, achieve quick mushrooms (ears) production, and early mushroom (ear) harvest, to reduce the chance of pest damage.

065 What are the physical prevention and control measures for pests and diseases of edible fungi?

(1) Barrier method. Use a 60-mesh insect-proof net for the doors, windows, and ventilation of cultivation sites to prevent adult insects such as mushroom flies and mushroom mosquitoes from entering and causing damage.

(2) High (low) temperature insecticidal method. When cultivating edible fungi, first pile up the culture material, let the material heat up to above 60 ℃, and maintain it for a certain period of time, which can kill a large number of pests in the material; drying mushrooms (ears) on a cement field can kill the pests in them; during storage, you can take out the mushroom body that has bugs and kill the pests in it in a cold storage or refrigerator.

(3) Lure and kill. Hang a special insect-killing lamp for edible fungi, or hang a yellow sticky insect board to lure and kill. For mites, you can spread gauze on the mushroom bed, sprinkle a layer of freshly fried rapeseed cake powder, the mites will gather on the gauze, then soak the gauze in lime water, and the mites will be killed. Do this several times, and the mite-killing effect can be over 90%.

(4) Water immersion method. Inject water into the cultivation bag (or bottle), or immerse in water and press it down, soak for 2-3 hours, and the larvae will suffocate and die.

066 What are the biological prevention and control measures for pests and diseases of edible fungi?

(1) Bacterial preparations. Bacillus thuringiensis and avermectin can control mites, flies, mosquitoes, and nematodes.

(2) Plant preparations. Pyrethrum ester, rotenone, nicotine, bitter vine, and tobacco

extracts have good prevention and control effects on various pests of edible fungi.

(3) Antibiotic agents. Streptomycin and chlortetracycline can prevent and control bacterial diseases of edible fungi with ideal results. Copper sulfate, Bordeaux mixture, etc. are sterile acaricides and fungicides.

067 How to carry out chemical prevention and control?

Especially during the fruiting period, the use of chemicals is not recommended. The cultivation cycle of edible fungi is short, and they are directly consumed, so pesticides are likely to remain in the fruiting body. During cultivation, when pesticides must be used, the following should be noted: Highly toxic pesticides are strictly prohibited, and those with a long residual period, difficult to decompose, or with an irritating odor should not be directly used on mushroom beds or bags; Try to choose high-efficiency, low-toxicity, low-residue agents that are harmless to humans, livestock, and edible fungi, and master the appropriate concentration and the right time for prevention and control.

REFERENCE

[1]赵奎华,王克,郑怀民. 葡萄病虫害防治图册[M]. 沈阳:辽宁科学技术出版社,1993.
[2]李淑珍,吴国兴,徐贵轩. 图说保护地桃葡萄栽培技术[M]. 北京:中国农业出版社,2000.
[3]高梅,潘自舒,武景和,等. 果树生产技术[M]. 北京:化学工业出版社,2009.
[4]董清华,姚允聪,高遐虹. 葡萄三高栽培技术[M]. 北京:中国农业大学出版社,1998.
[5]安新哲,武景和,张红. 葡萄优质丰产栽培掌中宝[M]. 北京:化学工业出版社,2012.
[6]李峰,温室葡萄栽培管理关键技术[M]. 北京:中国农业大学出版社,2019.
[7]刘军,龚林忠. 葡萄种植技术培训教材[M]. 北京:中国农业大学出版社,2015.
[8]陈俏彪. 食用菌栽培技术[M]. 北京:中国农业出版社,2015.
[9]黄毅. 食用菌工厂化栽培[M]. 福州:福建科学技术出版社,2014.
[10]李明,等. 食用菌病虫害防治关键技术[M]. 北京:中国三峡出版社,2006.
[11]杨桂梅,苏允平. 食用菌生产[M]. 北京:中国农业大学出版社,2014.
[12]李术臣,刘光东,杨文平,等. 北方反季层架立体栽培香菇关键技术[J]. 食用菌,2018(1)56-57.
[13]马雪梅,安玉森,李艳华,等. 小孔单片黑木耳栽培技术[J]. 黑龙江农业科学,2012(29):160-161.
[14]王延锋,戴元平,徐连堂,等. 黑木耳棚室立体吊袋栽培技术集成与示范[J]. 中国食用菌,2014(1):30-33.
[15]王勇,魏雪生,张志军,等. 食用菌污染袋白色链孢霉的发生及防治[J]. 北方园艺,2010(23).
[16]张功友. 承德地区双层拱棚栽培错季香菇新技术要点[J]. 中国农业文摘农业工程,2017(3)70-71.
[17]曹春英,孙日波. 花卉栽培[M]. 3 版. 北京:中国农业出版社,2014.
[18]包满珠. 花卉栽培[M]. 北京:中国农业出版社. 2001.
[19]张树宝. 花卉生产技术[M]. 重庆:重庆大学出版社,2006.
[20]胡惠蓉. 120 种花卉花期调控技术[M]. 北京化学工业出版社,2009.
[21]龙雅宜. 切花生产技术[M]. 北京:金盾出版社,1997.
[22]成海钟. 切花栽培手册[M]. 北京:中国农业出版社,2000.
[23]吴少华. 鲜切花栽培和保鲜技术[M]. 北京:科学技术文献出版社,1999.
[24]中国农业科学院蔬菜花卉研究所. 中国蔬菜栽培学[M]. 北京:中国农业出版社,2010.
[25]吴乾兴,刘勇,袁廷庆,等. 设施甜瓜 3 种授粉方式的效果比较[J]. 中国瓜菜,2013,

26(4) :31-32.
[26]闫德斌,牛庆生,常志光,等.蜜蜂为大棚甜瓜授粉试验报告[J].蜜蜂杂志,2012(7):1-2.
[27]吴会昌.春大棚薄皮甜瓜坐果期管理关键技术[J].中国蔬菜,2018(4):98.
[28]李惠明,赵康,赵胜荣,等.蔬菜病虫害诊断与防治实用手册[M].上海:上海科学技术出版社,2012.